Coumarins

Biology, Applications and Mode of Action

Coumarins

Biology, Applications and Mode of Action

Edited by

Richard O'Kennedy

School of Biological Sciences, Dublin City University, Ireland

and

R. Douglas Thornes

CK Byrnes Research Centre, Beaumont Hospital, Dublin, Ireland

JOHN WILEY & SONS

Chichester · New York · Weinheim · Brisbane · Singapore · Toronto

National 01243 779777
International (+44) 1243 779777
e-mail (for orders and customer service enquiries): cs-books@wiley.co.uk
Visit our Home Page on http://www.wiley.co.uk
or http://www.wiley.com

Other Wiley Editorial Offices

John Wiley & Sons, Inc., 605 Third Avenue,
New York, NY 10158-0012, USA

VCH Verlagsgesellschaft mth, Pappelallee 3,
D-69469 Weinheim, Germany

Jacaranda Wiley Ltd, 33 Park Road, Milton,
Queensland 4064, Australia

John Wiley & Sons (Asia) Pte Ltd, 2 Clementi Loop #02-01,
Jin Xing Distripark, Singapore 129809

John Wiley & Sons (Canada) Ltd, 22 Worcester Road,
Rexdale, Ontario M9W 1L1, Canada

Library of Congress Cataloging-in-Publication Data

Coumarins : biology, applications, and mode of action / edited by
R. O'Kennedy and R.D. Thornes.
p. cm.
Includes bibliographical references and index.
ISBN (invalid) 0-421-95997-4 (alk. paper)
1. Coumarins—Therapeutic use. 2. Coumarins—Physiological effect. I. O'Kennedy, R. (Richard) II. Thornes, R. D. (Robert Douglas)
[DNLM: 1. Coumarins. QV 193C855 1997]
RM666.C76C68 1997
615′.7—dc20
DNLM/DLC
for Library of Congress 96–31627
CIP

British Library Cataloguing in Publication Data

A catalogue record for this book is available from the British Library

ISBN 0-471-96997-4

Typeset in 10/11½pt Times from the author's disks by Techset Composition Ltd, Salisbury
Printed and bound by CPI Antony Rowe, Eastbourne
This book is printed on acid-free paper responsibly manufactured from sustainable forestation, for which at least two trees are planted for each one used for paper production.

Contents

Contributors

Declan P. Bogan	School of Biological Sciences, Dublin City University, Dublin 9, Ireland.
John R. Casley-Smith	The Henry Thomas Laboratory (Microcirculation Research), University of Adelaide, Adelaide, SA 5005, Australia.
Judith R. Casley-Smith	The Henry Thomas Laboratory (Microcirculation Research), University of Adelaide, Adelaide, SA 5005, Australia.
Deirdre Cooke	School of Biological Sciences, Dublin City University, Dublin 9, Ireland.
Ciaran F. Duffy	School of Biological Sciences, Dublin City University, Dublin 9, Ireland.
Scot W. Ebbinghaus	Division of Hematology/Oncology, University of Alabama at Birmingham, Alabama, USA.
Denise Egan	School of Science, Regional Technical College, Tallaght, Co. Dublin, Ireland.
Brian Fitzpatrick	School of Biological Sciences, Dublin City University, Dublin 9, Ireland.
John A. Freeman	Division of Urology and the University of North Carolina Lineberger Comprehensive Cancer Center, Chapel Hill, NC 27599-7235, USA.
Gary J. Keating	School of Biological Sciences, Dublin City University, Dublin 9, Ireland.
Matti A. Lang	International Agency for Research on Cancer, 150 Cours Albert Thomas, F-69372 Lyon Cedex 08, France.

Teresa McCormack	School of Biological Sciences, Dublin City University, Dublin 9, Ireland.
M. Ernest Marshall	Division of Medical Oncology, University of Texas Medical Branch, Galveston, Texas, USA.
James L. Mohler	Division of Urology and the University of North Carolina Lineberger Comprehensive Cancer Center, Chapel Hill, NC 27599-7235, USA.
Richard O'Kennedy	School of Biological Sciences, Dublin City University, Dublin 9, Ireland.
Markku Pasanen	Department of Pharmacology and Toxicology, University of Oulu, FIN-90220 Oulu, Finland.
Olavi Pelkonen	Department of Pharmacology and Toxicology, University of Oulu, FIN-90220 Oulu, Finland.
Neil B. Piller	Department of Public Health, School of Medicine, Flinders University of South Australia, GPO Box 2100, Adelaide, SA 5001, Australia.
Hannu Raunio	Department of Pharmacology and Toxicology, University of Oulu, FIN-90220 Oulu, Finland.
Arja Rautio	Department of Pharmacology and Toxicology, University of Oulu, FIN-90220 Oulu, Finland.
Heiko Reinartz	School of Biological Sciences, Dublin City University, Dublin 9, Ireland.
Barbara Seliger	University Hospital Mainz, III: Medical Clinic, Department of Hematology/Oncology, Langenbecksstr. 1, 55101 Mainz, Germany.
Malcolm R. Smyth	School of Chemical Sciences, Dublin City University, Dublin 9, Ireland.
R. Douglas Thornes	Colman K. Byrnes Research Centre, Beaumont Hospital, Dublin 9, Ireland.
Inge Weinmann	Heilwigstraße 160, 2000 Hamburg 20, Germany.
Bernadette T. Williams	Division of Urology and the University of North Carolina Lineberger Comprehensive Cancer Center, Chapel Hill, NC 27599-7235, USA.
Kurt S. Zänker	Institute of Immunology, University Witten/Herdecke, Stockumerstr. 10, 58448 Witten, Germany.
Gerhard J. Zlabinger	Institute of Immunology, University of Vienna, Borschkegasse 8a, A-1090 Vienna, Austria.

Preface

The greatest need in clinical medicine today is for the long-term control of cellular immunity. Coumarin, a simple molecule, and most of its derivatives, have been known for more than a century but their vital role in plant and animal biology has not been fully exploited. Coumarins prevent disease, modulate growth and maturation and defence systems and have anti-oxidant properties.

Murray, Mendez and Brown in their book *The Natural Coumarins* (John Wiley, 1982) detailed their occurrence, chemistry and biology. This present book concentrates on the biological and clinical actions of coumarin, 7-hydroxycoumarin and, to a lesser extent, 4-hydroxycoumarin.

John Wiley & Sons Ltd, the publishers, commissioned this book to highlight the potential of coumarins in medicine and to detail methods of laboratory assessment in an attempt to reveal their modes of action. Without their support this book would not have been possible.

Our thanks are due to all the contributors and to Barbara Drew and Jim Murray for their assistance.

The enthusiasm and interest of Mr Hans-Henning Schaper motivated many of the contributors. His sudden death in 1991 left a great void. This book is dedicated to his memory.

Richard O'Kennedy
R. Douglas Thornes

CHAPTER 1

History of the Development and Applications of Coumarin and Coumarin-related Compounds

Inge Weinmann
Heilwigstraße 160, 2000 Hamburg 20, Germany

1.1 THE CHEMOTAXONOMY OF COUMARIN IN PLANTS

In 1967 Duquenois[1] drew up a 'family tree' of the plant families and species thought to contain coumarin and coumarin derivatives as secondary plant products. A simplified version is shown in Figure 1.1, depicting just the families in which coumarin occurs free or in glycosidic form. More recent findings have confirmed the described distribution of coumarin and its derivatives throughout the plant kingdom[2–4]. Hegnauer's work on the chemotaxonomy of plants[4], now extending to 11 volumes, covers the distribution of natural compounds classified according to family, species and genus following a sophisticated schedule. This is the source of the most reliable and comprehensive information on the occurrence of coumarins. To date, naturally occurring coumarins have been isolated from over 800 species of plants and microorganisms, and more than 1000 coumarin derivatives have been described[5].

1.2 OCCURRENCE AND FUNCTION OF COUMARINS IN PLANTS

Coumarins are found free or as heterosides in many dicotyledonous families, including the Apiaceae (*Amni majus*), Asteraceae (*Trilisa odoratissima*), Fabiaceae (*Meli-*

Coumarins: Biology, Applications and Mode of Action. Edited by R. O'Kennedy and R.D. Thornes.

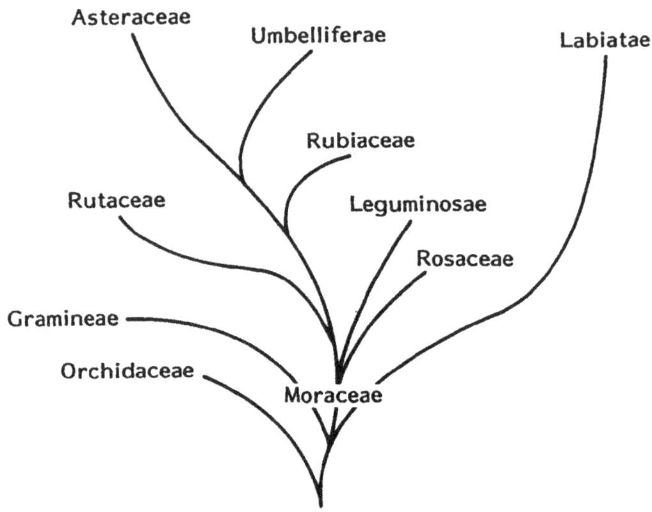

Figure 1.1. Plant families with coumarin-containing species.

lotus officinalis), Rosaceae (*Prunus mahaleb* L., known as coumarin plant), Rubiaceae (*Asperula odorata*), Rutaceae (*Ruta graveolens*), Solanaceae (*Atropa belladonna*) etc. (Species with high coumarin content are shown in brackets.) Many monocotyledonous plants, especially the Gramineae and orchids, also contain large amounts of coumarins. The compounds are found in various parts of plants including the trunk, stem, foliage, flowers, fruits and seeds and can be isolated by extraction with suitable solvents. Coumarin and dihydrocoumarin are frequently isolated by steam distillation[6].

The distribution of biologically active coumarins in a wide range of plants seems to correlate with their ability to inhibit the growth and sporulation of fungal plant pathogens and other ailments. Accumulation in grasses and orchids prevents infection (*Antoxanthum odoratum*, *Orchis militans*), while *Melilotus infesta* is avoided by certain beetles because of the presence of high concentrations of coumarin and its derivatives.

Coumarin and coumarin derivatives are commonly found in grass, corn and forage plants and appear to act as repellents against beetles and other terrestrial invertebrates. They are also formed as a response to traumatic injury, during the wilting process, by plant diseases or through drying. The enhanced production of phenylalanine metabolites is directly involved in hypersensitive resistance of plants to viruses[7].

1.3 BIOSYNTHESIS OF COUMARIN

The biosynthesis of the coumarin ring system[8] reveals, as a key step, the formation of the heterocyclic ring from an *o*-hydroxylated cinnamic acid. The latter is generated in the shikimic–phenylalanine pathway, which appears to be the common intermediate in the biosynthesis of coumarins in higher plants, as the polyketide pathway is the preferred synthesis in microorganisms. The hydroxycinnamate pathway is described as the metabolic sequence leading from the aromatic amino acid L-phenylalanine to various ring-substituted cinnamic acids, which are characterised by a free phenolic hydroxyl group at the *p*-position. In the biosynthesis of the numerous 7-hydroxylated coumarins the first step is the conversion of *trans*-cinnamic acid into *trans-p*-coumaric acid. 7-Hydroxycoumarin is generated from the *p*-coumaric acid[9]. It is noteworthy that so many biologically active coumarins possess a 7-hydroxycoumarin moiety. (See Chapter 2.)

By administering cinnamic acid to *Melilotus alba* cultures coumarin is formed by the action of cinnamic acid *cis-trans* isomerase, present in higher plants[2]. The route starts with *o*-coumaric acid leading to coumaric acid glucoside, which can be hydrolysed to *cis-o*-coumaric acid, which in turn cyclises spontaneously to coumarin. The cinnamic acids produced in this pathway represent important intermediates for the biosynthesis of a wide variety of phenolic natural products not only as the end-products of the general phenylpropanoid metabolism, but also as the most central intermediates of the entire metabolism in plants[10]. Mead *et al.*[11] proved in the 1950s that coumarin derivatives were trace metabolites of phenylalanine. Van Sumere *et al.*[12] detected traces of umbelliferone, aesculetin, ferulic acid, caffeic acid, *p*-hydroxyphenylacetic acid and kynurenine in normal human urine. Pathological urines (cancer and leukaemia patients), treated with prednisone, showed a surprisingly high content of umbelliferone and/or aesculetin. Six- to sevenfold increases above normal levels were found in patients suffering from phenylketonuria[13].

Thiede[14] improved the sensitivity of the test method for coumarin in blood and plasma to the picogram range; he found a regular endogenous coumarin level (about 2 ng/ml) in spite of a strictly coumarin-free diet during the study. Indeed, limited current biosynthesis of coumarin via the phenylalanine pathway is not impossible.

1.4 MEDICINAL PLANTS AND OFFICINAL HERBS CONTAINING COUMARIN

The *Dispensatory of the United States of America* Part I, 25th edn (1955), states under 'Uses of coumarin': '**Coumarin** has no recognized medicinal usage, having been employed only for its odorous qualities'. (One part of coumarin is approximately equivalent to three parts vanillin.) In Part II: page 1750, a note on **Melilotus** states that it 'was formerly employed as a supposed analgesic in neuralgias; there is no evidence however, of its value'.

In other pharmacopoeias the herbal drug has been officinal in the past: for example, in *DAB 6*—the German Pharmacopoeia 1926, 6th edn (reprint 1951:

334); the Polish pharmacopoeia (*Pharmacopoeia Polonica*, Editio III [1954]: 307–308 and *Farmacopea Polska* IV, Editio IV, 2: 278–279).

In 1990 the German Federal Institute of Drugs and Medicinal Products (BfArM) published a monograph on *Meliloti Herba*[15], in which the use of melilot is indicated for symptoms and signs in chronic venous insufficiency like pains and heaviness in the legs, calf cramps, itching and swelling; adjuvant treatment of thrombophlebitis, of the post-thrombotic syndrome, of haemorrhoids and lymphostasis. Further recommendations exist for the topical administration including dose references related to coumarin. Pharmacodynamic effects are described as anti-oedematous in inflammatory and stasis oedema by increase of venous reflux and increase of lymphokinesis. These particulars are adopted by Wagner and Wiesnauer[16] in a manual of phytotherapy. The monograph *Meliloti herba* (Melilot) in the new publication by Wichtl[17], *Herbal Drugs and Phytopharmaceuticals*, belongs to a very remarkable collection on herbal medicine.

A description of the use of coumarin drugs in empirical and traditional medicine is included in Madaus[18]. Melilotus is described as a remedy for wounds, inflammation, insomnia and pain in general. In the late 1930s Merck developed a mild hypnotic based on coumarin-3-carboxylic acid that had sedative effects in low doses and hypnotic properties in elevated doses[19]. In experimental animals high doses of coumarin in pharmacologic–toxicological trials produced a strong, but short-lasting sedative effect. Ritschel and Hardt[20] confirmed in the gerbil parallel effects of coumarin and blood levels in the brain. Here, it was also shown that coumarin passes through biological membranes and barriers. The less lipophilic 7-hydroxycoumarin did not enter the brain.

Hagers Handbook of Pharmaceutical Practice[21], published in 1938, gives a comprehensive botanical and pharmaceutical characterisation of the coumarin. It cites its external use as an emollient or as a herbal pad. In a supplement (1944), an essence containing 33% coumarin made from fresh leaves and flowers is described. In the *CRC Handbook of Medicinal Herbs*[22], yellow sweet clover is said to have astringent, anti-spasmodic, diuretic, emollient and many more effects. It was used in asthma cigarettes and as a decoction in lotions and enemas. The medical uses correspond mainly to those recommended nowadays. Weiss[23] includes *Melilotus officinalis* in the group of plants used for the treatment of venous disorders and varicose veins: ‘Melilotus is a lympho-venous angiolyticum’.

1.5 BIOPHARMACEUTICAL PROPERTIES

The chemical properties of coumarin and coumarin-related compounds are described in Chapter 2. Further details may also be obtained from Tietze[24].

The introduction of coumarin for clinical use required testing of its biopharmaceutical properties[25]. In respect to coumarin and 7-hydroxycoumarin their low solubility in water warrants caution regarding bioavailability upon oral administration.

The relatively high apparent partition coefficient is favourable for fast absorption by passive diffusion, once the drugs are in aqueous solution. Pharmacokinetic data in humans[26–28] indicate a widespread distribution within the body. Intestinal and renal absorption are not influenced by varying gastrointestinal or urinary pH because coumarin is non-charged, and 7-hydroxycoumarin has a pK_a of 7.77. The extent of protein binding is well below the critical value of > 80% (being 35% and 47% for coumarin and 7-hydroxycoumarin, respectively). Both substances are bound to erythrocytes *in vitro* (27% and 21%, respectively). There is a pronounced haemorheological effect on erythrocytes (i.e. a decrease in aggregation and increase in flexibility), which is regarded as responsible for the positive influence of coumarin and 7-hydroxycoumarin on microcirculation and capillary endothelium function[29]. Hence, its value for use in venous disorders.

Coumarins are active in plant metabolism, taking part in growth regulation. They antagonise the auxines through their ability to inhibit seed germination and root growth[30,31]. The powerful UV absorption of umbelliferone and derivatives protects young plants from sun radiation[32].

It is possible to draw inferences from knowledge of the biological activity of coumarins in plants as to their effects in animal and human organisms regarding, for example, bacteriostatic, virostatic and antitumour effects[33]. The natural coumarins, like other unsaturated lactones, may exert various effects on living organisms. The range of physiological effects, both in plants and in animals, is indeed extensive[30,34–36]. Table 1.1, based on data collected from Soine[34], lists a whole series of observed pharmacological activities.

Central effects, for example sedative, hypnotic and antipyretic effects, have also been described[37–39]. Kovach *et al.* showed a marked effect on the heart and circulation, as well as an increase in capillary blood flow in organs, and a favourable effect on myocardial ischemia in the dog[40]. Wound-healing was also favourably influenced by coumarin in dogs[43]. Of general interest is a dose-dependent effect on the vessel wall. The effect was named as adrenolytic and spasmolytic[41], but Laven[42] interprets it as a musculotropic action of the peripheral nervous system. In the rabbit peripheral effects of coumarin have been described with an indirect analgesic produced by the abolition of oedema and inflammation[42–44]. Oedema paw models and peritonitis models in rats have also been used and demonstrated the effects of coumarin[45–48].

Koh in 1979, as part of his MD thesis[49], compared coumarin action with steroidal and non-steroidal drugs in mice, rats and rabbits in models of inflammation. The results varied and were species-dependent.

It is also interesting to note that coumarin and 7-hydroxycoumarin inhibit the biosynthesis of prostaglandins and leucotrienes in comparable order of magnitude to acetylsalicylic acid[50].

The most important effect, in particular in respect to the underlying pathological condition, is the anti-oedematous activity, closely followed by endothelial protection, lymphokinetic and haemorheological properties. These are described in detail in Chapters 7 and 8. The reader is also referred to work by Földi[51], Casley-Smith[52,53], Piller[54] and Knight[55].

Table 1.1. Some pharmacological and physiological properties of coumarins.

Pharmacological activities of coumarin	References
Plant growth regulation	31,32,158
Cytological properties	34,158
Spasmolytic action	32,37
Narcotic effect	37
Sedative and hypnotic effect	19,20
Respiratory stimulation	34,37
Uricosuric action	35
Bacteriostatic, tuberculostatic, antifungal action	30,32,33,100
Anthelminthic action	158
Piscidal action	158
Analgesic action	160,161
Antiphlogistic effect	45,49,50
Antipyretic effect	37–39
Anti-oedematous effect	48,52,55,118
Vasodilator effect	34,40,42
Choleretic action	34,162
Diuretic action	159,163
Hypothermal action	34
Oestrogenic action	34
Endothel protection	164,165

1.6 TOXICOLOGY

The toxicological profile of coumarin was described as enigmatic in the past[56], but now the species-dependent metabolism, hepatotoxicity and genotoxicity in causal connection with single metabolites and intermediates have been explored.

Conventional toxicity studies with coumarin have been summarised recently in a Technical Report of the National Toxicology Program[5]. Coumarin was nominated by the US Food and Drug Administration and the National Cancer Institute for toxicity and carcinogenicity studies because 'it is widely distributed in perfumes, soaps, and other common household products; because of continued interest in using coumarin products as flavor-enhancing agents for foods, and because complete comparative toxicity and carcinogenicity studies in rats and mice had not been performed'. Toxicity studies (GLP-conform) with coumarin show that it is a very well-investigated compound[57–64]. In rodents, chronic lesions and tumorigenesis might be seen after overdosing of the compound for months to years. An expert opinion to the NTP Report states, for instance: 'Organ-specific toxicity occurs in species and strains only, that metabolize coumarin qualitatively and quantitatively different from man. The tumour-inducing effects of coumarin have to be seen—in sensitive species—as chronic organ damage, implicating carcinoma as sequelae of unspecific lesions because of a considerable exceeding of the MTD-range' and 'Animal strains revealing high spontaneous tumour rates do not reflect the situation in man'[65]. (The NTP studies showed the well-known reactions of Fisher rats 344 and $B6C3F_1$ mice, currently used in the NTP program, under the conditions of two-year gavage studies. There was some evidence of carcinogenic activity in male rats but no evidence of carcinogenicity in female rats receiving 25, 50, or 100 mg/kg. There was some evidence of carcinogenic activity in male mice and clear evidence of carcinogenic activity in female mice. Genotoxic effects were rare; coumarin induced gene mutations in *Salmonella typhimurium* strain TA 100 in the presence of S9 activation; no other mutagenic activity was noted[66–69].) 7-Hydroxycoumarin did not induce gene mutations in *Salmonella typhimurium* strains TA 98, TA 100, TA 1537, with and without S9, at doses up to 2 mg/ml, but it was reported to induce mutations in *Klebsiella pneumoniae* without metabolic activation[70]. Chromosomal effects were seen in Chinese hamster ovary cells[71].

Moreover coumarin can cause chromosome and DNA breakage in plant cells[30], but also exhibits antimutagenic effects against genotoxicity generated by 4-nitroquinoline-1-oxide or UV radiation in *E. coli*[72], and inhibits the induction of sister chromatid exchange in ovary cells (Chinese hamster) by mitomycin[73].

The induction of detoxifying and the inhibition of toxifying enzymes by coumarin and 7-hydroxycoumarin has been shown in further animal models *in vivo*[74,75]. These include an increase in NAD(P)H : chinonreductase and activity enhancement of glutathione-*S*-transferase[76]. Coumarin was shown to induce detoxifying CYP450 isoenzymes so that the benzopyrene 7,8-epoxide, a precursor of the ultimate carcinogen in the rat liver, could not develop[77].

The seeming ambivalence of the toxic effects of coumarin is connected with a pronounced species-dependent metabolism (see table 2 from ref. 78). The procedure of extrapolation from animal studies to the human situation, therefore, requires extraordinary care and precision[79]. The introduction of alternative procedures, using for instance liver preparations (cryopreserved hepatocytes, liver microsomes or 'precision-cut liver slices' techniques) has shown promising results. The direct comparison of the biotransformation of test substances in liver preparations from conventional test species and other animals might allow us to choose the species which metabolises compounds most similarly to humans. Specimens of human liver have been tested and compared with preparations of monkey, rat, mouse, hamster, guinea-pig and other mammals[80–85]. The great differences in the patterns of coumarin metabolites in humans, primates and several mice strains and gerbils on the one hand, and rodents, hamster, guinea-pig, rabbit and dog on the other hand clearly indicate the polymorphism of the metabolising enzymes' activity[86].

Discussion about a potential risk of coumarin for use in humans resulting from the generation of 3-hydroxycoumarin via 3,4-epoxide, presumes a biotransformation as in the rat. It is likely that the normal first step is the hydroxylation in C7 and C3 with succeeding conjugation. By opening the α-pyrone ring moiety the cascade of *o*-hydroxyphenylcarbonic compounds can derive from 3-hydroxycoumarin, which is excreted in the urine, partly as a 3-hydroxycoumarin conjugate.

1.7 ANTICARCINOGENIC EFFECTS

Feuer and Kellen[87] and Wattenberg[88] showed that coumarin had anticarcinogenic effects; Wattenberg[89] categorised coumarin and 7-hydroxycoumarin as chemopreventive or blocking agents. The ability of coumarin to prevent cells from entering mitosis and to arrest cells in the G_0/G_1 phase, so that less cells can reach the phase of synthesis (S phase) of the cell cycle, was shown, in plants, by Scheel[33] and in rats by Tseng *et al.*[90]. In various tumour models of test animals (prostate cancer of the rat, Maucher *et al.*[91], or in the model of transgenic mice, bearing the oncogene MMTV-H-*ras* and especially relevant in respect to human tumour pathology[92]) coumarin inhibits the malignant transformations. It is very likely the inhibition takes place via the DNA-sited poly(ADP-ribose) polymerase, leading to rapid normalisation of cell morphology[93].

1.8 HUMAN EXPOSURE IN RELATION TO COUMARINS

Egan *et al.*[36] gave a good review of the history of coumarins including a time table that begins with the first isolation of the compound from Tonka beans (*Dipterex odoratus* L) in 1822, to its manifold use in medicine, industry and laboratory applications. Natural sources never played a role in the isolation of coumarin, because the synthesis is simple and inexpensive. The first synthetic procedure was published in 1868 by Perkins. The industrial use was rather intensive as a flavouring substance and fixative in perfumery, in the cosmetic and toilet industry, in general, and as essential ingredients (light-protecting 7-hydroxycoumarin derivatives) in sun oil.

In 1953 a report of a private pharmacological testing laboratory stated that coumarin was toxic in animals, and this resulted in the voluntary withdrawal of coumarin from the market and, also, a federal prohibition of its use as a compound in food flavours, notably in chocolate products. In a statement appearing in the US Federal Register of 1954 under the Federal Food, Drug and Cosmetic Act, foods containing coumarin are legally regarded as adulterated[94]. The UK authorities recommended its withdrawal in 1965 (Food Standards Committee [1965])[95]. Germany maintains the so-called 'Höchstmengenverordnung' (maximum permissive concentration), published [1988][96] in the Amtsblatt der Europäischen Gemeinschaften: 'Coumarin is allowed in food 2 mg/kg, in beverages 2 mg/kg, 10 mg/kg in selected caramel sweets, 50 mg/kg in chewing gum and 10 mg/kg in alcoholic beverages. It is not allowed to be added, as such, to foods or flavours. It is allowed, if either naturally occurring in foods or flavours, which are obtained from natural sources'.

The 'opinion on coumarin' (a constituent of natural flavouring materials limited by Annex II of Flavourings Directive 881 3881 EEC) was expressed on 16 December 1994 at the 95th meeting of the Scientific Committee for Food in Brussels, and the text will appear in the 36th Series of Reports of the Scientific Committee for Food.

Coumarin has been used since the 1900s, and its use in fragrances in the USA amounts to about 250 000 lb a year[97]. In a survey conducted from 1981 to 1983, the National Institute of Occupational Safety and Health estimated that approximately 100 000 workers might come in contact with coumarin[95]. It should be borne in mind that coumarin is absorbed by the skin far more easily than it is taken up after oral intake, and that blood levels are maintained for several hours following topical administration[28,98,99].

Typical concentrations of coumarin found in common household products range from 0.03 to 0.2% in soaps, 0.003 to 0.02% in detergents, 0.015 to 0.1% in creams and lotions, and 0.3 to 0.8% in perfumes.

1.9 GENERAL CLINICAL APPLICATIONS OF COUMARINS

The interest in coumarin as an investigator drug is still valid, though the parent molecule has been submitted to a thorough derivatisation and screening of the derivatives between 1950 and 1970. Except for the group of anticoagulants derived from 4-hydroxycoumarin, other coumarin derivatives such as the antibiotic novobiocin (complex coumarin from two *Streptomyces* species) or hymecromen, a spasmolytic agent for bilious affections and complaints, did not attain clinical use or a true medical importance, respectively. The furocoumarins, however, are playing a role in the treatment of psoriasis and other dermatological diseases (PUVA—psoralen ultraviolet A range therapy).

Coumarin and 7-hydroxycoumarin in cell cultures exert dose-dependent cytostasis and cytotoxicity which could not be aligned with *in vivo* effects in the same order of magnitude. Substance screening for fungicidal, virostatic, bacteriostatic activity is recorded[100], and experiments with tumour cell lines have been carried out[101–107].

Coumarin and 7-hydroxycoumarin have been patented[108,109] in Germany and in Australia and Europe[110]. Patents are for the scavenging of free radicals and the resolution of high-protein oedemas.

Paya *et al.*[111] tested 16 plant-derived or synthetic coumarins for their ability to inhibit lipid peroxidation and to scavenge free oxygen radicals. Free radicals generate endothelial lesions of the cell membrane and spasms of blood vessels and lymphatics[112,113]. They also provoke the proliferation of fibroblasts[114]. Coumarin and 7-hydroxycoumarin are active scavengers of the OH′ radical. The inhibition and reduction of those processes are the basis of the defence and protection of the terminal vascular bed[111,115,116].

The effect and efficacy in the resolution of high-protein oedemas in animals and humans is the main indication of oral and topical coumarin monoproducts and Venalot® depot (a combination of 15 mg coumarin and 90 mg troxerutin in one coated tablet) (see Chapter 7). Of medical interest are the anti-oedematous and lymphokinetic effects of coumarin and 7-hydroxycoumarin as well as the additive effect of oxyrutins[55,117–121]. The indication of 'lymphoedemas of different origins' is registered in Switzerland; in Switzerland and Germany the clinical administration is approved for chronic venous insufficiency stage I, II (venolymphous disorders with tension, heaviness, itching and cramps), the reduction or hindering of fibrotic degeneration, and the treatment of venous ulcers. 'Complex physical therapy' as first-line therapy of the impaired macrocirculation of the affected veins has to be supplied by phlebological pharmacotherapy to treat the languished microcirculation; combined physical and pharmacotherapy has proven to be beneficial.

This therapeutic use goes back to the experience of practitioners and clinicians over 40 years (including the early formulations of coumarin standardised herbal medicine, such as melilotus products). Controlled clinical trials and open controlled studies and field tests are available in more than 300 studies between 1970 and today[52,122].

The lymphoedemas of different origins, in particular the long-term therapy of post-mastectomy oedema or filaritic oedema and elephantiasis, form a most important and extended field of application for coumarin[123–129]. A further indication is wound-healing, including post-operative and post-traumatic oedema[130–134].

A new field of growing interest is the adjuvant use of coumarin and 7-hydroxycoumarin in oncology[135]. The inhibition of mitosis in plants stimulated first investigations of a possible antitumour effect[31] (see Chapter 11). The interest in the coumarins in oncology therapy was probably sparked off by observations that warfarin and other coumarin anticoagulants initiated fibrinolysis in the surroundings of tumours and stimulated the early infiltration of macrophages of tumours. In the 1960s and 1970s in Ireland[136] and in Germany[137] systematic research, based on the contingency of clotting, thrombosis and cancer, tried to inhibit the process of the spread of carcinoma. But—as Hilgard put it—'Despite an attractive and experimentally well-documented hypothesis, warfarin therapy of human malignant diseases remains of unproven benefit for the patient.'

Studies on blocking tumour dissemination by modifying coagulation were successful in lung cancer but without benefit in colorectal cancer and other malignancies[138].

Promising findings in more than 700 patients suffering from renal cell carcinoma have been available for at least eight years. Cancers of the prostate were treated after primary therapy by tumour resection and/or nephrectomy with varying doses of coumarin or 7-hydroxycoumarin (sustained release tablets) over different periods of time or until regression of the disease respectively. Malignant melanomas results were also treated. This work is described in detail in Chapters 9, 10 and 11.

The International Society of Coumarin Investigators (ISCI) files coumarin(s) in the group of biological response modifiers with influence on the immune system (immunomodulators). Zlabinger *et al.*[139] tried to define the coumarin effect on the immune system, and have deduced from their results of binding studies that binding of coumarin is characterised by high-capacity (ubiquitous intracellular receptor proteins) but low affinity conditions. The lack of knowledge about the 'cascade of interference' with the immune system requires experiments on the intracellular reaction of coumarin with tumour cells[140,141].

1.10 PHARMACOKINETICS

The bioavailability of coumarin and its metabolites has been measured following oral and topical administration[26–28,98,142,143]. Coumarin is characterised as a substance with low oral availability (0 to < 10%) because of a pronounced first-pass effect[142]. The oral bioavailability has to be extended to that of the 'bioactive principle', i.e. coumarin plus 7-hydroxycoumarin. The metabolite has almost identical AUC values after intravenous and oral application of coumarin (AUC after oral administration runs to 89% of the AUC after i.v. administration). The free 7-hydroxycoumarin is detected in very small amounts together with its conjugates in blood and urine.

The conjugates might give rise to free 7-hydroxycoumarin by a recycling process (enterohepatic circulation). While in animals the excretion via faeces has been determined (see below), no analyses of unchanged coumarin or metabolites are available in humans. The gut flora (bacterial enzymes) are likely to be involved. Tests with a radiolabelled compound in the baboon[144], mice[145] and rats[146] and with 7-hydroxycoumarin in guinea-pigs[147] prove the distribution throughout the body to nearly all organs and tissues.

In humans, 7-hydroxycoumarin is excreted in conjugated form, almost completely as glucuronide in the urine. The excreted amount of 7-hydroxycoumarin reflects—at a given time—the hepatic level of CYP2A6 (coumarin-7-hydroxylase). The equipment of liver microsomes with enzyme activity is individually established by genetic mutation. High and poor, fast and slow metabolisers have been detected in several populations (Table 1.2)[148–151] (see Chapter 3).

With many drugs and their active phase I metabolites, conjugation with glucuronic acid causes an accelerated hydrophilisation or detoxification, respectively. Selected glucuronides contribute to the pharmacological effect of the parent molecule or even cause it[152]. The resulting effect depends on the affinity to single macromolecules as well as the concentration–time profile of the biologically active substance. 7-Hydroxycoumarin-glucuronide was not active over a wide dose range in tests with tumour cells[153,154], while coumarin and 7-hydroxycoumarin clearly demonstrated 'direct antiproliferative effects' in seven cell lines established in the Institute of Pathology, University of Düsseldorf. The original tumours were typical representatives of the three different types of renal cell carcinoma. An *in vitro* growth inhibition test was performed, in which the tumour cells were exposed to coumarin, 7-hydroxycoumarin and 7-hydroxycoumarin-glucuronide at concentrations of 2, 20 and 200 μg/ml. The growth inhibitory effects were exclusively observed at concentrations of 200 μg/ml for coumarin and its main metabolite. Responsiveness to both was observed in all histologic types of human renal cell carcinoma, the extent of

Table 1.2. Coumarin metabolism of different species.

Species	Urinary excretion as % of administered coumarin			References
	7-Hydroxycoumarin	Other metabolites	Unchanged coumarin	
Human	94–98.6	4: *o*-HPLA	3.4	142
	68–92	1-6: *o*-HPAA		166
	63.4			167
	ave. 50			149
	ave. 86			151,168
	36–40	0.002–0.13: 3-hydroxycoumarin, 4-6: *o*-HPAA	1.3	143,169
Baboon	60			170
Marmoset	0.5	0.7: 3-hydroxycoumarin; 61.1: *o*-HPAA	1.8	171
Dog	3			170
Beagle dog	6–12		45	78
Cat	19			170
Rabbit	10–16	> 4 to > 7: 3-, 4-, 5-, 6- and 8-hydroxycoumarin. 20.7–25.6: *o*-HPLA and *o*-HPAA	12.3–16.7	172
Squirrel monkey	1			170
Guinea-pig	1			170
Hamster	5; traces (Syrian hamster)			170,175
Rat	0.3–0.5	2, 3–2, 8: 3-, 4-, 6-, 8-hydroxycoumarin. 13.1–28.1: *o*-HPLA and *o*-HPAA; up to 44.7	3.1–7.4	172,175
Mouse	3			170,173
Various strains	0.9–7.5			
DBA/2/lac mouse	11.3–12.5			170
Pig	12			170
Chicken	84	8: *o*-HPLA and HPAA; unidentified metabolites		174

o-HPLA, *o*-hydroxyphenyl lactic acid; *o*-HPAA, *o*-hydroxyphenyl acetic acid.

response differed from cell line to cell line. 7-Hydroxycoumarin was more effective than coumarin and significantly ($P < 0.05$) inhibited the proliferation in all cell lines of the study. Siegers[154] used human colon carcinoma cell lines (Caco-2), human hepatoma cells (HepG2) and acute lymphatic leukaemia cells (CCRF CEM) to measure the antiproliferative effect of coumarin and 7-hydroxycoumarin when used at the therapeutic blood levels achieved in oncological studies. In human colon carcinoma cell in culture and in hepatoma cell lines the inhibitory concentrations are in the microgram or even the milligram range (400–500 μg/ml Caco-2, 162–312 μg/ml for HepG2). Such concentrations of free coumarin or 7-hydroxycoumarin could, indeed, never be detected in patients[28,143,155].

Schäfer[155] records values of 365 ng/ml (determination in the serum) after a single dose of 1200 mg, 803 ng/ml after 1600 mg, 4.1 μg/ml after oral application of 4–4.5 g, 26.3 μg/ml after 7 g coumarin. The concentrations of the phase I metabolite varied between about 50 ng/ml and 2000 ng/ml over the entire range of dosages from 400 mg to 7 g[156], of the phase II metabolite between about 8 μg/ml and 255 μg/ml. Sharifi[143] could determine 100–2400 ng/ml coumarin after oral application of 2 g coumarin. Frohlich concluded from the AUC values following the peroral administration of 100 μg coumarin that 5.78% of the dose was recovered in the blood plasma. The maximal concentration—3.4 μg/ml—was measured 3.8 h after dosing. To achieve cytostatic effects *in vivo* the interaction of both substances would be needed and is indeed probable through similar kinetic behaviour of 7-hydroxycoumarin and its pro-drug character[36]. However, it is not clear how the effective drug concentration can be reached, because the blood concentration of both substances after therapeutic doses is only in the nanogram range. It is likely, on the other hand, that 7-hydroxycoumarin can be set free from the considerable amounts of 7-hydroxycoumarin conjugate and coumarin from deep tissue compartments.

To estimate a possible hepatopathogenic risk of coumarin, the side-effects were compiled from clinical studies[156,157,158], and all notifications that were made to the WHO and the National Institutes of Health. The data have been analysed and evaluated according to the ADR categorisation of Karck and Lasagna[159].

Based on the number of patients treated with coumarin as a single substance between 1971 and 1996 for a duration of six weeks, the resulting prevalence of ADRs (adverse drug reactions) is 1:7000 (70:500 000). In no case of the registered adverse events could a causal relationship between the treatment with coumarin and the adverse event be categorised as 'definite' or 'certain'.

The prevalence of hepatopathogenic adversial events and adversial drug reactions, respectively, during or due to coumarin-containing formulations (Venalot® Depot) based on patient treatment figures (3 × 2 coated tablets/day for six weeks) between 1971 and 1996, the resulting prevalence of ARDs is 1:1.23 million (6:7.375 millions) (personal communication, I. Weinmann).

Coumarin metabolism in humans, almost exclusively involving hydroxylation through CYP2A6, does not lead to toxic metabolites and intermediates. The interindividual differences in specific enzyme activity discussed in Chapter 3 did not lead to hepatotoxic reactions.

1.11 CONCLUSIONS

The beneficial effects of the medicinal administration of coumarins at present appears to be due to coumarin itself, but the more interesting development will probably be the further exploration of coumarin derivatives and 7-hydroxycoumarin derivatives containing functional groups with known effects.

REFERENCES

1 Duquenois P. Coumarines et dérivés: Répartition dans le règne végétal et biosynthèse. *Pharm Weekbl [Sci]* 1967; **102**: 443–55.
2 Bell EA, Charlwood BV. *Secondary Plant Products*. Berlin: Julius Springer, 1980: 334–40.
3 Murray RDH, Mendez J, Brown SA. *The Natural Coumarins*. Chichester: John Wiley, 1982.
4 Hegnauer R. *Chemotaxonomie der Pflanzen*, vols 1–11A. Basel & Stuttgart: Birkhäuser, 1962–1994.
5 National Toxicology Program Technical Report on the Toxicology and Carcinogenesis of Coumarin in F344/N Rats and $B6C3F_1$ Mice (Gavage Studies). US Department of Health and Human Services, NIH Publication No. 92-3153, 1992.
6 Steinegger E, Hänsel R. *Pharmakognosie*. Berlin: Julius Springer, 1992: 376–83.
7 Fritig B, Hirth L, Ourisson G. Biosynthesis of the coumarins. *Phytochemistry* 1963; 9: 1963.
8 Bernfeld P. *Biogenesis of Natural Compounds*. Oxford: Pergamon Press, 1967: 788–97.
9 Nuth P. Naturstoffchemie. *Mikrobielle, pflanzliche und tierische Naturstoffe*. Stuttgart: Wissenschaftliche Verlagsanstalt, 1990: 523–30.
10 Swain T, Harborne JB, Van Sumere Ch F. *Biochemistry of Plant Phenolics*. New York: Plenum Press, 1979.
11 Mead JAR, Smith JN, Williams RT. Studies in detoxication 72. The metabolism of coumarin and of *o*-coumaric acid. *Biochem J* 1957; **68**: 67–74.
12 Van Sumere CF, Teuchy H, Massart L. Coumarins in normal and pathological human urines. *Clin Chim Acta* 1959; **4**: 590–3.
13 Hattori M, Onisawa J, Goto M. Increased excretion of umbelliferone in phenylketonuria. *J Biochem (Tokyo)* 1969; **65**: 655–7.
14 Thiede M. Abschlußbericht: Bestimmung von Cumarin und seinen Metaboliten in Plasma und Urin. Methodenentwicklung und Validierung. [Determination of coumarin and its metabolites in plasma and urine. Development and validation of analytical methods. Investigational report (unpublished)]. Berlin 1993: 1–30.
15 *Meliloti herba, Steinkleekraut*. Aufbereitungsmonographie Berlin: Bundesanzeiger, 1990.
16 Wagner H, Wiesnauer M. Phytotherapie. In: *Phytopharmaka und pflanzliche Homöopathika*. Stuttgart & Jena: Gustav Fischer, 1995.
17 Wichtl M. *Herbal Drugs and Phytopharmaceuticals*. Stuttgart: Wissenschaftliche Verlagsanstalt, 1994.
18 Madaus G. *Lehrbuch der biologischen Heilmittel*. Part 1: *Heilpflanzen*, vol. 2. Leipzig: Georg Thieme, 1938: 1862–5.
19 v. Werder F. Über Abkömmlinge der Cumarin-3-carbonsäure, eine neue Klasse synthetischer Heilmittel. *Merck's Jahresberichte Pharm* 1937: 88–101.

20 Ritschel WA, Hardt TJ. Pharmacokinetics of coumarin, 7-hydroxycoumarin and 7-hydroxycoumar-inglucuronide in the blood and brain of gerbils following intraperitoneal administration of coumarin. *Arzneim-Forsch/Drug Res* 1983; **33**: 1254–8.

21 *Hagers Handbuch der pharmazeutischen Praxis*. Berlin: Julius Springer, 1938: Vol. 2. Melilotus: 155–156; Supplement 1944: 792.

22 Duke JA. *CRC Handbook of Medicinal Herbs*. Boca Raton, FL: CRC Press, 1987: 305.

23 Weiss RF. *Lehrbuch der Phytotherapie*. Stuttgart: Hippokrates, 1991: 251ff.

24 Tietze LF. Expertise on the physical properties, chemical transformations, and stability of coumarin. Institut für Organische Chemie der Universität Göttingen, 1990: pp 1–7.

25 Ritschel WA, Grummich KW, Kaul S, Hardt TJ. Biopharmaceutical parameters of coumarin and 7-hydroxycoumarin. *Pharmaz Ind* 1981; **43**: 271–6.

26 Ritschel WA, Brady ME, Tan HSI, Hoffmann KA, Yiu IM, Grummich KW. Pharmacokinetics of coumarin and its 7-hydroxycoumarin-metabolites upon intravenous and peroral administration of coumarin in man. *Eur J Clin Pharmacol* 1977; **12**: 457–61.

27 Ritschel WA, Hoffmann KA. Pilot study on bioavailability of coumarin and 7-hydroxycoumarin upon peroral administration in a sustained-release dosage form. *J Clin Pharmacol* 1981; **21**: 294–300.

28 Frölich JC. Studie zur Untersuchung der absoluten Bioverfügbarkeit von oral verabreichtem sowie topisch appliziertem Cumarin bei gesunden Probanden. [On the bioavailability of coumarin following oral or topical administration of coumarin in healthy volunteers. Investigational report (unpublished)]. Med. Hochschule, Hannover: 1995; pp 1–244.

29 Radtke H, Kiesewetter H, Schneider R. Die rheologische Wirksamkeit von Cumarin und Troxerutin. *Therapiewoche* 1983; **33**: 6662–6.

30 Grigg GW. Genetic effects of coumarins. *Mutat Res* 1977/78; **47**: 161–81.

31 Riordan DJ, Daly I. The growth-inhibiting properties of lactones. *Ir J Med Sci* 1954; **6**: 157–61.

32 Chadzaj Jal. Die biologischen Eigenschaften und die pharmakologische Wirkung der Cumarine und Furocumarine. *Terpenoidy i kumariny*. Moscow & Leningrad: GV Pigulevskij, 1965: 25–30.

33 Scheel LD. The biological action of the coumarins. *Microbiol Toxins* 1972; **8**: 47–66.

34 Soine TO. Naturally occurring coumarins and related physiological activities. *J Pharm Sci* 1964; **53**: 231–64.

35 Nakajima S, Kawazu K. Coumarin and euponin, two inhibitors for insect development from leaves of *Eupatorium japonicum*. *Agric Biol Chem* 1980; **44**: 2893–9.

36 Egan D, O'Kennedy R, Moran E, Cox D, Prosser E, Thornes RD. The pharmacology, metabolism, analysis, and applications of coumarin and coumarin-related compounds. *Drug Metab Rev* 1990; **22**: 503–29.

37 Kitagawa H, Iwaki R. Coumarin derivatives for medicinal purposes. XVII Pharmacological studies on coumarin derivatives having biological activity. *J Pharm Soc Japan* 1963; **83**: 1124–8.

38 Ritschel WA, Alcorn GJ, Ritschel-Beurlin G. Antipyretic and pharmacokinetic evaluation of coumarin in the rabbit after endotoxin administration. *Methods Find Exp Clin Pharmacol* 1982; **4**: 407–11.

39 Ritschel WA, Alcorn GJ, Ritschel-Beurlin G. Therapeutic concentration of coumarin as antipyretic. *Methods Find Exp Clin Pharmacol* 1984; **6**: 363–5.

40 Kovach AGB, Hamar J, Dora Ö, Marton I *et al*. Die Wirkung von Cumarin aus *Melilotus officinalis* am Kreislauf des Hundes. *Arzneim-Forsch/Drug Res* 1970; **20**: 1630–33.

41 Wezler K, Schlüter F. Dehnungsversuche an isolierten kleinen Arterien des Pferdemesenteriums. Kaninchenohrpräparat nach Krawkow-Pissemski. *Ringelheimer Biologische Umschau* 1966; **21**: 383–91.

42 Laven R. der Einfluß von 5, 6-Benzo-a-pyron(Cumarin) auf den Tonus der glatten Gefäßmuskulatur. Berichtsband der 3. *Gemeinsamen Jahrestagung der Deutschen, Österreichischen und Schweizerischen Gesellschaften für Angiologie*. Bern: H. Huber, 1984: 296–7.

43 Hopf G, Kaeßmann HJ, Pekker I. Schäfer EA, Weber HG. Experimentelle Untersuchungen der Beeinflussung des postoperativen Extremitätenödems am Hunde mit einem Cumarinpräparat. *Arzneim-Forsch/Drug Res* 1971; **21**: 5.

44 Casley-Smith JR, Clodius L, Piller NB, Földi M. A model of lymphoedema in the rabbit's ear. *Chirurgia plastica* 1977; **4**: 5–14.

45 Földi-Börscök E, Bedall FK, Rahlfs VW. Die antiphlogistische und ödemhemmende Wirkung von Cumarin aus *Melilotus officinalis*. *Arzneim-Forsch Drug Res* 1971; **21**: 2025–30.

46 Földi-Börscök E. Wirkung von Benzopyron auf die durch Rinderserum-Albumin herbeigeführte exsudative Peritonitis bei der Ratte. *Arzneim-Forsch/Drug Res* 1977; **27**: 378–9.

47 Uhlig G. Antiphlogistische Wirksamkeit verschiedener Pharmaka Vergleichende tierexperimentelle Untersuchungen über die antiphlogistische Wirksamkeit verschiedener Pharmaka auf das posttraumatische Ödem. *Zeitschrift für Allgemeinmedizin* 1981; **57**: 44–8.

48 Dauber KE. Vergleichende tierexperimentelle Untersuchung verschiedener peroral verabreichter antiphlogistisch wirksamer Medikamente am posttraumatischen Rattenpfotenödem. Dissertation, Freiburg, 1982.

49 Koh MMS. Modification of the inflammatory response by therapeutic agents. Thesis, University of London, 1979: 1–240.

50 Lee RE, Bykadi G, Ritschel WA. Inhibition of prostaglandin biosynthesis by coumarin, 4-hydroxycoumarin, and 7-hydroxycoumarin. *Arzneim-Forsch Drug Res* 1981; **31**: 640–2.

51 Földi M. Postmastektomisches Lymphödem: Pathophysiologie und Klinik. *Diagnostik* 1978; **11**: 411–14.

52 Casley-Smith JR. Die Wirkung der Benzopyrone beim Lymphödem und bei anderen eiweißreichen Ödemen. *Vasomed* 1994; **6**: 232–4.

53 Gaffney RM, Casley-Smith JR. Excess plasma proteins as a cause of chronic inflammation and lymphoedema: Biochemical estimations. *J Pathol* 1981; **133**: 229–42.

54 Piller NB, Clodius L. Lymphoedema of the rabbit ear following partial and complete lymphatic blockade; its effects on fibrotic development, enzyme types and their activity levels. *Br J Exp Pathol* 1978; **59**: 319–26.

55 Knight KR, Khazanchi RK, Pederson WCH, McCann JJ, Coe SA, O'Brien B. Coumarin and 7-hydroxycoumarin treatment of canine obstructive lymphoedema. *Clin Sci* 1989; **77**: 69–76.

56 Van Sumere CF, Teuchy H. The coumarin enigma. *Arch Int Physiol Biochim* 1974; **79**: 665.

57 Bär F, Griepentrog F. Die Situation in der gesundheitlichen Beurteilung der Aromatisierungsmittel für Lebensmittel. *Medizin und Ernährung* 1967; **8**: 244–51.

58 Griepentrog F. Pathologisch-anatomische Befunde zur karzinogenen Wirkung von Cumarin im Tierversuch. *Toxicology* 1973; **1**: 93–102.

59 Evans JG, Gaunt IF, Lake BG. Two-year toxicity study on coumarin in the baboon. *Food Cosmet Toxicol* 1979; **17**: 187–93.

60 Brune H, Deutsch-Wenzel R. *Toxikologische Prüfung von Cumarin im 30 Monate Fütterungsversuch an Ratten*. Hamburg: Board for Preventive Medicine and Environmental Protection Ltd, 1984: 1–359.

61 Preuss-Ueberschär C, Ueberschär S. Licht- und elektronenmikroskopische Untersuchungen zur Hepatotoxizität von Benzopyronen in Abhängigkeit von Dosis und Anwendungsdauer. *Arzneim-Forsch/Drug Res* 1988; **38**: 1318–26.

62 Ueno I, Hirono I. Non-carcinogenic response to coumarin in Syrian Golden hamsters. *Food Cosmet Toxicol* 1981; **19**: 353–5.

63 Hunter B, Pulsford AM, Heywood R, Gibson WA, Street AE, Raed R, Gopinath C, Anderson A. Coumarin – potential tumourigenic effects in prolonged administration to mice. Final report No. 82493, Huntingdon Res. Centre, Contract Research (confidential): 1983.

64 Hunter B, Warren S, Heywood R, Gopinath C, Gibson W, Anderson A, Gregson RI. Coumarin – tumourigenic and toxic effects in prolonged administration to rats following in utero exposure – main phase. Final report No. 83651, Huntingdon Res. Centre, Contract Research (confidential): 1984.

65 Siegers CP. STELLUNGNAHME ZU NTP-Report No. 422 Cumarin (Expert Opinion), 1993.

66 Stoltz DR, Scott PM. Mutagenicity of coumarin and related compounds to *Salmonella typhimurium* (Abstract). *Can J Genet Cytol* 1980; **22**: 679.

67 Norman RL, Wood AW. o-Hydroxyphenylethanol, a novel lactone ring-opened metabolite of coumarin. *Drug Metab Dispos* 1984; **12**: 543–9.

68 Haworth St, Lawlor T, Mortelmans K, Speck W, Zeiger E. Salmonella mutagenicity test results for 250 chemicals. *Environ Mutagen Suppl* 1983; **1**: 3–142.

69 Yoon JS, Mason JM, Valencia R, Woodruff RC, Zimmering S. Chemical mutagenesis testing in *Drosophila*. IV. Results of 45 coded compounds tested for the National Toxicology Program. *Environ Mutagen* 1985; **7**: 349–67.

70 Voogd CE, van der Stel JJ, Jacobs JAA. On the mutagenesis action of some enzyme immunoassay substrates. *J Immunol Methods* 1980; **36**: 55–61.

71 Galloway SM, Armstrong MJ, Reuben C *et al.* Environmental and molecular mutagenesis. *J Environ Mutagen Soc* 1987; **10** (Suppl 10): 1–149.

72 Ohta T, Watanabe K, Moriya M, Shirasu Y, Kada T. Anti-mutagenic effects of coumarin and umbelliferone on mutagenesis induced by 4-nitroquinoline 1-oxide or UV irradiation in *E. coli. Mutat Res* 1983; **117**: 135–8.

73 Sasaki YF, Imanishi H, Ohta T, Shirasu Y. Effects of anti-mutagenic flavourings on SCEs induced by chemical mutagens in cultured Chinese hamster cells. *Mutat Res* 1987; **189**: 313–18.

74 Benson AM, Hunkeler MJ, Talalay P. Increase in NAD(P)H : quinone reductase by dietary antioxidants: Possible role in protection against carcinogenesis and toxicity. *Proc Natl Acad Sci USA* 1980; **77**: 5216–5220.

75 DeLong M, Procharska HJ, Talalay P. Induction of NAD(P)H : quinone reductase in murine hepatoma cells by phenolic antioxidants, azo dyes, and other chemoprotectors: A model system for the study of anticarcinogens. *Proc Natl Acad Sci USA* 1986; **83**: 787–91.

76 Sparnins VL, Venegas PL, Wattenberg LW. Glutathione *S*-transferase activity: enhancement by compounds inhibiting chemical carcinogenesis and by dietary constituents. *J Nat Cancer Inst* 1982; **68**: 493–6.

77 Teikemeier G, Bolsen K, Merk HF, Khan WA, Goerz G. Influence of coumarin on chemical carcinogenesis. Düsseldorf: Department of Dermatology of the University; Köln: Department of Dermatology of the University; Cleveland, Ohio: Department of Dermatology of the Case Western Reserve University and Veterans Administration Medical Center; 1986: Study Report, unpublished data.

78 Ritschel WA, Grummich KW. Pharmacokinetics of coumarin and 7-hydroxycoumarin upon i.v. and p.o. administration in the euthyroid and hypothyroid Beagle dog. *Arzneim-Forsch/Drug Res* 1981; **31**: 643–9.

79 Greim H, Filser JG. Risikoabschätzung in der Toxikologie. Voraussetzungen für die Übertragbarkeit von Daten aus dem Tierversuch auf den Menschen. In: Dengler HJ, Mutschler E eds. *Fremdstoffmetabolismus und Klinische Pharmakologie*. Stuttgart & Jena: Gustav Fischer, 1994: 141–51.

80 Lake BG, Gray TJB, Evans JG *et al.* Studies on the mechanism of coumarin-induced toxicity in rat hepatocytes: Comparison with dihydrocoumarin and other coumarin metabolites. *Toxicol Appl Pharmacol* 1989; **97**: 311–23.

81 Lake BG, Gaudin H, Price RJ, Walters DG. Metabolism of [3-^{14}C]coumarin to polar and covalently bound products by hepatic microsomes from the rat, Syrian hamster, gerbil and humans. *Fd Chem Toxic* 1992; **30**: 105–15.

82 Fentem JH, Fry JR. Metabolism of coumarin by rat, gerbil and human liver microsomes. *Xenobiotica* 1992; **22**: 357–67.

83 Fentem JH, Fry JR. Species differences in the metabolism and hepatotoxicity of coumarin. *Comp Biochem Physiol* 1993; **104C**: 1–8.

84 Van Iersel MLPS, Henderson CJ, Walters DG, Price RJ, Wolf CR, Lake BG. Metabolism of (3-C^{14}]coumarin by human liver microsomes. *Xenobiotica* 1994; **24**: 795–803.

85 Steensma A, Beamand JA, Walters DG, Price RJ, Lake BG. Metabolism of coumarin and 7-ethoxycoumarin by rat, mouse, guinea pig, Cynomolgus monkey and human 'precision cut liver slices'. *Xenobiotica* 1994; **24**: 357–63.

86 Pearce R, Greenway D, Parkinson A. Species differences and interindividual variation in liver microsomal cytochrome P 450 2A enzymes: Effects on coumarin, dicumarol, and testosterone oxidation. *Arch Biochem Biophys* 1992; **298**: 211–25.

87 Feuer G, Kellen JA. Link between carcinogenicity and hepatic metabolism of 7,12-dimethylbenz(a)anthracene. *Oncology* 1974; **30**: 499–508.

88 Wattenberg LW, Lam LKT, Fladmoe AV. Inhibition of chemical carcinogen-induced neoplasia by coumarins and α-angelicalactone. *Cancer Res* 1979; **39**: 1651–54.

89 Wattenberg LW. Chemoprevention of cancer. *Cancer Res* 1985; **45**: 1–8.

90 Tseng A, Jr. Lee WMF, Kirsten E *et al.* Prevention of tumourigenesis of oncogene-transformed rat fibroblasts with DNA site inhibitors of poly(ADP-ribose) polymerase. *Proc Natl Acad Sci USA* 1987; **84**: 1107–11.

91 Maucher A, Kager M, von Angerer E. Evaluation of the antitumour activity of coumarin in prostate cancer models. *J Cancer Res Clin Oncol* 1993; **119**: 150–4.

92 Kahn J, Preis P, Waldman F, Tseng A, Jr. Coumarin modulates the cell-cycle progression of an MTV-EJras cell line. *J Cancer Res Clin Oncol* 1994; **120**: Suppl 19–22.

93 Nakayasu M, Shima H, Aonuma Sh *et al.* Deletion of transfected oncogenes from NIH 3T3 transformants by inhibitors of poly(ADP-ribose)polymerase. *Proc Natl Acad Sci USA* 1988; **85**: 9066–70.

94 Bedoukian PZ. Coumarin. In: *Perfumery and Flavoring Synthetics*. Amsterdam: Elsevier, 1967: 131–40.

95 Cohen AJ. Critical review of the toxicology of coumarin with special reference to interspecies differences in metabolism and hepatotoxic response and their significance to man. *Fd Cosmet Toxicol* 1979; **17**: 277–89.

96 Amtsblatt der Europäischen Gemeinschaften. Richtlinie des Rates zur Angleichung der Rechtsvorschriften der Mitgliedsstaaten über Aromen zur Verwendung in Lebensmitteln und über Ausgangsstoffe für ihre Herstellung. Official Gazette L 184, 15 July 1988.

97 Opdyke DLJ. Monographs on fragrance raw materials. Coumarin. *Fd Cosmet Toxicol* 1974; **12**: 385–8.

98 Laven R, Schäfer EA. *Pharmakokinetik von Cumarin beim Menschen nach Applikation verschiedener Darreichungsformen.* Salzgitter: Fa Schaper & Brümmer, Chem-Pharmaz. Fabrik 1988.

99 Ritschel WA, Hussain AS. Transdermal absorption and topical availability of coumarin. *Methods Find Exp Clin Pharmacol* 1988; **10**: 165–9.

100 Hamburger H, Hostettmann K. Bioactivity in plants: The link between phytochemistry and medicine. *Phytochemistry* 1991; **30**: 3864–74.

101 Moran E, Prosser E, O'Kennedy R, Thornes ED. The effect of coumarin and 7-hydroxycoumarin on the growth of human tumour cell lines. *J Irish Coll Phys Surg* 1993; **22**(Suppl 1): 41–3.

102 Siegers CP, Bostelmann HC. Effect of coumarin on cell proliferation in human tumour cell lines. *J Irish Coll Phys Surg* 1993; **22**(Suppl 1): 47–8.

103 Marshall ME, Albainey-Jenei S, Umerani A, Zhao Q. Direct cytotoxic effects of coumarin and 7-hydroxycoumarin on human malignant cell lines *in vitro*. *J Irish Coll Phys Surg* 1993; (Suppl 1): 71–2.

104 Marshall ME, Kervin K, Benefield C *et al.* Growth-inhibitory effects of coumarin(1,2-benzopyrone) and 7-hydroxycoumarin on human malignant cell lines in vitro. *J Cancer Res Clin Oncol* 1994; **120** (Suppl): 3–10.

105 Myers RB, Parker M, Grizzle WE. The effects of coumarin and suramin on the growth of malignant renal and prostate cell lines. *J Cancer Res Clin Oncol* 1994; **120** (Suppl): 11–13.

106 Seliger B, Petterson H. 7-Hydroxycoumarin inhibits oncogene-induced transformation of murine fibroblasts. *J Cancer Res Clin Oncol* 1994; **120** (Suppl): 23–7.

107 Seliger B. 7-Hydroxycoumarin mediated induction of differentiation of human malignant glioblastoma cells. *J Irish Coll Phys Surg* 1993; **22** (Suppl 1): 44–6.

108 Elstner EF, Schuff-Werner P, Seidel D. Verwendung von Cumarin und Cumarin-Derivaten. Deutsches Patentamt Nr. DE 37 15 990A1; Offenlegungstag: 08. 12. 1988.

109 Schuff-Werner P. Einfluß von 1, 2-benzopyron(Cumarin) auf den oxidativen Metabolismus von Phagozyten. *Antiphlogistische Wirkung durch Beeinflussung der Radikalbildung*. [Influence of 1,2-benzopyrone(coumarin) on the oxidative metabolism of phagocytes. Antiphlogistic effect by the influence on the formation of free radicals. Investigational report (unpublished)] Göttingen: Abteilung Klinische Chemie Zentrum Innere Medizin Universitätsklinikum 1987.

110 Casley-Smith JR. Australian Patent Abridgment. The method of treating high protein oedemas in humans or animals, by administering to the body the compound 7-hydroxycoumarin, 1987.

111 Paya M, Halliwell B, Hoult JRS. Interactions of series of coumarins with reactive oxygen species. *Biochem Pharmacol* 1992; **44**: 205–14.

112 Hladovec J. Protective effect of oxygen-derived free radical scavengers on the endothelium in vivo. *Physiol Bohemoslovaca* 1986; **35**: 97–103.

113 Zawieja DC, Greiner ST, Davis KL, Hinds WM, Granger HJ. Reactive oxygen metabolites inhibit spontaneous lymphatic contractions. *Am J Physiol* 1991; **260**: 1935–43.

114 Murell GAC, Francis MJO, Bromley L. Modulation of fibroblast proliferation by oxygen free radicals. *Biochem J* 1990; **265**: 659–65.

115 Borzeix MG, Agignard J, Dedieu F *et al.* Effect of a combination of coumarin derivatives and rutoside on venous and lymphatic circulations during severe constriction of the caudal vena cava in rabbits. *Arzneim-Forsch/Drug Res* 1995; **45**: 261–6.

116 Marshall ME, Conley D, Hollingworth P, Brown S, Thompson JS. Effects of coumarin (1,2-benzopyrone) on lymphocyte, natural killer cell, and monocyte functions in vitro. *J Biol Response Mod* 1989; **8**: 70–85.

117 Piller NB. An integration of the modes of action of coumarin. *Arzneim-Forsch/Drug Res* 1977; **27**: 1135–8.

118 Piller NB. The induction of controlled proteolysis in high protein oedemas by coumarin. *Z Lymphol* 1977, **1**: 39–43, 1978, **2**: 30–4, 1979, **3**: 111–7.

119 Casley-Smith JR, Casley-Smith Judith R. *High-protein Oedemas and the Benzo-Pyrones*. Sydney & Baltimore: JB Lippincott, 1986.

120 Mislin H. Die Wirkung von Cumarin aus *Melilotus officinalis* auf die Funktion des Lymphangions. *Arzneim-Forsch/Drug Res* 1971; **21**: 852–3.

121 Földi M, Kovach AGB, Varga L, Zoltan ÖT. Die Wirkung des Melilotus-Präparates Esberiven auf die Lymphströmung. *Ärztliche Forschung* 1962; **16**: 99–102.

122 Wüstenberg P. *Klinik und Therapie der chronischen venösen Insuffizienz* Marshall M (ed.). Karlsruhe: Braun, 1994.

123 Desprez-Curely JP, Cluzan R, Pecking A. Benzopyrones and post-mastectomy lymphoedemas. Double-blind trial placebo versus sustained release coumarin with trioxyethylrutin. In: Casley-Smith JR, Piller NB eds. *Progress in Lymphology X. Proceedings of the Xth International Congress of Lymphology.* Adelaide: University of Adelaide Press, 1985.

124 Clodius L, Piller NB. The conservative treatment of postmastectomy lymphoedema patients with coumarin results in a marked continuous reduction in arm swelling. In: Bartos V, Davidson JW eds. *Advances in Lymphology. Proceedings of the 8th International Congress of Lymphology.* Prague: Avicenum, 1981: 471–6.

125 Cluzan R, Pecking A. Benzopyrone (Lysedem®) double blind crossing over study in patients with secondary upper limb edemas. In: Nishi M, Uchino S, Yabuki S eds. *Progress in Lymphology XII.* Amsterdam: Elsevier, 1990: 453–4.

126 Jamal S, Casley-Smith JR, Casley-Smith Judith R. The effects of 5,6-benzo-[a]-pyrone(coumarin) and DEC on filaritic lymphoedema and elephantiasis in India. Preliminary results. *Ann Trop Med Parasitol* 1989; **83**: 287–90.

127 Casley-Smith JR, Jamal S, Casley-Smith Judith R. Reduction of filiaritic lymphoedema and elephantiasis by 5,6-benzo-a-pyrone (coumarin) and the effects of diethylcarbamazine (DCE). *Ann Trop Med Parasitol* 1993; **87**: 247–58.

128 Casley-Smith JR, Wang CT, Casley-Smith Judith R, Cui Zi-Hai. Treatment of filarial lymphoedema and elephantiasis with 5,6-benzo-a-pyrone (coumarin). *Br Med J* 1993; **307**: 1037–41.

129 Chang Ti-Seng, Gan Ji-Liang, Huang Wen-yi. The use of 5,6-benzo-[a]-pyrone (coumarin) and heating by microwaves in the treatment of chronic leg lymphoedema. (in press) *Lymphology* 1993.

130 Becker HM, Niedermaier G, Orend KH. Benzopyrone in der Therapie des postrekonstruktiven Ödems. *Fortschr Med* 1985; **103**: 593–6.

131 Clodius L. Antibiotikaprophylaxe in der plastischen Wiederherstellungschirurgie? *Helv Chir Acta* 1978; **45**: 505–10.

132 Pethö A. Die Wirksamkeit von Benzopyronen bei der posttraumatischen Entzündung. *Arzneim-Forsch/Drug Res* 1981; **31**: 1303–7.

133 Pethö A. Nachbehandlung von Laparatomien. *Münchner Medizinische Wochenschrift* 1984; **126**: 943–5.

134 Pethö A. Nachbehandlung von vaginalen Hysterektomien mit einem Benzopyron-Präparat. *Fortschr Med* 1984; **102**: 493–6.

135 Thornes RD, Daly L, Lynch G, Brown H, Tanner A, Keane F *et al.* Prevention of early recurrence of high risk malignant melanoma by coumarin. *Eur J Surg Oncol* 1989; **15**: 431–5.

136 Thornes RD. Coumarins, melanoma and cellular immunity. *Protective Agents in Cancer* 1983: 43–56.

137 Hilgart P. Coumarin anticoagulants and cancer: a personal review of old ideas and new hypotheses. *J Irish Coll Phys Surg* 1993; **22** (Suppl 1): 62–3.

138 Zacharski LR. Warfarin therapy for cancer. *J Irish Coll Phys Surg* 1993; **22** (Suppl 1): 20–3.

139 Zlabinger GJ, Nöhammer C, Böhmig GA, Menzel JE. Mode of action of coumarin in immune cells. *J Cancer Res Clin Oncol* 1994; **120** (Suppl): 17–18.

140 Stuhlmeyer K, Theyer G, Baumgartner G, Zlabinger GL. Synergistic effect of coumarin (1,2-benzopyrone) and endotoxin in the induction of human interleukin-1. *Clin Exp Immunol* 1991; **84**: 317–23.

141 Zänker KS. Identification of members of the coumarin family as cancer chemopreventive agents: Evaluation in in-vitro models and human clinical trials. In: Sydney E Salmon ed. *Adjuvant Therapy of Cancer*, vol. VII. Philadelphia: JB Lippincott, 1993: 405–10.

142 Ritschel WA, Brady ME, Tan HSI. First-pass effect of coumarin in man. *Int J Clin Pharmacol Biopharm* 1979; **17**: 99–103.

143 Sharifi S. Pharmakokinetische Untersuchungen von Cumarin und seinen Metaboliten nach oraler und intravenöser Applikation bei gesunden Probanden. Dissertation im Fachbereich Medizin der Universität; Zentrum Pharmakologie und Toxikologie, Abteilung Klinische Pharmakologie, Göttingen, 1990.

144 Chasseaud LF, Taylor T, Waller A, Biggs S. The metabolic fate of ^{14}C-coumarin after single oral doses, either alone or admixed with troxerutin, to baboons. Huntingdon Research Centre Reports SCB7 & SCB16/79574, Huntingdon, 1979.

145 Chasseaud LF, Taylor T, Waller A. The metabolism of ^{14}C-coumarin by DBA/2J mice and comparison of hepatic covalent binding of radioactivity in DBA/2J, CFLP and CDI mice following intraperitoneal doses of ^{14}C-coumarin. Huntingdon Research Centre Report SCB9/78859, Huntingdon, 1978.

146 Piller NB. Tissue levels of (3-^{14}C) coumarin in the rat: Distribution and excretion. *Br J Exp Pathol* 1977; **58**: 28–34.

147 Greib E, Haag-Berrurier M. Nouvelle recherche sur l'élimination de l'ombelliférone et sa répartition dans quelques organes du lapin. *Thérapie* 1972; **27**: 25–34.

148 Sotaniemi EA, Rautio A, Bäckström M, Arvela P, Pelkonen O. AYP3A4 and CYP2A6 activities marked by the metabolism of lignocaine and coumarin in patients with liver and kidney diseases and epileptic patients. *Br J Clin Pharmacol* 1995; **39**: 71–6.

149 Rautio A, Kraul H, Kojo A, Salmela E, Pelkonen O. Interindividual variability of coumarin-7-hydroxylation in healthy volunteers. *Pharmacogenetics* 1992; **2**: 227–33.

150 Pelkonen O, Sotaniemi EA, Ahokas JT. Coumarin 7-hydroxylase in human liver microsomes. Properties of the enzyme and interspecies comparisons. *Br J Clin Pharmacol* 1985; **19**: 59–66.

151 Iscan M, Rostami H, Iscan M *et al.* Interindividual variability of coumarin 7-hydroxylation in a Turkish population. *Eur J Clin Pharmacol* 1994; **47**: 315–18.

152 Spahn-Langguth H, Benet LZ, Zia-Amirhosseini P *et al.* Aktive und reaktive Glucuronide. Reversible und irreversible Interaktionen von Glucuronsäurekonjugaten mit Proteinen. In: Dengler HJ, Mutschler E eds. *Fremdstoffmetabolismus und Klinische Pharmakologie*. Stuttgart: Gustav Fischer, 1994; 29–33.

153 Gerharz CD, Reinicke P, Gabbert HE. Analysis of the growth-inhibitory effects of coumarin, 7-hydroxycoumarin and 7-hydroxycoumaringlucuronide in the clear cell, chromophilic, and chromophobic types of renal cell carcinoma in vitro. Investigational report (unpublished), Düsseldorf: Universität, Institut für Pathologie, 1994: 1–10.

154 Siegers CP. Einfluß von Cumarin, 7-Hydroxycumarin und 7-Hydroxycumaringlucuronid auf die Zellproliferation menschlicher Tumorzellinien [Influence of coumarin, 7-hydroxycoumarin and 7-hydroxycoumaringlucuronide on the cell proliferation of human tumor cell lines. Investigational report (unpublished)], Lübek: Medizinische Universität Institut für Toxikologie, 1994: 1–35.

155 Schäfer EA. Bestimmung der Konzentration von Cumarin, freiem und konjugiertem 7-Hydroxycumarin in Patientenseren nach Gabe von 0, 6–7g Cumarin. Salzgitter-Ringelheim: Forschungsbericht aus der Biochemischen Abteilung der Firma Schaper & Brümmer, 1991: 1–25.

156 Marshall ME, Butler K, Fried A. Phase I evaluation of coumarin (1,2-benzopyrone) and cimetidine in patients with advanced malignancies. *Mol Biother* 1991; **3**: 170–8.

157 Cox D, O'Kennedy R, Thornes RD. The rarity of liver toxicity in patients treated with coumarin (1,2-benzopyrone). *Human Toxicol* 1989; **8**: 501–6.

158 Bose PK. On some biochemical properties of natural coumarins. *J Indian Chem Soc* 1958; **58**: 367–75.

159 Karch FE, Lasagna L. Adverse drug reactions. A critical review. *JAMA* 1975; **234**: 1236–41.

160 Laven R, Hess R. Reizschwellenerhöhung im Zahnpulpatest an Kaninchen durch Cumarin: Unspezifischer (lokaler) Effekt oder analgetische Wirkung? *Methods Find Exp Clin Pharmacol* 1982; **4**: 551–3.

161 Stern P, Dezelic A, Kosak R. Über die analgetische und antipyretische Wirkung von k-Vitamin und Dicumarol mit besonderer Berücksichtigung des 4-Hydroxycumarins. *Arch exp Pathol Pharmakol* 1959; **230**: 356–9.

162 Fontaine L, Belleville M, Lechevin IC, Tete R. Activités cholérétiques et spasmolytiques. Pharmacologie générale de la méthyl-4-ombelliféron. *Thérapie* 1968; **23**: 52–62.

163 Pflug JJ, Taylor LM. An experimental assessment of the action of a drug for the swollen leg. *Folia Angiologica* 1976; **24**: 72–7.

164 Hladovec J. Vasotrope Arzneistoffe – ein Überblick auf der Basis eines vereinheitlichenden Konzepts ihres Wirkungsmechanismus. *Arzneim-Forsch/Drug Res* 1977; **27**: 1073–76.

165 Hladovec J. The effect of digoxin on endothelaemia in rats and its modification by a combination of troxerutin and coumarin. *Arzneim-Forsch/Drug Res* 1978; **28**: 982–3.

166 Shilling WH, Crampton RF, Longland RC. Metabolism of coumarin in man. *Nature* (Lond.) 1969; **221**: 664–5.

167 Moran E, O'Kennedy R, Thornes RD. Analysis of coumarin and its urinary metabolites by high-performance liquid chromatography. *J Chromatogr* 1987; **416**: 165–9.

168 Cholerton S, Idle ME, Vas A. Comparison of a novel thin-layer chromatographic-fluorescence detection method with a spectrofluorometric method for the determination of 7-hydroxycoumarin in human urine. *J Chromatogr* 1992; **575**: 325–30.

169 Meinecke I. Ergebnisbericht zur analytischen Bestimmung von 2-Phenylessigsäure im Urin von gesunden Probanden nach peroraler Einnahme von 1 und 2g Cumarin. Experimental Report (unpublished), Universität Göttingen 1995: 1–21.

170 Gangolli ShD, Shilling WH, Grasso P, Gaunt IF. Studies on the metabolism and hepatotoxicity of coumarin in the baboon. *Biochem Soc Trans* 1974; **2**: 310–12.

171 Lake BG, Walters DG, Gangolli ShD. Comparison of the metabolism and disposition of [3-^{14}C]coumarin in the rat and marmoset (*Callithrix jacchus*). *Toxicol Lett* 1989; **45**: 299–306.

172 Feuer G. The metabolism and biological actions of coumarin. *Prog Med Chem* 1974; **10**: 85–158.

173 Lush IE, Andrews KM. Genetic variation between mice in their metabolism of coumarin and its derivatives. *Genet Res* 1978; **31**: 177–86.

174 Cacini W, Ritschel WA. The excretion of coumarin and hydroxycoumarins by the avian kidney *in vivo*. *Arch Int Pharmacodyn Ther* 1980; **243**: 197–208.

175 Lake BG, Evans JG, Walter DG, Price RJ. Studies on the hepatotoxicity and metabolism of coumarin in the rat and Syrian hamster. *Human Exp Toxicol* 1990; Oral communication: 356.

CHAPTER 2

The Chemistry and Occurrence of Coumarins

Gary J. Keating and Richard O'Kennedy
School of Biological Sciences, Dublin City University, Dublin, Ireland

2.1 INTRODUCTION

The coumarins belong to a group of compounds known as the benzopyrones, all of which consist of a benzene ring joined to a pyrone—a six-membered heterocyclic ring containing one oxygen atom and five sp^2-hybridised carbons. Both γ- and α-pyrones exist, with the prefixes referring to the position at which the oxygen atom is found. The structure of α-pyrone is shown below:

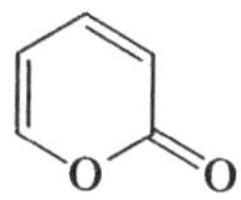

Coumarin (shown below) and the other members of the coumarin family are benzo-α-pyrones, while the other main members of the group—the flavonoids—contain the γ-pyrone group.

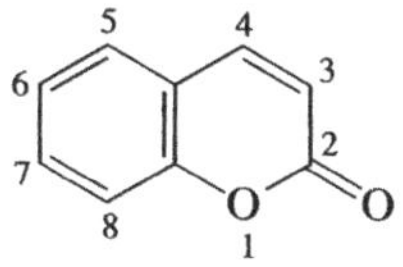

Coumarins: Biology, Applications and Mode of Action. Edited by R. O'Kennedy and R.D. Thornes.

Coumarins have well-defined absorption spectra (see Chapter 12), and display a characteristic fluorescence upon excitation with ultraviolet light, in addition to having a large range of physiological activity.

Many coumarins are comprised of isoprenoid chains joined to one of the carbons of the nucleus, or to an oxygen atom, for example, a phenolic oxygen. The isoprenoid chains may consist of one, two or three units, and many variations of this type of structure are to be found among the coumarins. Interactions of the prenyl group with an *ortho*-phenolic group can lead to the formation of another heterocyclic ring. Oxidative interaction may also occur. As the vast majority of the coumarins are oxygenated at C-7, 7-hydroxycoumarin is frequently considered to be the parent compound for the more structurally complex members of the family. Coumarins may also be found in nature in combination with sugars, as glycosides. A variety of chromatographic and spectroscopic methods for the identification of coumarins exist (see Chapter 12). The coumarins can be roughly categorised as follows:

(a) *Simple*. These members of the coumarin family are the hydroxylated, alkoxylated and alkylated derivatives of the parent compound, coumarin, along with their glycosides.

7-Hydroxycoumarin

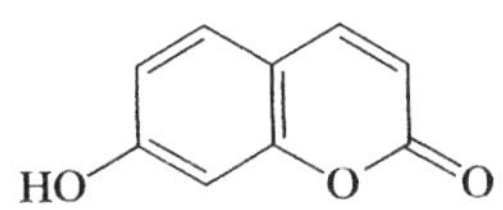

(b) *Furanocoumarins*. These compounds consist of a five-membered furan ring attached to the coumarin nucleus. Most of the members of this group are derivatives of the linear furanocoumarin psoralen, or its more stable angular isomer, angelicin. (The terms 'linear' and 'angular' refer to the orientation of the furan ring with respect to the coumarin nucleus.) The dihydrofuranocoumarins and dihydropyranocoumarins are also members of this group.

Psoralen

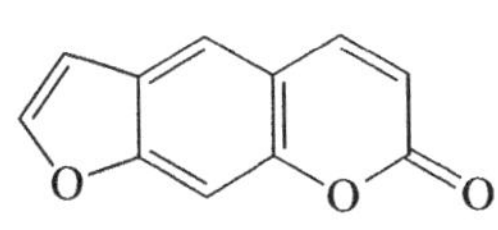

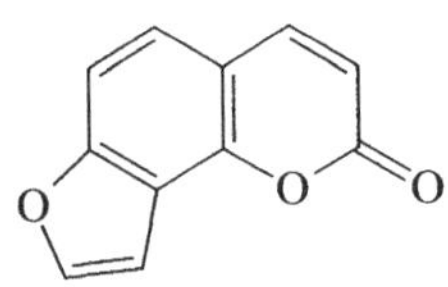

Angelicin

(c) *Pyranocoumarins*. Members of this group are analogous to the furanocoumarins, and contain a six-membered ring. As with the furanocoumarins, linear and angular types exist:

Xanthyletin

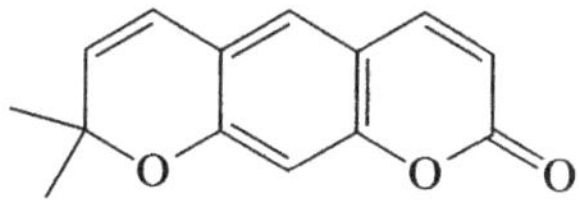

Seselin

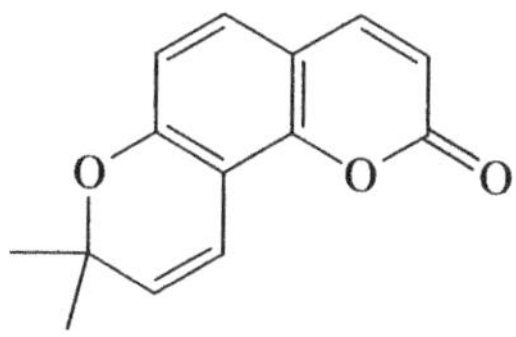

(d) *Coumarins substituted in the pyrone ring*. Examples of this group include 4-hydroxycoumarin and the 3-methyl-4-hydroxycoumarins, such as autumnariol:

4-Hydroxycoumarin

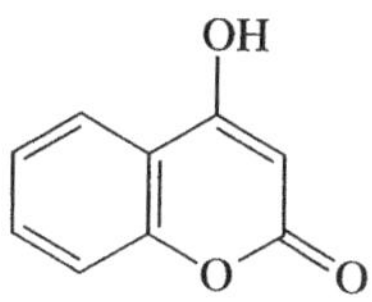

Autumnariol

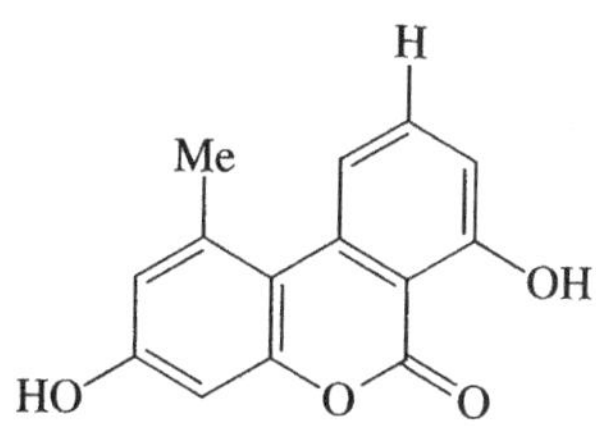

The synthetic compound warfarin—widely used both as a rodenticide and an oral anticoagulant—contains a 4-hydroxy group:

Warfarin

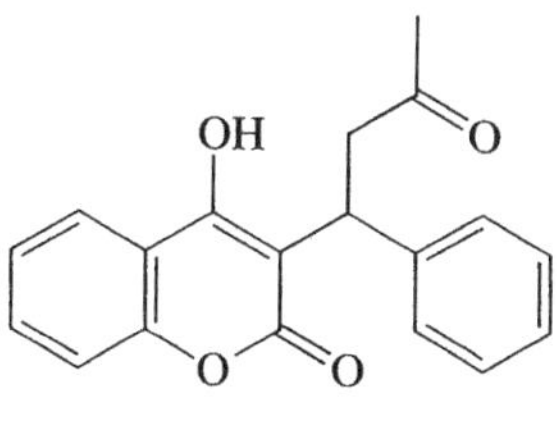

Another coumarin containing a 4-hydroxy group, phenprocoumon, has been found to possess antiviral activity and significantly inhibit the HIV-1 protease responsible for the maturation of the causative virus of acquired immunodeficiency syndrome (AIDS)[1]:

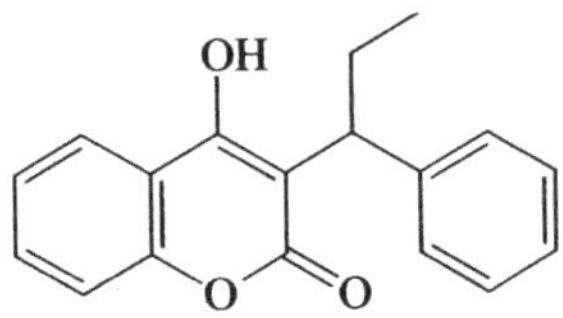

Phenprocoumon

Members of this group of coumarins have also been widely used as fluorophores. The 4-methylumbelliferones are commonly employed to generate a fluorescent signal in assays for the detection of bacterial contamination of water supplies[2]. Peptide derivatives of 7-amino-4-methyl coumarin (AMC) have been extensively utilised for the investigation of protease activity[3] (see Chapter 13).

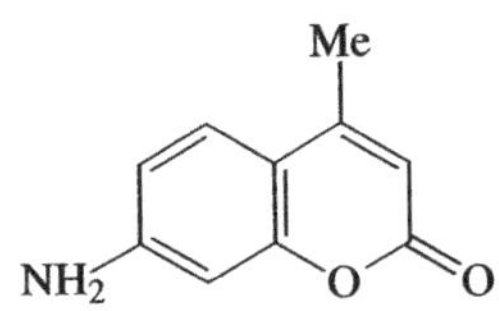

7-Amino-4-methyl coumarin

The coumestans—or coumaroncoumarins—which occur naturally in alfalfa and ladino clover, are all based on the structure of coumestan:

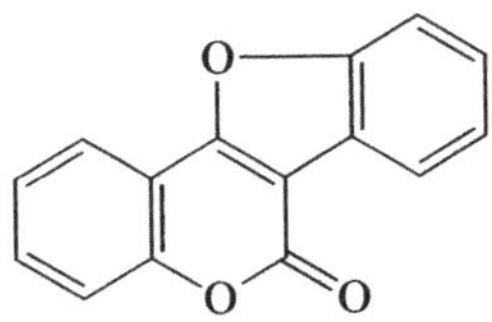

Coumestan

One important member of this group is coumestrol, which has been shown to display an oestrogenic activity.

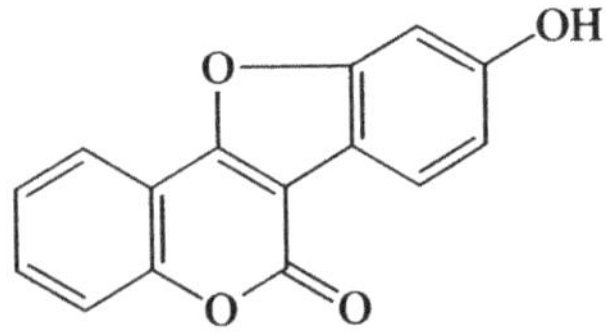

Coumestrol

The coumarinolignans consist of a phenylpropanoid and a coumarin unit linked together by means of a dioxane bridge. Cleomiscosin A is one example of this group:

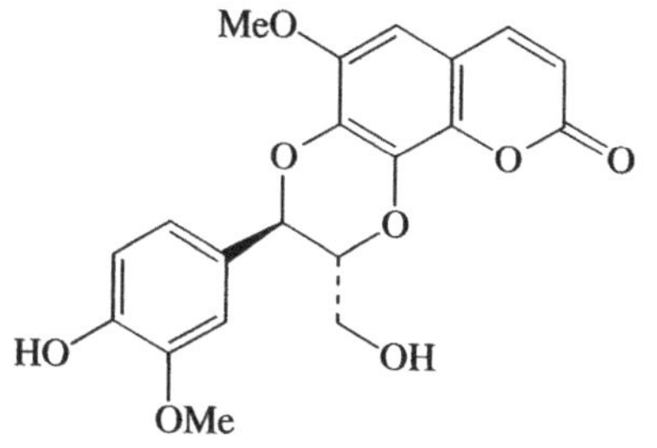

Cleomiscosin

The biscoumarins—such as demethyldaphnoretin, which is found in *Daphne gnidiodes*—consist of two coumarin units linked together:

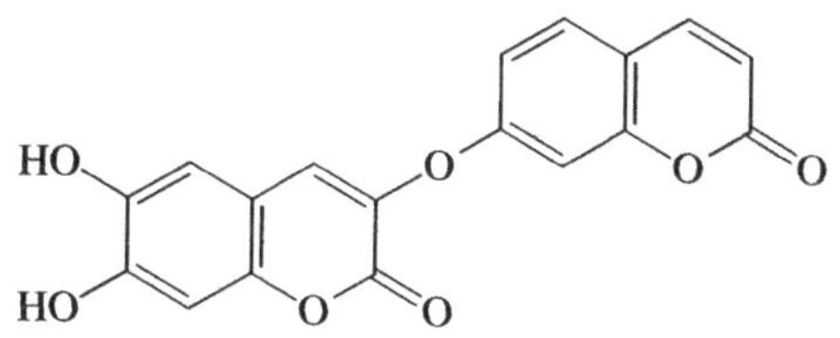

Demethyldaphnoretin

2.2 NOMENCLATURE

The nomenclature of coumarin compounds is frequently problematic. The parent compound, coumarin, is variously known as 2-oxo-1,2-benzo-pyran, 5,6-benzo-α-pyrone, 5,6-benzo-2-pyrone, 1,2-benzopyrone, 2*H*-1-benzopyran-2-one, *cis-o*-coumaric acid lactone, 2*H*-benzo[*b*]pyran-2-one, coumarinic anhydride and 2*H*-chromen-2-one. Members of the coumarin family have been isolated from hundreds of species of plants and microorganisms, giving rise to another problem in the naming of the compounds. Many of the coumarins that have been isolated are known by common names that are generally derived from the Latin or colloquial name of the plant in which they were discovered, or from the place where the plant grows, resulting in the same compound having several different trivial names. Frequently, the trivial name assigned to such a compound gives little or no information as to its structural composition, and in some cases may even be misleading. For example, 7-hydroxycoumarin is also known as umbelliferone (due to its occurrence in plants of the Umbelliferae), skimmetin and hydrangin.

2.3 OCCURRENCE

Although most of the natural coumarins in existence have been isolated from the higher plants, some members of the family have been discovered in microorganism and animal sources. Yeasts of the genus *Streptomyces* produce antibiotics like novobiocin, chartreusin, and coumermycin A_1:

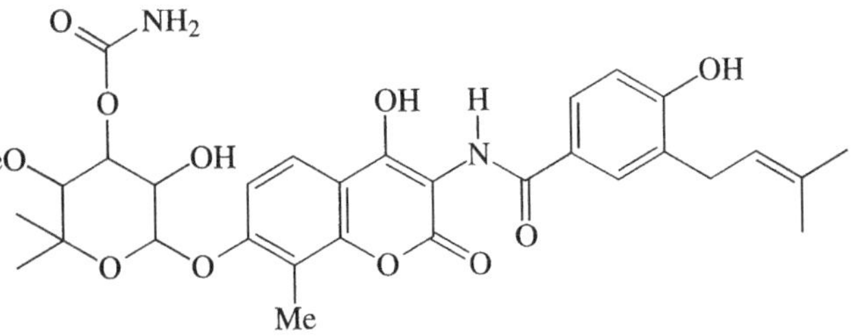

Novobiocin

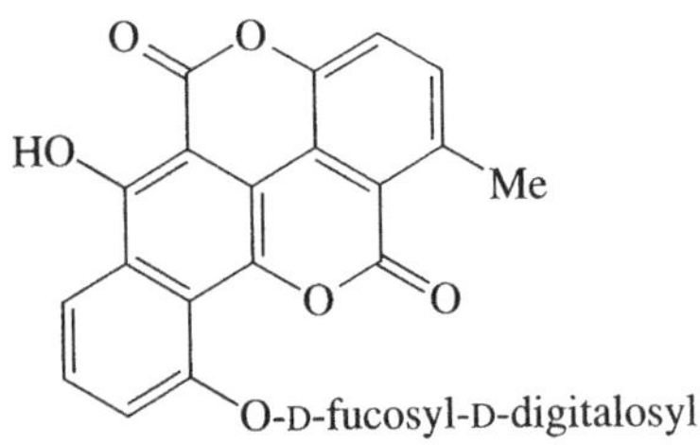

Chartreusin

The aflatoxins are a group of highly toxic fungal metabolites that occur in the fungus *Aspergillus* sp. The most commonly occurring member of the group is aflatoxin B_1 (AFB_1), produced by *Aspergillus flavus* and *A. parasiticus*:

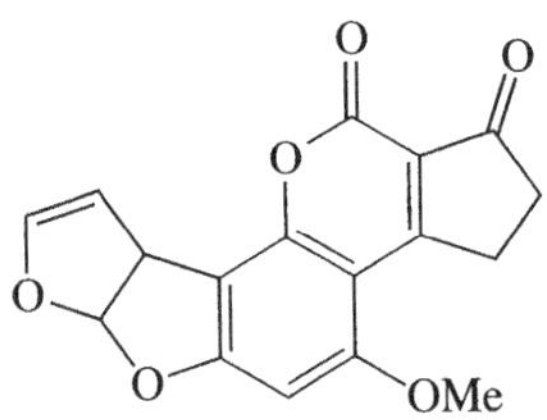

Aflatoxin B_1

A range of metabolites of AFB_1 have been isolated. AFM_1 (or 4-hydroxy-AFB_1) is a major metabolite of AFB_1 in rats and cows, among other species[4]. AFQ_1 and AFP_1 have been shown to be major metabolites of AFB_1 in the mouse and Rhesus monkey[5,6]:

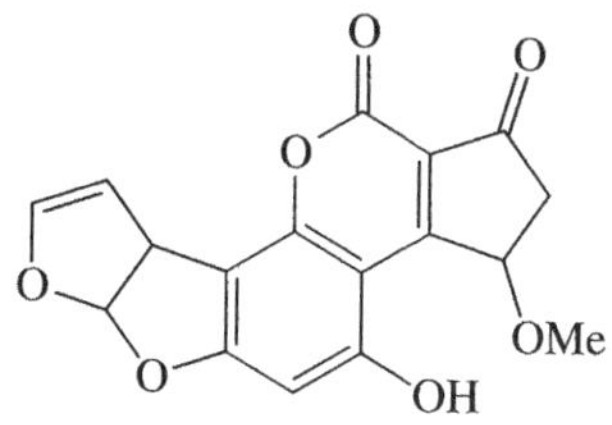

Aflatoxin Q_1

Aflatoxicol H is another metabolite of AFB_1 in primates[7]. AFB_{2a} and AFG_{2a} are metabolic products of AFB_1 and AFG_1, respectively[8]. Other metabolites of AFB_1 include aflatoxicol and deshydroaflatoxin D_1.

Dasycladus vermicularis—a member of the green algal family Dasycladaceae—has been shown to produce 3,6,7-trihydroxycoumarin[9]:

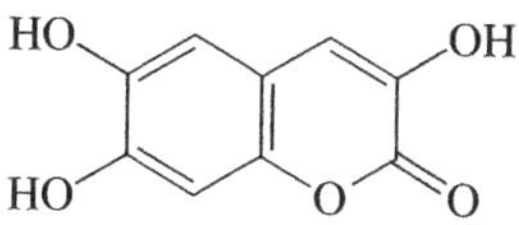

3,6,7-Trihydroxycoumarin

The mould *Emericella desertorum* produces desertorin A, B and C[10]:

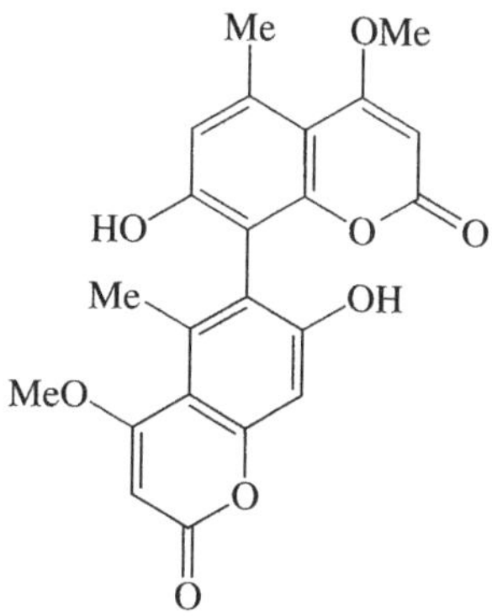

Desertorin A

Necatorin has been extracted from the wild mushroom *Lactarius necator*[11]:

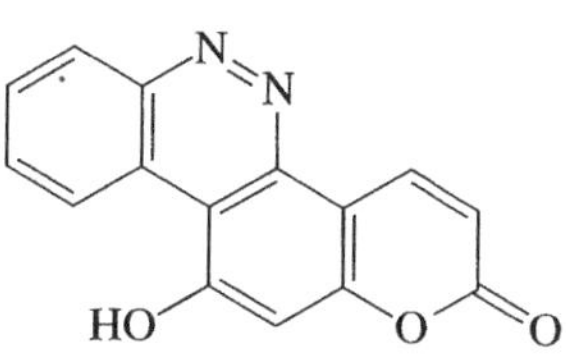

Necatorin

Lamillarin A, B, C and D are metabolites of the mollusc *Lamellaria* sp.[12]:

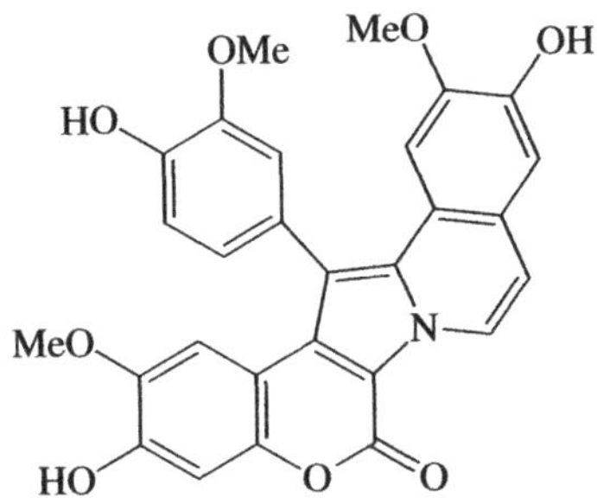

Lamillarin D

The castoreum pigments are also of animal origin, having been extracted from secretions of the beaver scent gland[13]:

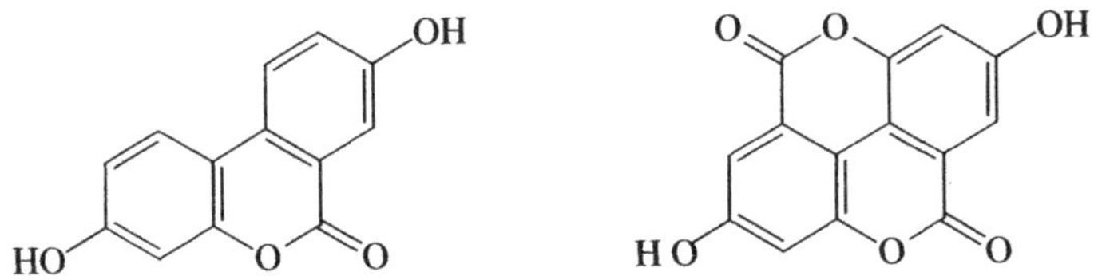

Castoreum pigments

As stated above, though, most coumarins occur in the higher plants, with the richest sources being the Rutaceae and Umbelliferae, in which all parts of the plant have been discovered to contain coumarins. The coumarins belong to a group known as the secondary products of plant metabolism, the exact roles of which in plant physiology are not clearly understood. Although distributed throughout all parts of the plant, the coumarins occur at the highest levels in the fruits, followed—in order of decreasing occurrence—by the roots, stems and leaves. In addition, seasonal changes and environmental conditions may affect the occurrence in various parts of the plant.

Murray *et al.*[14] have given a comprehensive overview of the chemistry and occurrence of natural coumarins, including a table of the botanical sources of all of the coumarin compounds known at time of publication. A subsequent review article[15] covers the natural coumarins isolated from mid-1980 to mid-1988. Table 2.1 lists the new coumarins that have been extracted from natural sources since this time.

Table 2.1. Naturally occurring coumarins.

Compound	Source	References
Acetoxyaurapten	*Zanthoxylum schinifolium*	16
14-Acetoxybadrakemin	*Heptaptera anisoptera*	17
12-Acetoxycalanolide A,B	*Calophyllum lanigerum*	18
14-Acetoxycolladonin	*Heptaptera anisoptera*	17
7-Acetoxy-5,6-dimethoxycoumarin	*Pelargonium sidoides*	19
10′-Acetoxy-1′-epilycoserone	*Aphyllocladus denticulatus*	20
(*E*)-ω-Acetoxyferprenin	*Ferula communis*	21
(*Z*)-ω-Acetoxyferprenin	*Ferula communis*	20
(*E*)-ω-Acetoxyferulenol	*Ferula communis*	21
(*Z*)-ω-Acetoxyferulenol	*Ferula communis*	21
8-Acetoxy-5-hydroxyumbelliprenin	*Asafetida* sp.	22
cis-2-Acetoxy-5-methoxy-6-geranyloxymellein	*Bunium paucifolium*	23
5-*O*-(6″-Acetyl)-β-galactopyranosyl-3′,4′-dihydroxy-7-methoxy-4-phenylcoumarin	*Exostema caribaeum*	24
2″-*O*-Acetyloxypeucedanin hydrate	*Angelica glauca*	25
(+)-*trans*-4′-Acetyl-3′-tigloylkhellactone	*Peucedanum japonicum*	26
Acrimarine A,B,C	*Citrus funadoko*	27
Aesculetin-6-*O*-β-D-apiofuranosyl-(1→6)-*O*-β-glucopyranoside	*Lonicera gracilipes* var. *glandulosa*	28
Agrimonolide-6-*O*-β-D-glucopyranoside	*Agrimonia pilosa*	29
Albiflorin I	*Boenninghausenia albiflora*	30
Albiflorin II,III	*Boenninghausenia albiflora*	31
Aleuritin	*Aleurites fordii*	32
Alloxanthoxyletol	*Pilocarpus goudotianus*	33
Angelol K,L	*Angelica pubescens*	34
2′-Angeloyl-3′-isovaleryl vaginate	*Angelica archangelica archangelica*	35
Anhydronotoptoloxide	*Notopterygium incisum*	36
Anisocoumarin A,B,C,D	*Clausena anisata*	37
Anisocoumarin E,F,G,H	*Clausena anisata*	38
Anisocoumarin I,J	*Clausena anisata*	39
Aphyllocladone	*Aphyllocladus denticulatus*	20
Aphyllodenticulide	*Aphyllocladus denticulatus*	20
7-*O*-[β-D-Apiofuranosyl-(1→6)-β-D-glucopyranosyl]-6-methoxycoumarin	*Xeomphis spinosa*	40
Assafoetidin	*Ferula assafoetida*	41
cis-Avicennin	*Eriostemon apiculatus*	42
Bakuchinin	*Psoralea corylifolia*	43
Bergaptol-*O*-β-D-glucopyranoside	*Notopterygium forbesii*	44
(*E*)-Betulin-3-*p*-coumarate	*Phebalium tubercolosum* subsp. *megaphyllum*	45
(*Z*)-Betulin-3-*p*-coumarate	*Phebalium tubercolosum* subsp. *megaphyllum*	45
Bishassanidin	*Citrus hassaku*	46

(*continues*)

Table 2.1. (*continued*)

Compound	Source	References
Bisnorponcitrin	*Citrus hassaku*	46
Bisosthenon	*Citrus* sp.	47
Bisparasin	*Citrus* sp.	48
Buntansin	*Citrus grandis buntan*	49
Buntansin B,C	*Citrus grandis buntan*	50
Calanolide A,B,C	*Calophyllum lanigerum*	18
Calanone	*Calophyllum teysmannii*	51
Capensin	*Gochniata argentina*	52
cis-Casegravol	*Citrus* sp.	53
Casegravol isovalerate	*Citrus* sp.	54
Chamaejasme	*Stellera chamaejasme*	55
Chloculol	*Murraya paniculata*	56
Citrubuntin	*Citrus grandis*	57
Citrumarin A,B,C,D	*Citrus hassaku*	58
Citrusarin A,B	*Citrus* sp.	53
Claudimarin A	*Citrus hassaku*	59
Cleomiscosin B	*Salsola laricifolia*	60
Cleomiscosin D	*Cleome viscosa*	61
Columbianetin-β-D-glucopyranoside	*Angelica pubescens*	62
Costatolide	*Calophyllum teysmannii* var. *inophylloide*	63
Cyclohoehnelia coumarin	*Ethulia vernonioides*	64
Daphneside	*Daphne arisanensis*	65
10′,11′-Dehydrocyclolycoserone	*Aphyllocladus denticulatus*	20
10′,11′-Dehydro-1′βH-lycoserone	*Aphyllocladus denticulatus*	20
5-Deoxyprotobruceol-I-regioisomer	*Boronia lanceolata*	66
5-Deoxyprotobruceol-II-regioisomer	*Boronia lanceolata*	66
5-Deoxyprotobruceol-III-regioisomer	*Boronia lanceolata*	66
6,7-Di-(3,3-dimethylallyloxy)-5-hydroxycoumarin	*Drumondita hasselli*	67
6,7-Di-(3,3-dimethylallyloxy)-5-methoxycoumarin	*Drumondita hasselli*	67
5-[[3-(4,5-Dihydro-5,5-dimethyl-4-oxo-2-furanyl)-2-butenyl]oxy]-7*H*-furo[3,2-g][1]benzopyran-7-one	*Dorstenia brasiliensis*	68
5-[3-(4,5-Dihydro-5,5-dimethyl-4-oxo-2-furanyl)butoxy]-7*H*-furo[3,2-g][1]benzopyran-7-one	*Dorstenia brasiliensis*	68
(−)-2,3-Dihydro-9-*O*-β-glucosyloxy-2-isopropenyl-7*H*-furo[3,2-g][1]benzopyran-7-one	*Apium graveolens*	69
2,3-Dihydro-9-hydroxy-2-[1-(6-feruloyl)-β-D-glucosyloxy-1-methylethyl]-7*H*-furo[3,2-g][1]benzopyran-7-one	*Skimmia japonica*	70

Table 2.1. (*continued*)

Compound	Source	References
(+)-2,3-Dihydro-9-hydroxy-2[1-(6-sinapinoyl)β-D-glucosyloxy-1-methylethyl]-7*H*-furo[3,2-*g*][1]benzopyran-7-one	*Apium graveolens*	69
4-[6,10-Dihydroxy-7,9-dehydro-6,7-dihydrogeranyloxy]-5-methyl coumarin	*Mutisia orbignyana*	71
6,8-Dihydroxy-5,7-dimethoxycoumarin	*Pelargonium sidoides*	19
5,3′-Dihydroxy-7,4′-dimethoxy-4-phenylcoumarin	*Exostema caribaeum*	24
(−)-7-(6,7-Dihydroxy-3,7-dimethyl-5-oxooctanoxy)-coumarin	*Phebalium squameum*	72
4-[4,10-Dihydroxygeranyloxy]-5-methyl coumarin	*Mutisia orbignyana*	71
4-[5,10-Dihydroxygeranyloxy]-5-methyl coumarin	*Mutisia orbignyana*	71
5,8-Dihydroxy-7-(4-hydroxy-5-methylcoumarin-3)-coumarin	*Gerbera anandria*	73
3,8-Dihydroxy-4-methoxycoumarin	*Gerbera anandria*	73
6,8-Dihydroxy-7-methoxycoumarin	*Pelargonium sidoides*	19
6,7-Dihydroxy-5-methoxycoumarin 6-β-D-glucopyranoside	*Prunus prostrata*	74
7-(2,3-Dihydroxy-3-methylbutyloxy)-5-hydroxy-6-methoxycoumarin	*Pterocaulon alopecuroides*	75
7-(2,3-Dihydroxy-3-methylbutyloxy)-6-methoxycoumarin	*Pterocaulon alopecuroides*	75
11,12-Dihydroxy-5-methylcoumestan	*Mutisia acuminata* var. *hirsuta*	76
7,4′-Dimethoxy-5′-hydroxy-4-phenyl-5,2′-oxidocoumarin	*Exostema caribaeum*	77
8-(χ-χ-Dimethyl-allyl)-1-methoxycoumestrol	*Lotus creticus*	78
7-(3,3-Dimethylallyloxy)-6-hydroxy-5-methoxycoumarin	*Drumondita hasselli*	67
4-[6,10-Dioxo-neryloxy]-5-methyl coumarin	*Mutisia orbignyana*	71
Donatin	*Pilocarpus goudotianus*	33
Edgeworin	*Edgeworthia chrysantha*	79
Edgeworoside A,B,C	*Edgeworthia chrysantha*	79
Edulisin III,IV,V	*Angelica edulis*	80
(+)-Elisin	*Pilocarpus goudotianus*	33
1′,2′,6′,7′,8′-Epicyclolycoserone	*Aphyllocladus denticulatus*	20
2′-Epiisotriptiliocoumarin	*Triptilion benaventei*	81
6′,7′-Epoxyauraptene	*Aegle marmelos*	82
Epoxycollinin	*Zanthoxylum schinifolium*	16
7,10-Epoxy-7,14-dihydronassauvirevolutin	*Nassauvia argentea*	83
5-(3,4-Epoxy-2,7-dimethyl-6,7-octenoyl) psoralen	*Dorstenia contrajerva*	83
(+)-7-(6,7-Epoxy-3,7-dimethyl-5-oxooctanoxy)-coumarin	*Phebalium squameum*	72

(*continues*)

Table 2.1. (*continued*)

Compound	Source	References
4-[5,10-Epoxygeranyloxy]-5-methyl coumarin	*Mutisia orbignyana*	71
4-[4β-7-Epoxy-6β-hydroxy-6,7-dihydroxygeranyloxy]-5-methyl coumarin	*Mutisia orbignyana*	72
3-[4,10′-Epoxylinalyl]-5-methyl coumarin	*Mutisia orbignyana*	71
1′,2′-Epoxy-6-methoxycasegravol-4′-(3-methyl) butanoate	*Phebalium elatius* subsp. *beckleri*	85
(−)-2′β-(2ζ,3-Epoxy-3-methyl-1-oxobutyl)-2′-deoxybruceol	*Eriostemon myoporoides*	86
Ethylnotopterol	*Notopterygium incisum*	36
Euchretin A	*Euchresta japonica*	87
Euchretin B,C	*Euchresta formosana*	88
Fernolin	*Feronia limonia*	89
Ferocolicin	*Ferula assafoetida*	41
6′-*O*-*trans*-Feruloylnodakenin	*Notopterygium forbesii*	45
4-[5-Formyl-3-methyl-pent-2*E*,4*E*-dien-1-yl]-5-methyl coumarin	*Mutisia orbignyana*	71
6-Formylumbelliferone	*Citrus* sp.	90
Frachinoside	*Fraxinus chinensis*	91
Fraxetin	*Gochniata argentina*	52
Fraxidin	*Gochniata argentina*	52
Funadonin	*Citrus funadoko*	92
2′4′ : 4,5-Furocoumarin	*Mutisia acuminata* var. *hirsuta*	76
5-*O*-β-Galactopyranosyl-3′,4′-dihydroxy-7-methoxy-4-phenylcoumarin	*Exostema caribaeum*	24
5-*O*-β-Galactopyranosyl-7,4′-dimethoxy-4-phenylcoumarin	*Exostema caribaeum*	24
5-*O*-β-D-Galactosyl-7-methoxy-3′,4′-dihydroxy-4-phenylcoumarin	*Exostema caribaeum*	77
8-Geranyl-5,7-dihydroxycoumarin	*Eriostemon myoporoides*	86
8-Geranyl-7-hydroxycoumarin	*Eriostema tomentellus*	93
(*R*)-6-*O*-(4-Geranyloxy-2-hydroxy) cinnamoylmarmin	*Citrus hassaku*	94
4-Geranyloxy-5-methyl coumarin	*Mutisia orbignyana*	71
Gigasol	*Angelica gigas*	95
cis-Grandmarin	*Citrus* sp.	90
trans-Grandmarin isovalerate	*Citrus* sp.	90
Hassanidin	*Citrus* sp.	96
cis-Hassanin	*Citrus* sp.	96
trans-Hassanin	*Citrus* sp.	96
Hassanon	*Citrus* sp.	90
Hassmarin	*Citrus hassaku*	97
Hemidesmin 1,2	*Hemidesmus indicus*	98
Hemidesminine	*Hemidesmus indicus*	99
Hoehnelia coumarin	*Ethulia vernonioides*	64

Table 2.1. (*continued*)

Compound	Source	References
(*E*)-6-(7-Hydroperoxy-3,7-dimethylocta-2,5-dienyl)-7-hydroxycoumarin	*Eriostemon tomentellus*	93
(*E*)-7-(6-Hydroperoxy-3,7-dimethylocta-2,7-dienyloxy)coumarin	*Phebalium tuberculosum* subsp. *megaphyllum*	45
(*E*)-8-(6-Hydroperoxy-3,7-dimethylocta-2,7-dienyloxy)psoralen	*Phebalium tuberculosum* subsp. *megaphyllum*	45
(−)-2β-(2ζ-Hydroperoxy-3-methylbut-3-enyl)-2′-deoxybruceol	*Eriostemon myoporoides*	86
(−)-2β-{(*E*)-3-Hydroperoxy-3-methylbut-1-enyl}-2′-deoxybruceol	*Eriostemon myoporoides*	86
14-Hydroxycolladonin	*Heptaptera anisoptera*	17
4-Hydroxycoumestrol	*Erythrina sigmoidea*	100
10′-Hydroxycyclolycoserone	*Aphyllocladus denticulatus*	20
5-Hydroxy-8,8-dimethyl-10-(3′,7′-dimethylocta-1′6′-dien-3′-yl)-2*H*,8*H*,benzo[1,2-*b* : 5,4-*b*′]dipyran-2-one	*Paramignya monophylla*	101
6-(6-Hydroxy-3,7-dimethylocta-2,7-dienyl)-7-hydroxycoumarin	*Eriostema tomentellus*	93
(*E*)-6-(7-Hydroxy-3,7-dimethylocta-2,5-dienyl)-7-hydroxycoumarin	*Eriostema tomentellus*	93
7-(6*R*-Hydroxy-3,7-dimethyl-2*E*,7-octadienyloxy) coumarin	*Citrus hassaku*	94
(*E*,*E*)-8-(7-Hydroxy-3,7-dimethylocta-2,5-dienyloxy) psoralen	*Phebalium tuberculosum* subsp. *megaphyllum*	45
4-[4-Hydroxy-6-*epi*-hydroperoxy-7,9-dehydro-6,7-dihydrogeranyloxy]-5-methyl coumarin	*Mutisia orbignyana*	71
5α′-Hydroxy-6′,7′-epoxytriptiliocoumarin	*Triptilion benaventei*	81
4′β-Hydroxyeriobrucinol	*Eriostemon brucei cinerus*	102
(*E*)-ω-Hydroxyferprenin	*Ferula communis*	21
(*Z*)-ω-Hydroxyferprenin	*Ferula communis*	21
(*E*)-ω-Hydroxyferulenol	*Ferula communis*	21
(*Z*)-ω-Hydroxyferulenol	*Ferula communis*	21
4-[4-Hydroxygeranyloxy]-5-methyl coumarin	*Mutisia orbignyana*	71
4-[4-Hydroxygeranyloxy]-5-methyl coumarin acetate	*Mutisia orbignyana*	71
4-[10-Hydroxygeranyloxy]-5-methyl coumarin	*Mutisia orbignyana*	71
4-[10-Hydroxygeranyloxy]-5-methyl coumarin acetate	*Mutisia orbignyana*	71
7-(10′-Hydroxygeranyl)scopoletin-10′-*O*-β-D-fucopyranosyl-(6→1)-β-D-glucopyranoside	*Coptis trifolia*	103
7-(10′-Hydroxygeranyl)scopoletin-10′-*O*-β-D-glucopyranoside	*Coptis trifolia*	103

(*continues*)

Table 2.1. (*continued*)

Compound	Source	References
4-[4-Hydroxy-6-hydroperoxy-7,9-dehydro-6,7-dihydrogeranyloxy]-5-methyl coumarin	*Mutisia orbignyana*	71
4-[4-Hydroxy-7-hydroperoxy-5,6*E*-dehydro-6,7-dihydrogeranyloxy]-5-methyl coumarin	*Mutisia orbignyana*	71
4-[10-Hydroxy-6-hydroperoxy-7,9-dehydro-6,7-dihydrogeranyloxy]-5-methyl coumarin	*Mutisia orbignyana*	71
4-[10-Hydroxy-7-hydroperoxy-5,6*E*-dehydro-6,7-dihydrogeranyloxy]-5-methyl coumarin	*Mutisia orbignyana*	71
7-Hydroxy-6-(2-(*R*)-hydroxy-3-methylbut-3-enyl)-coumarin	*Boronia lanceolata*	66
10′-Hydroxyisolycoserone	*Aphyllocladus denticulatus*	20
(±)-7-Hydroxy-6-linalycoumarin	*Citrus hassaku*	94
10′-Hydroxy-1′*βH*-lycoserone	*Aphyllocladus denticulatus*	20
4-[10-Hydroxy-oxo-7,9-dehydro-6,7-dihydrogeranyloxyl-5-methyl coumarin	*Mutisia orbignyana*	71
(+)-(2′*S*,3′*R*)-3-Hydroxymarmesin	*Brosimum gaudichaudii*	104
(2*R*)-2′-Hydroxymarmesin 2′-*O*-*β*-D-glucopyranoside	*Diplolophium buchananii*	105
5-Hydroxy-7-methoxycoumarin-8-*O*-*β*-D-glucoside	*Daphne pseudo-mezereum*	106
6-Hydroxymethylherniarin	*Citrus funadoko*	92
(−)-2′*β*-{(*E*)-3-Hydroxy-3-methylbut-1-enyl}-2′-deoxybruceol	*Eriostemon myoporoides*	86
(−)-2′*β*-(2ζ-Hydroxy-3-methyl-1-oxobutyl)-2′-deoxybruceol	*Eriostemon myoporoides*	86
4-Hydroxy-5-methyl-3-[8′-oxo-farnesyl]-coumarin	*Aphyllocladus denticulatus*	20
4-Hydroxy-5-methyl-3-[8′-oxo-9′-hydroxyfarnesyl]-coumarin	*Aphyllocladus denticulatus*	20
3-(3-Hydroxy-3-methyl-*trans*-but-1-enyl)-7-methoxycoumarin	*Asterolasia squamuligera*	107
4-[10-Hydroxyneryloxy]-5-methyl coumarin	*Mutisia orbignyana*	71
4-[10-Hydroxyneryloxy]-5-methyl coumarin acetate	*Mutisia orbignyana*	71
5-Hydroxyseselin	*Citrus* sp.	90
5′-Hydroxytriptiliocoumarin	*Triptilion benaventei*	81
8-Hydroxytriptiliocoumarin	*Triptilion benaventei*	81
5-Hydroxyumbelliprenin	*Asafetida* sp.	22
6-Hydroxyumbelliprenin	*Asafetida* sp.	22
Isoarnottinin	*Ammi majus*	108
Isobyakangelicin angelate	*Peucedanum palustre*	109
Isolycoserone	*Aphyllocladus denticulatus*	20
Isomurralonginol isovalerate	*Murraya paniculata*	56
Isotriptiliocoumarin	*Triptilion benaventei*	80

Table 2.1. (*continued*)

Compound	Source	References
Jayantinin	*Boenninghausenia albiflora*	110
(+)-*trans*-Khellactone	*Peucedanum japonicum*	26
Khelmarin A	*Citrus hassaku*	111
Khelmarin B	*Citrus canariculata*	112
Khelmarin C	*Citrus hassaku*	46
Kinocoumarin	*Citrus* sp.	54
Kiyomal	*Citrus* sp.	54
Licoarylcoumarin	*Glycyrrhiza* sp.	112
Licopyranocoumarin	*Glycyrrhiza* sp.	112
Merrillin	*Merrillia caloxylon*	113
7-Methoxy-8-(2-acetoxy-3-methyl-1-oxobut-2-enyl) coumarin	*Phebalum ralstonii*	114
12-Methoxycalanolide A,B	*Calophyllum lanigerum*	18
6-Methoxy-3′,4′-dehydromurranganon	*Phebalium elatius* subsp. *beckleri*	115
6-Methoxy-3′,4′-dehydromurranganon-2-methylbutanoate	*Phebalium elatius* subsp. *beckleri*	115
6-Methoxy-3′,4′-dehydromurranganon-2-methylpropanoate	*Phebalium elatius* subsp. *beckleri*	115
6-Methoxy-3′,4′-dehydromurranganon propanoate	*Phebalium elatius* subsp. *beckleri*	115
cis-5-Methoxydecursidinol	*Citrus* sp.	96
trans-5-Methoxydecursidinol	*Citrus* sp.	96
6-Methoxy-3′,4′-dihydroisomicrominutin	*Phebalium elatius* subsp. *beckleri*	85
2-Methoxy-3,9-dihydroxycoumestone	*Tephrosia hamiltonii*	116
5-Methoxy-8,8-dimethyl-10-(3′,7′-dimethylocta-1′6′-dien-3′-yl)-2*H*,8*H*,benzo[1,2-*b* : 5,4-*b*′]dipyran-2-one	*Paramignya monophylla*	101
5-Methoxy-6-geranyloxymellein	*Bunium paucifolium*	23
5-Methoxy-8-*O*-*β*-D-glucosyloxy-psoralen	*Apium graveolens*	69
(+)-6-Methoxyisomurralonginol acetate	*Phebalium elatius* subsp. *beckleri*	85
5-Methoxy-6,7-methylenedioxycoumarin	*Simsia croquistii*	117
6-Methoxymicrominutinin	*Micromelum minutum*	118
(+)-6-Methoxymurranganon	*Phebalium elatius* subsp. *beckleri*	85
6-Methoxymurraol	*Phebalium elatius* subsp. *beckleri*	85
(+)-6-Methoxyphebalosin	*Phebalium elatius* subsp. *beckleri*	85
(+)-7-(3-Methylbut-2-enyloxy)-8-(2-hydroxy-3-methylbut-3-enyl)coumarin	*Eriostemon spicatus*	119
(+)-7-(3-Methylbut-2-enyloxy)-8-(4-hydroxy-3-methylbut-2-enyl)coumarin	*Eriostemon spicatus*	119
5-(3-Methyl-2-butenyloxy)-6,7-methylenedioxycoumarin	*Pterocaulon virgatum*	120
6,7-Methylenedioxy-8-methoxycoumarin	*Asterolasia trymalioide*	121
trans-*O*-Methylgrandin	*Citrus* sp.	54

(*continues*)

Table 2.1. (*continued*)

Compound	Source	References
(−)-2′β-(3-Methyl-1-oxobut-2-enyl)-2′-deoxybruceol	*Eriostemon myoporoides*	86
(*E*)-Methylsuberenol	*Citrus funadoko*	92
(*Z*)-Methylsuberenol	*Citrus funadoko*	92
(−)-2′β-(3-Methyl-1ζ,2ζ,3-trihydroxybutyl)-2′-deoxybruceol	*Eriostemon myoporoides*	86
Microminutinin	*Micromelum minutum*	122
Minumicrolin isovalerate	*Murraya paniculata*	56
Mirificoumestan glycol	*Pueraria mirifica*	123
Mirificoumestan hydrate	*Pueraria mirifica*	123
Murracarpin	*Murraya paniculata omphalocarpa*	124
Murraculatin	*Murraya paniculata*	125
Murralonginol isovalerate	*Murraya paniculata*	56
Murrangatin isovalerate	*Murraya paniculata*	56
Murrayacarpin-A	*Murraya paniculata omphalocarpa*	124
Murrayacarpin-B	*Murraya paniculata omphalocarpa*	124
Murrayanone	*Murraya paniculata*	125
Mutisifuro coumarin diacetate	*Mutisia orbignyana*	71
Nassauvirevolutin A,B,C	*Nassauvia* sp.	126
2′-*epi*-Nassauvirevolutin C	*Nassauvia argentea*	83
Neoacrimarine A,B	*Citrus* sp.	127
Neoacrimarine C,D	*Citrus hassaku*	128
Nordenletin	*Citrus hassaku*	129
Notoptolide	*Notopterygium incisum*	36
Obtusifolin	*Haplophyllum obtusifolium*	130
Omphalocarpin	*Murraya paniculata omphalocarpa*	124
Oxaclausarin	*Citrus hassaku*	129
Oxanordentatin	*Citrus hassaku*	129
(*E*)-ω-Oxoferprenin	*Ferula communis*	21
(*E*)-ω-Oxoferulenol	*Ferula communis*	21
4-[10-Oxo-geranyloxyl]-5-methyl coumarin	*Mutisia orbignyana*	71
4-[10-Oxo-neryloxy]-5-methyl coumarin	*Mutisia orbignyana*	71
Palustroside	*Ledium palustre*	131
Paniculonol isovalerate	*Murraya paniculata*	132
Peroxymurraol	*Murraya exotica*	132
Peroxytamarin	*Citrus* sp.	53
Peucedanocoumarin I,II,III	*Peucedanum praeruptorum*	133
Peujaponiside	*Peucedanum japonicum*	134
Peujaponisinol A,B	*Peucedanum japonicum*	135
Plicadin	*Psoralea plicata*	136

Table 2.1. (*continued*)

Compound	Source	References
Praealtin A,B,C,D	*Aster praealtus*	137
Praeroside I,II,III,IV,V	*Peucedanum praeruptorum*	138
6-Prenyloxy-1-methoxycoumarin	*Carduus tenuiflorus*	139
Pseudobruceol I,II	*Eriostemon brucei cinerus*	102
Ptilosin	*Haplophyllum ptilostylum*	140
Ptilostol	*Haplophyllum ptilostylum*	140
Puerarostan	*Pueraria tuberosa*	141
Pummeloquinone	*Citrus* sp.	142
Purpurasol	*Pterocaulon purpurascens*	143
Pygmaeoherin	*Pygmaeopremna herbacea*	144
Schinicoumarin	*Zanthoxylum schinifolium*	16
Schinilenol	*Zanthoxylum schinifolium*	16
Schininallylol	*Zanthoxylum schinifolium*	16
Schinindiol	*Zanthoxylum schinifolium*	16
3′-Senecioylxanthyletin	*Peucedanum praeruptorum*	145
Seselinol	*Citrus hassaku*	129
Seselinol isovalerate	*Citrus hassaku*	129
Setarin	*Setaria italica*	146
(*Z*)-Suberenol	*Citrus funadoko*	92
(9*S*,10*R*)-9,10,11,5′-Tetrahydroxyumbelliprenin	*Heptaptera anisoptera*	147
Toddacoumalone	*Toddalia asiatica*	148
Tortuoside	*Seseli tortuosum*	149
(9*S*,10*R*)-9,10,5′-Triacetoxy-11-hydroxyumbelliprenin	*Heptaptera anisoptera*	147
6,7,8-Trihydroxycoumarin	*Pelargonium sidoides*	19
7,4′,5′-Trihydroxy-4-phenyl-5,2′-oxidocoumarin	*Exostema caribaeum*	77
(9*S*,10*R*)-9,10,11-Trihydroxyumbelliprenin	*Heptaptera anisoptera*	147
10,11,5′-Trihydroxyumbelliprenin	*Heptaptera anisoptera*	147
Triptiliocoumarin	*Triptilion benaventei*	81
Triptispinocoumarin	*Triptilion spinosum*	81
Xeroboside	*Xeromphis obovata*	150

2.4 BIOSYNTHESIS OF COUMARINS

In the higher plants, coumarins are generally produced via the shikimate–chorismate biosynthetic pathway as a derivative of cinnamic acid, which is also a precursor of many other natural products, such as methyl salicylate, cinnamic aldehyde and amygdalin. The shikimate–chorismate pathway is responsible for the formation of the aromatic amino acids. Phosphoryl group transfer from ATP followed by nucleophilic displacement results in the formation of 3-enolpyruvyl-shikimate-5-phosphate from shikimate, and the elimination of phosphate gives chorismate:

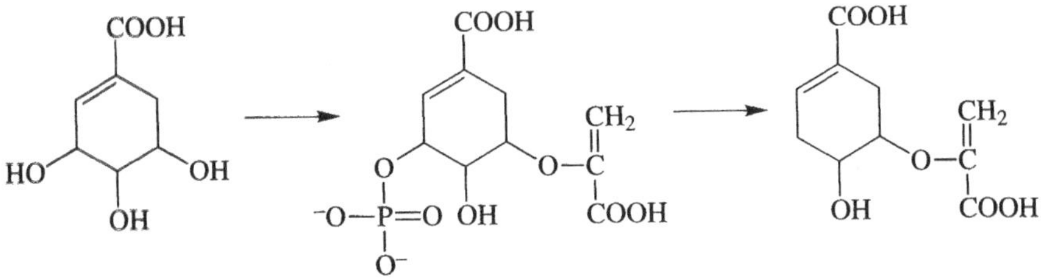

Shikimate 3-Enolpyruvyl-shikimate-5-phosphate Chorismate

The enzyme chorismate mutase then catalyses the rearrangement of chorismate to prephenate, which is converted to phenylpyruvate by means of a 1,4 elimination. A glutamate-dependent transamination results in the formation of phenylalanine:

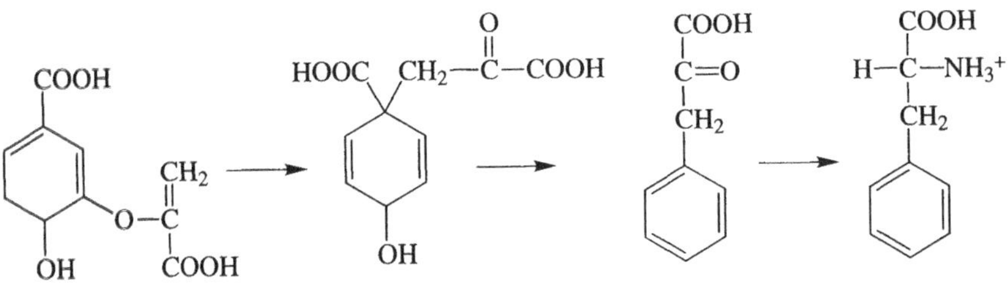

Chorismate Prephenate Phenylpyruvate Phenylalanine

Phenylalanine is then converted to *trans*-cinnamic acid by the action of the enzyme phenylalanine ammonia-lyase, which eliminates NH_3^+:

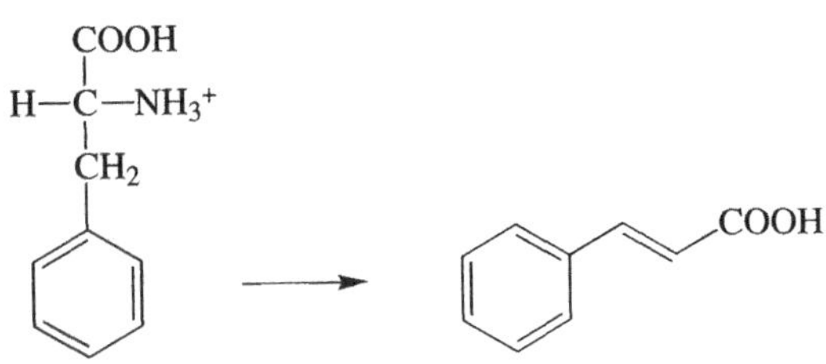

Phenylalanine *trans*-Cinnamic acid

From this point, simple coumarins which are oxygenated at C-7 follow a biosynthetic route different to those which are not oxygenated at this position. In the latter case, the *trans*-cinnamic acid is first 2′-hydroxylated, then glucosylated to give *trans*-2′-glucosyloxycinnamic acid. This is then converted to the *cis*-isomer:

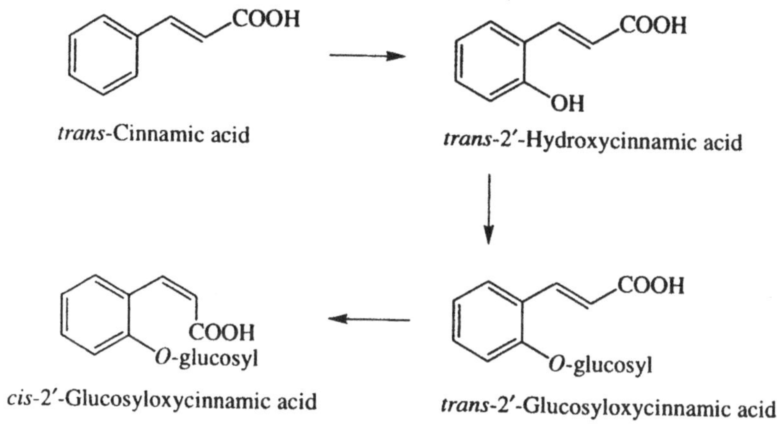

cis-2′-Glucosyloxycinnamic acid is also known as coumarinyl glucoside, and is the bound form in which coumarin exists in plants. The hydroxylation and glucosylation of the *trans*-cinnamic acid are enzyme-catalysed reactions, and both enzymes involved have been identified[151,152]. The *trans-cis* isomerisation, however, is mediated by UV light[153].

The 7-oxygenated simple coumarins arise as a result of *para*-hydroxylation, rather than the *ortho*-hydroxylation which is the first step in the formation of the non-oxygenated compounds. As in the previous reaction scheme, a glucosylation reaction is also involved, followed by UV-dependent *trans-cis* isomerisation[154]. The scheme below shows the formation of *cis*-2′-glucosyl-4-methoxycinnamic acid, the bound form of herniarin (7-methoxycoumarin, found in *Lavandula officinalis*):

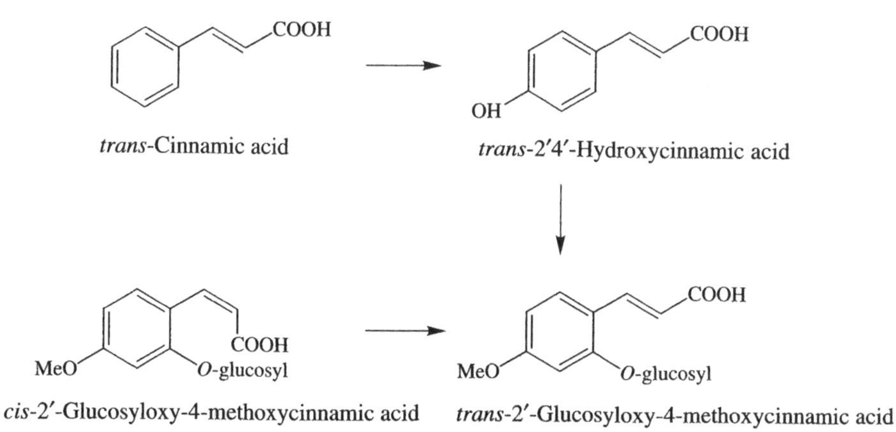

The structure of the unbound compound, herniarin, is shown below:

Herniarin 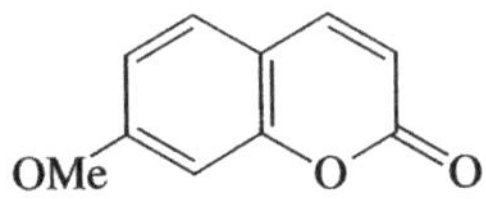

The furanocoumarins (both linear and angular) are produced using 7-hydroxycoumarin as the parent compound, with the other two carbons of the furan ring coming from mevalonic acid. The mevalonate derivative dimethylallylpyrophosphate is involved in the prenylation of position 6 of 7-hydroxycoumarin to form the furanocoumarin precursor demethylsuberosin (DMS).

Dimethylallylpyrophosphate 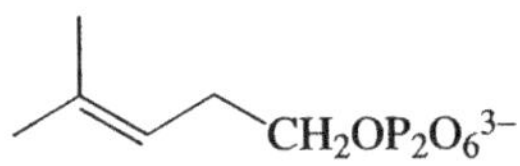

It is thought that the double bond of the prenyl group in DMS undergoes epoxidation[154] prior to cyclisation to a dihydrofuranocoumarin, which is then converted into a furanocoumarin, as in the synthesis of psoralen:

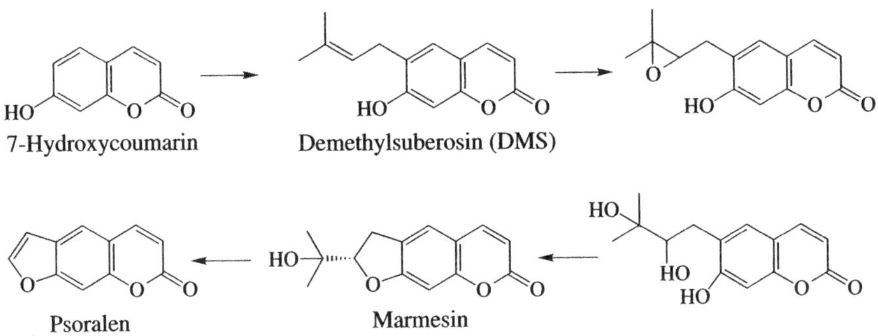

The formation of angular furanocoumarins proceeds by an analogous series of reactions, with, for example, angelicin being formed from 7-hydroxycoumarin via colombianetin[156], following prenylation at position 8 to give osthenol. In the case of furanocoumarins which possess hydroxyl or methyl ether groups, *O*-methylation and hydroxylation appear to occur after psoralen or angelicin have been fully formed.

It seems most likely that the linear and angular pyranocoumarins are formed by a biosynthetic route similar to that of the furanocoumarins, differing only in the last cyclisation step. Although the shikimate–chorismate pathway is the predominant route for the biosynthesis of coumarin compounds, including the 4-hydroxycoumarins, exceptions do exist. The phenylcoumarins, such as coumestrol, have been found to originate partly from the shikimate–chorismate pathway and partly from the polyketide pathway[157], while the aflatoxins are synthesised entirely from acetate via polyketide intermediates like averufin[158] and versicolorin A[159].

2.5 CHEMICAL SYNTHESIS

The classical reaction for the formation of coumarin—the Perkin reaction—involves heating *o*-hydroxybenzaldehyde with sodium acetate and acetic anhydride at 180 °C:

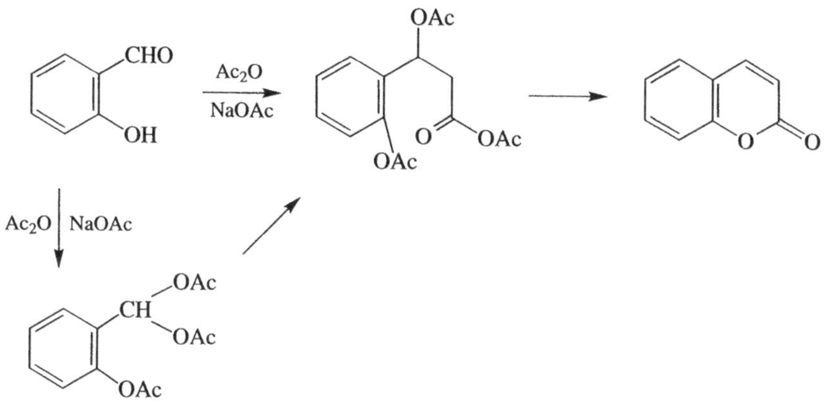

Since this reaction was first used for the preparation of coumarin in 1868, many of the natural coumarins have been chemically synthesised. The synthesis of simple coumarins with methoxy or hydroxy groups is generally still carried out by means of the Perkin reaction, despite the fact that yields are frequently poor.

As the formation of the pyrone ring is the most important step in many of the chemical syntheses of coumarin, several methods have been developed to introduce functional groups into this structure. One strategy is to prepare a phenol which contains the desired substituents of the coumarin before the pyrone ring is formed. Alternatively the coumarin nucleus may be synthesised first, and a process such as *C*- or *O*-alkylation or nuclear oxygenation employed to produce the required compound.

Chemical syntheses have often been used as an aid in elucidating the structures of newly isolated coumarin compounds. Table 2.2 lists the coumarins that have been synthesised since 1988.

2.5.1 Synthesis of Coumarin–Protein Conjugates

For some applications, for example, immobilisation of coumarins on sensor surfaces or immunisation of animals, it is necessary to produce protein conjugates of coumarin compounds. A 7-hydroxycoumarin conjugate of bovine serum albumin (BSA), which has been used in biosensor-based analytical methods, was prepared by the formation of 3-acetylamino-7-acetoxycoumarin, followed by its conversion to 3-amino-7-hydroxycoumarin[267] and the generation of a diazonium ion and conjugation

Table 2.2 Chemically synthesised coumarins

Compound	Authors	Reference
Acrimarins	Reisch *et al.*, 1991	160
6-Acetyl-3,4-dimethyl-7-hydroxy-2*H*-1-benzopyran-2-one	Nicolaides *et al.*, 1994	161
6-Acetyl-7-hydroxy-4-methyl-2*H*-1-benzopyran-2-one	Nicolaides *et al.*, 1994	161
Aflatoxin B_1 8,9-epoxide	Iyer and Harris, 1993	162
3-Alkyl-4-hydroxycoumarins	Appendino *et al.*, 1991	163
Aminomethylangelicins	Rao and Trivedi, 1991	164
4-Aminomethyl-dihydroangelicins	Rao and Trivedi, 1991	165
4-Aminomethyl-dihydropsoralens	Rao and Trivedi, 1991	165
Aminomethylpsoralens	Rao and Trivedi, 1991	164
Angelicin	Wulff *et al.*, 1988	166
	Ahluwalia *et al.*, 1988	167
	Zubia *et al.*, 1992	168
	Lee, 1995	169
Apetalolide	Palmer and Josephs, 1994	170
Arnottinin	Banerji *et al.*, 1988	171
3-Aroyl-2-aryl-4*H*-1-benzopyran-4-ones	Chawla and Sharma, 1990	172
3-*C*-Aroyl-2-aryl-4*H*-1-benzopyran-4-ones	Chawla and Sharma, 1990	173
6-Arylazocoumarins	Rajagopal and Seshadri, 1990	174
4-Arylcoumarin	Wattansin, 1988	175
4-Arylethenyl-6,7-dihydropyranocoumarins	Ahluwalia *et al.*, 1989	176
4-Arylethenylpyranocoumarins	Ahluwalia *et al.*, 1989	176
2-Arylidene-coumaran-3-ones	Chawla and Sharma, 1990	177
8-Aryl-7-phenyl-9*H*-benzofuro[3,2-*g*][1]benzopyran-9-ones	Ahluwalia and Khanduei, 1989	178
6-Aryl-5-phenyl-7*H*-furo[3,2-*g*]benzopyran-7-ones	Ahluwalia and Khanduri, 1989	178
Ayapin	Soman and Trivedi, 1993	179
Balsamiferone	Cairns *et al.*, 1994	180
5,6-Benzocoumarin	Harvey *et al.*, 1988	181
7,8-Benzocoumarin	Harvey *et al.*, 1988	181
5,6-Benzocoumarin	Skripskaya *et al.*, 1992	182
3-(Benzothiazol-2′-ylthio)-2*H*-1-benzopyran-2-ones	Ahluwalia *et al.*, 1989	183
3-Benzoylamino-5-oxo-5,6,7,8-tetrahydrocoumarins	Kocevar *et al.*, 1989	184
6-Benzylated coumarins	Kondedesmukah and Paradkar, 1993	185
3-Benzylcoumarins	Britto *et al.*, 1989	186
3-Benzyl-4-substituted coumarins	Britto *et al.*, 1989	186
Bergapten	Danheiser and Trova, 1995	187
Bromocoumarin	Harayama *et al.*, 1994	188
5-Bromoesculetin	Borges and Pinto, 1992	189
4-Butylcoumarin	Clinet, 1988	190

Table 2.2 (*continued*)

Compound	Authors	Reference
Calophyllolide	Palmer and Josephs, 1994	170
6-Chloroesculetin	Borges and Pinto, 1992	189
6-Chlorocoumarin	Harvey *et al.*, 1988	181
Citrusarin B	Ito *et al.*, 1991	53
Coumarin	Harvey *et al.*, 1988	181
Coumarin 3-*O*-acylisoureas	Bonsignore *et al.*, 1995	191
Coumarin azomethines	Tkach *et al.*, 1990	192
Coumarin-3-carboxylic acid	Shirokova *et al.*, 1988	193
Coumarin-containing phosphonates	Vishnyakova and Tarakanova, 1991	194
Coumarin crown ethers	Bulut and Erk, 1992	195
Coumarin 8-C-glycoside	Mahling and Schmidt, 1995	196
Coumarin-6-sulphonyl chloride	Al-Kindy and Miller, 1989	197
Coumestan	Pandit and Gadre, 1988	198
	Laschober and Kappe, 1990	199
Coumestans	Bhalerao *et al.*, 1989	200
	Pandey *et al.*, 1989	201
	Mali and Tilve, 1990	202
	Litinas and Stampelos, 1992	203
Coumestrol	Laschober and Kappe, 1990	199
Coumestrol methyl ether	Pandit and Gadre, 1988	198
Cyclobisuberodiene	Reisch *et al.*, 1990	204
Dehydrogeijerin	Reisch and Bathe, 1988	205
	Cairns *et al.*, 1992	206
trans-Dehydroosthol	Murray and Zeghdi, 1989	207
Demethylsuberosin	Cairns *et al.*, 1994	180
7-Diethylamino-3-(2-arylethenyl) coumarins	Tkach *et al.*, 1992	208
Diethyl coumarin-3-phosphonates	Bouyssou and Chenault, 1991	209
8-(Dihydrocinnamoyl) dihydrocoumarins	Ahluwalia *et al.*, 1988	210
Dihydrofurocoumarins	Rao and Trivedi, 1992	211
3,4-Dihydro-4-hydroxy pyrano coumarins	Nandaiah and Krupadanam, 1993	212
Dihydrosuberenol	Reisch *et al.*, 1991	213
4,7-Dihydroxycoumarin	Mizuno *et al.*, 1988	214
6,7-Dimethoxycoumarin-3-carboxylic acid	Hormi *et al.*, 1991	215
5,7-Dimethoxy-7′,6-oxydicoumarin	Reisch *et al.*, 1989	216
6,7-Dimethoxy-7′,5-oxydicoumarin	Reisch *et al.*, 1989	216
3-(1′,1′-Dimethylallyl)seselin	Hernandez Galan *et al.*, 1989	217
3,4-Dimethyldihydropyranocoumarins	Ahluwalia *et al.*, 1988	218
3,4-Dimethylpsoralen	Ahluwalia *et al.*, 1988	167
4,6-Dimethylpyrano[3,2-g]chromene-2,8-dione	Nicolaides *et al.*, 1994	161
1,5-Dioxano coumarins	Prashant and Krupadanam, 1993	219
Diphenyl coumarin-3-phosphinyloxide	Bouyssou and Chenault, 1991	209
2,3-Diphenyl-9-methyl-7*H*-furo[3,2-*g*]benzopyran-7-ones	Ahluwalia *et al.*, 1988	220

(*continues*)

Table 2.2 (*continued*)

Compound	Authors	Reference
DMACGO	Baburaj *et al.*, 1994	221
Esculetin	Borges and Pinto, 1992	189
7-Ethoxy-3,4-dimethylcoumarin	Ahluwalia *et al.*, 1988	222
4-Ethylcoumarin	Clinet, 1988	190
	Patra and Misra, 1990	223
Ethyl *E*-3-(7-Ethoxy-4-methyl-2-oxo-2*H*-1-benzopyran-6-yl)-2-butenoate	Nicolaides *et al.*, 1994	161
Ethyl *E*-3-(7-Hydroxy-4-methyl-2-oxo-2*H*-1-benzopyran-6-yl)-2-butenoate	Nicolaides *et al.*, 1994	161
(*E*)-*O*-Ethylsuberenol	Reisch *et al.*, 1991	213
8-Ethyltrimethoxy-6*H*-benzo[*d*]naphtho-[1,2-*b*]pyran-6-ones	Hua *et al.*, 1992	224
Fatagarine	Reisch *et al.*, 1988	225
3-(2-Formylphenoxymethyl)coumarin-2-one	Bode *et al.*, 1992	226
Furanocoumarins	Rao *et al.*, 1991	227
Furocoumarins	Hager *et al.*, 1990	228
	Soman and Trivedi, 1992	229
	Atta *et al.*, 1993	230
	Rao *et al.*, 1994	231
Furo-coumestans	Soman and Trivedi, 1993	232
Furodihydroisocoumarin	Bhide *et al.*, 1992	233
Furoisocoumestans	Soman and Trivedi, 1991	234
3,4-Fused coumarins	Minami *et al.*, 1992	235
Geijerin	Cairns *et al.*, 1992	206
Gleinadiene	Reisch *et al.*, 1990	236
Gleinene	Reisch *et al.*, 1990	236
Gravelliferone	Cairns *et al.*, 1994	180
Heratomin	Wulff *et al.*, 1988	166
4-(Herniarin-3-yl)-2-methyl-3-buten-2-ol	Reisch and Zappel, 1992	237
Heterocyclic [*c*]-fused coumarins	Steinfuehrer *et al.*, 1992	238
HOCGO	Baburaj *et al.*, 1994	221
3-(2-Hydroxyaryl)coumarins	Litinas and Stampelos, 1992	203
7-Hydroxycoumarin-3-carboxylic acid	Shirokova *et al.*, 1988	193
Hydroxycoumarins	Harayama *et al.*, 1994	239
4-Hydroxy-7-methoxycoumarin	Mizuno *et al.*, 1988	87
(+)-8-Hydroxy-3-methyl-3,4-dihydroisocoumarins	Mali *et al.*, 1992	240
8-Hydroxy-1-methyl-3*H*-naphtho [2,1-*b*]pyran-3-one	Oyman and Gunaydin, 1994	241
8-Hydroxy-4-methyl-2*H*-naphtho [2,3-*b*]pyran-2-one	Oyman and Gunaydin, 1994	241
2(*N*-*R*-Imino) benzopyrones	Zubkov *et al.*, 1994	242
Inophyllum E,C	Palmer and Josephs, 1994	170
6-Iodoumbelliferone	Zubia *et al.*, 1992	168
Isomurralonginol	Allison *et al.*, 1991	243

Table 2.2 (*continued*)

Compound	Authors	Reference
Isomurralonginol acetate	Allison *et al.*, 1991	243
Isophellopterin	Reed and Moore, 1988	244
3-Isoprenyl-4-hydroxycoumarins	Appendino *et al.*, 1992	245
Menthadienediylbis(coumarin)	Reisch *et al.*, 1991	213
Methoxycarbonylcoumarin	Harayama *et al.*, 1994	188
7-Methoxycoumarin-3-carboxylic acid	Shirokova *et al.*, 1988	193
Methoxycoumarins	Harayama *et al.*, 1994	239
5-Methoxycoumarin-3-carboxylic acid	Hormi *et al.*, 1991	215
6-Methoxycoumarin-3-carboxylic acid	Hormi *et al.*, 1991	215
7-Methoxycoumarin-3-carboxylic acid	Hormi *et al.*, 1991	215
8-Methoxycoumarin-3-carboxylic acid	Hormi *et al.*, 1991	215
5-Methoxy-4-methylcoumarin	Ahluwalia *et al.*, 1988	222
7-Methoxy-4-methylcoumarin-6-sulphonylamino acids	Ibrahim *et al.*, 1994	246
6/7-Methoxy-8/9-methylcycloalkeno[*c*] psoralens	Soman and Trivedi, 1992	247
3-Methyl-4-benzocoumarin	Clinet, 1988	190
3-Methylcoumarin	Clinet, 1988	190
4-Methylcoumarin	Clinet, 1988	190
	Patra and Misra, 1990	223
	Biswas *et al.*, 1992	248
6-Methylcoumarin	Harvey *et al.*, 1988	181
3-Methyl-3,4-dihydro-5-hydroxyisocoumarin	Bhide *et al.*, 1992	233
Methylfuroquinolinones	Guiotto *et al.*, 1989	249
(*E*)-*O*-Methylsuberenol	Reisch *et al.*, 1991	213
Methyl-substituted benzofurocoumarins	Ahluwalia *et al.*, 1988	250
Methyltriazolocoumarins	Rodighiero *et al.*, 1990	251
Microminutin	Allison *et al.*, 1991	243
Murralongin	Allison *et al.*, 1991	243
Murraol	Reisch and Bathe, 1988	205
	Murray and Zeghdi, 1989	207
6-Naphthoherniarin	Reisch and Bathe, 1988	205
8-Naphthoherniarin	Reisch and Bathe, 1988	205
Nieshoutin	Mali *et al.*, 1992	252
Nitrocoumarins	Harayama *et al.*, 1994	253
Oblongulide	Palmer and Josephs, 1994	170
Oreojasmine	Reisch *et al.*, 1988	225
Oroselol	Lee, 1995	169
Oroselone	Lee, 1995	169
Ostruthin	Cairns *et al.*, 1994	180
3-Perfluoroalkylated coumarins	Matsui *et al.*, 1991	254
	Huang *et al.*, 1994	255
4-Phenylcoumarin	Patra and Misra, 1990	223
6-Phenylcoumarin	Harvey *et al.*, 1988	181
4-(Phenylethyl)-3-(1-propenyl)-coumarin	Castellani *et al.*, 1994	256

(*continues*)

Table 2.2 (*continued*)

Compound	Authors	Reference
4-Phenylethyl-3-styryl-coumarin	Castellani *et al.*, 1994	256
Pimpinellin	Reed and Moore, 1988	244
4-Propylcoumarin	Patra and Misra, 1990	223
Psoralen	Zubia *et al.*, 1992	168
Pyranocoumarins	Rao *et al.*, 1991	227
	Atta *et al.*, 1993	230
	Rao *et al.*, 1994	231
Pyrano[3,2-*c*]coumarins	Appendino *et al.*, 1994	257
Pyrano-isocoumestans	Soman and Trivedi, 1993	232
Sphondin	Wulff *et al.*, 1988	166
(*E*)-Suberene	Reisch *et al.*, 1990	204
(*E*)-Suberenol	Reisch *et al.*, 1990	204
(*E*)-Suberodiene	Reisch *et al.*, 1990	204
Suberosin	Cairns *et al.*, 1994	180
3,4-Substituted coumarins	Gaon *et al.*, 1991	258
4,5-Substituted coumarins	Ishii *et al.*, 1991	259
4′-Substituted furocoumarins	Koenig and Otsuki, 1991	260
Swietenocoumarin G	Murray and Zeghdi, 1989	207
8-*tert*-Butyl-6-hydroxy-4-methyl coumarin	El-Farargy *et al.*, 1991	261
Thiocoumarins	Huang *et al.*, 1994	255
Thiosphondin	Wulff *et al.*, 1988	166
3-Thiosubstituted coumarins	Ahluwalia *et al.*, 1989	262
Toddacoumaquinone	Ishikawa *et al.*, 1995	263
Tormentolide A,B	Palmer and Josephs, 1994	170
3,5,8-Trimethylcoumarin	Rautenstrauch *et al.*, 1989	264
3,3′-Umbelliferone	Reisch and Zappel, 1992	265
3-Ureido coumarins	Tawada *et al.*, 1995	266

to protein[268,269]. A human serum albumin (HSA) conjugate of warfarin for immunisation has also been prepared[270].

2.6 CHARACTERISTIC CHEMICAL REACTIONS

The natural coumarins undergo a wide range of different chemical reactions, many of which have been used to elucidate the structure of newly isolated compounds. All members of the family will react with alkali, which can result in the lactone ring being fragmented or opened (depending on the relative vigour of the treatment). Reaction of coumarin with dilute sodium hydroxide leads to ring-opening and the formation of sodium coumarinate.

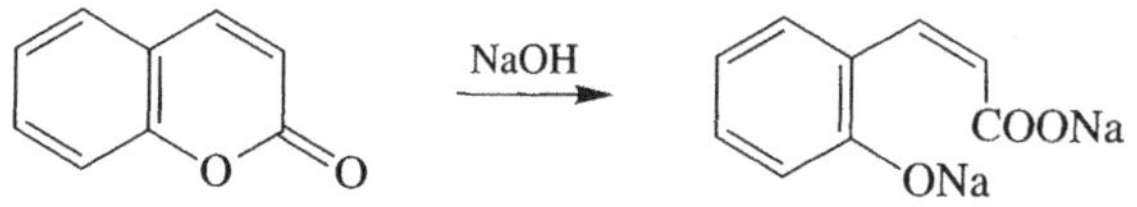

Harsher reaction conditions with alkali result in the production of acetic acid, along with fission products of the lactone ring. In the following reaction, mammeisin is heated at reflux together with 15% aqueous potassium hydroxide for 67 h, producing 3-methylbutanoic acid, isopentylphloroglucinol and acetophenone[271]:

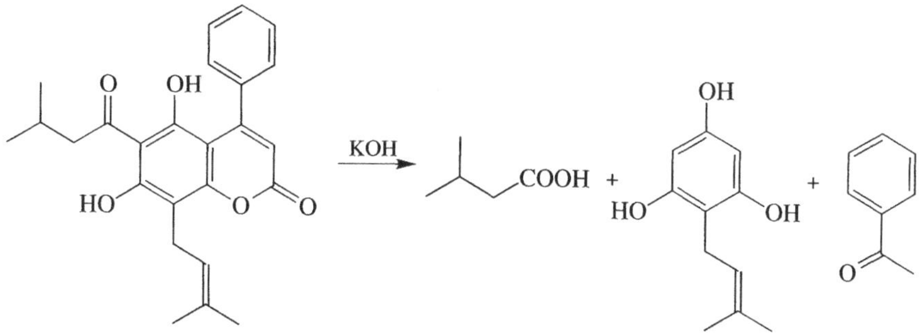

Side-chains of the compounds may also be degraded as a result of reaction with alkali. One example of this is the degradation of the seneciocyl side-chain of glabralactone by the action of hot 30% potassium hydroxide[272]:

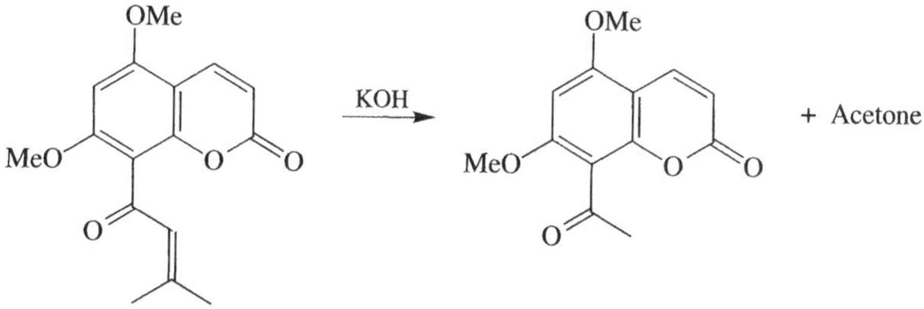

Ester hydrolysis may also occur.

Reactions of coumarins with acid are also known. Naturally occurring vinyl ethers can be hydrolysed; for example, the hydrolysis of peucedanin to give oreoselone:

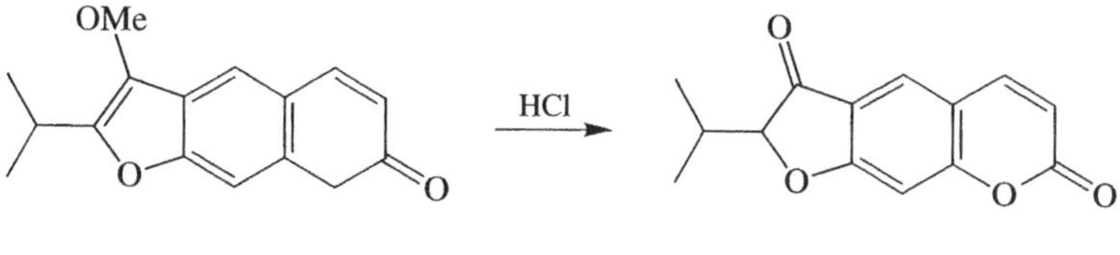

Many other ether residues in coumarin compounds will undergo hydrolytic cleavage. Coumarins containing an unsaturated side-chain *ortho* to a phenolic group will undergo cyclisation under mildly acidic conditions. Treatment of obliquetin with hydrogen bromide, for example, yields cycloobliquetin[273]:

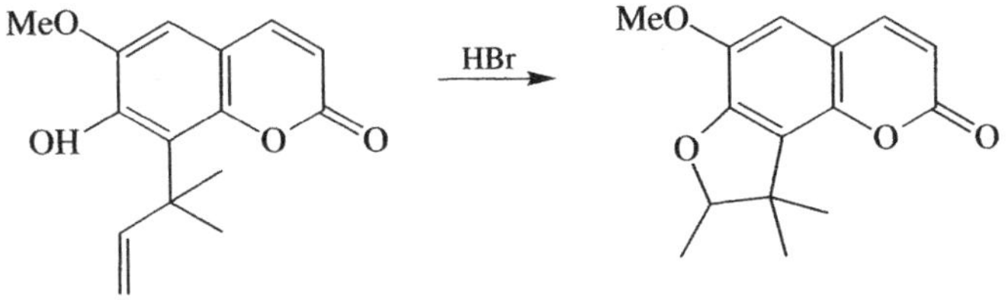

Deacylation and dealkylation may also be observed. One example of a dealkylation reaction is afforded by the reaction between furopinnin and acetic acid with sulphuric acid at 100 °C, resulting in the production of bergapten[274]:

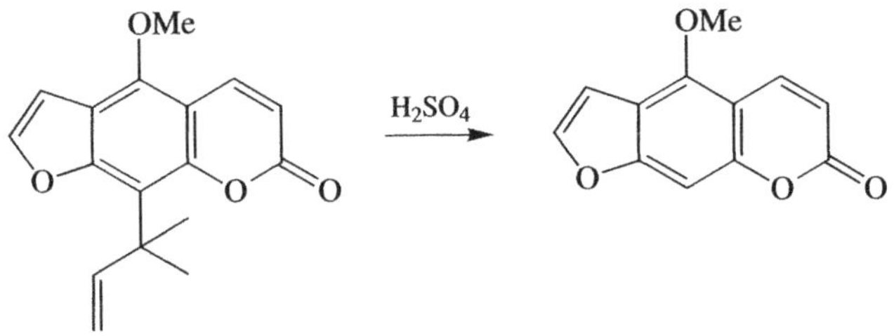

The conversion of leptophyllidin to leptophyllin (by treating with a 2 : 1 mixture of acetic acid and hydrochloric acid), shows that coumarins can undergo alkene hydration[275]:

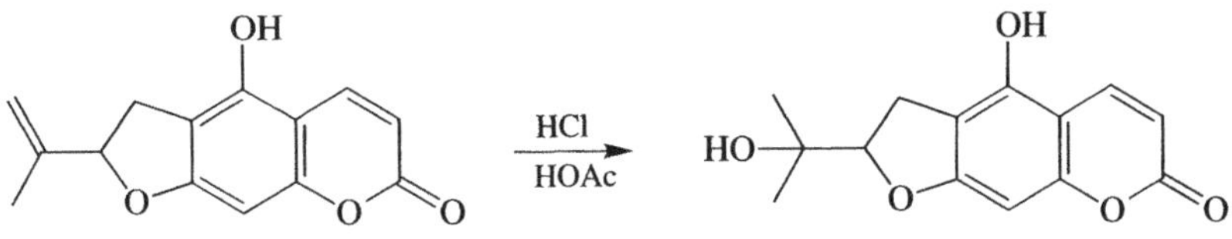

(*R*)-(+)-epoxyaurapten is converted to (*R*)-(+)-marmin by hydrolysis of the epoxide group. This involves reaction with sulphuric acid in tetrahydrofuran[276]:

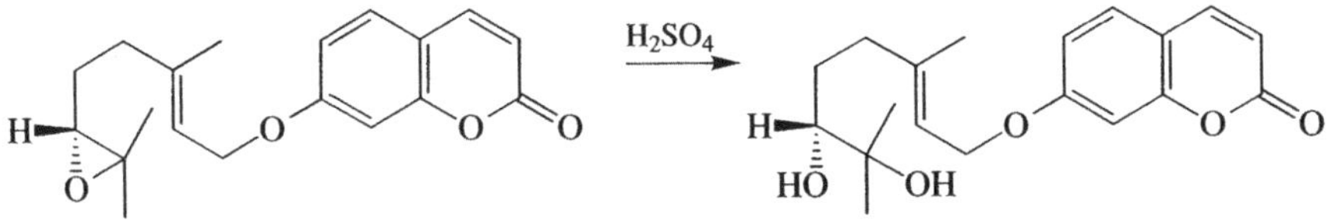

Treatment of apterin with 4% sulphuric acid results in dehydration and the formation of oroselone[277]:

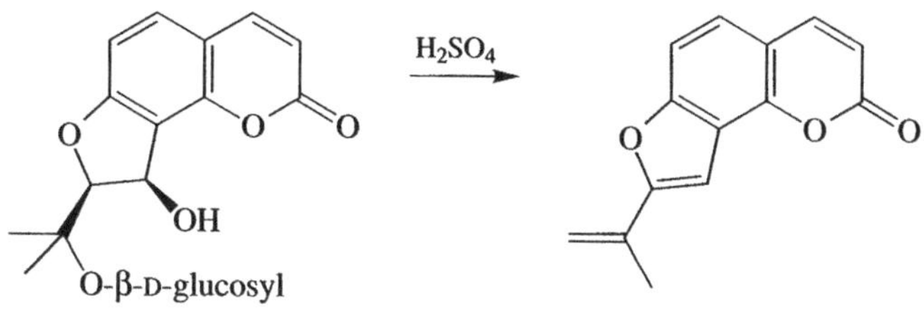

Oroselone can also be formed by the elimination of the ester group of athamantin, following treatment with 1 M HCl, displaying another characteristic acidic reaction[278]:

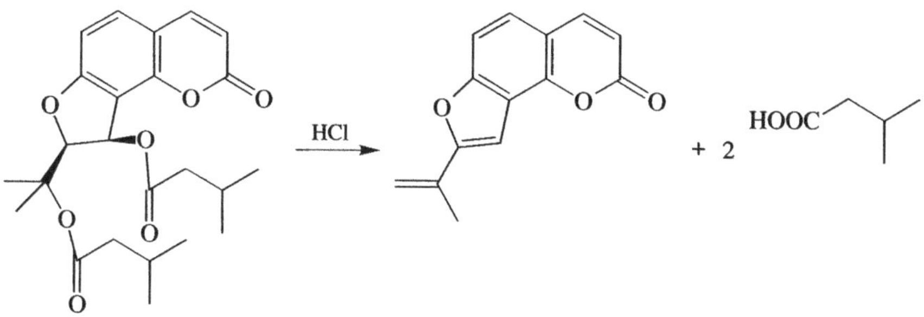

Isomerisation of allylic alcohols and epoxides also occurs upon treatment of coumarins with acid.

Coumarins may also undergo reduction and oxidation reactions. Reduction methods which have been investigated include catalytic hydrogenation, hydrogenolysis and the use of metal-ammonia, sodium amalgam and sodium borohydride.

Furropinnarin has been converted to tetrahydrofuropinnarin using palladium-charcoal as the catalyst[279]:

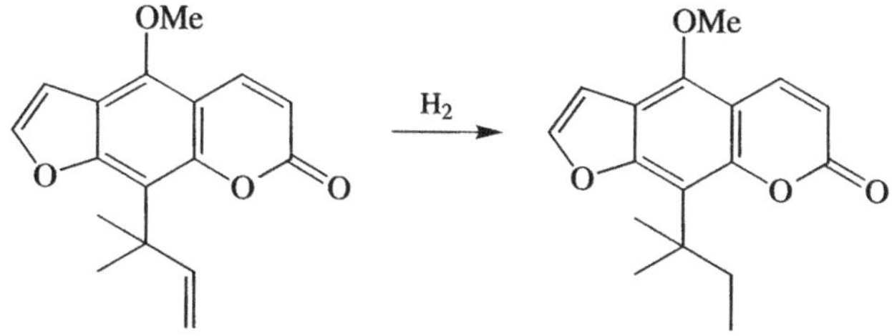

The production of columbianetin 3-methylbutanoate from peucedanidin, achieved by hydrogenation over platinum oxide, is an example of hydrogenolysis[280]:

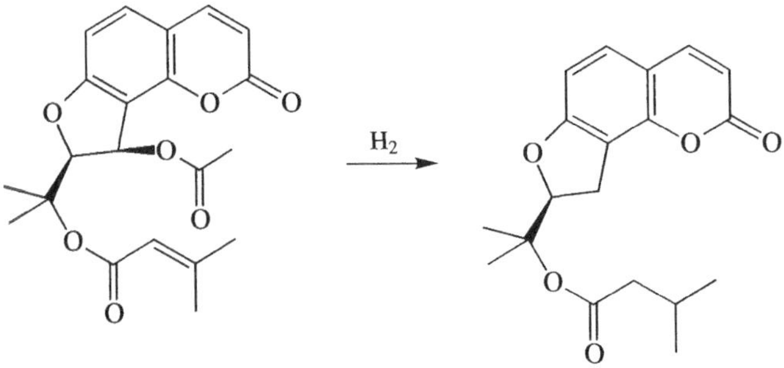

Oxidation of coumarin compounds has been achieved by means of potassium permanganate, chromium trioxide, nitric acid, ozone, hydrogen peroxide and other agents. Heliettin has been oxidised using potassium permanganate to give 3,3-dimethylpentan-2-onoic acid[281]:

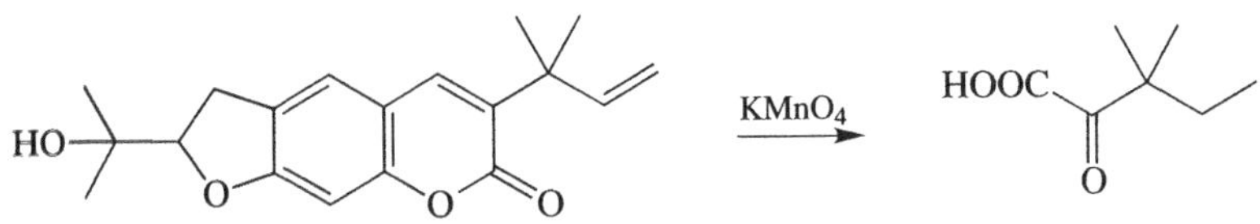

Hydrogen peroxide oxidation of sphondin leads to the production of furan-2,3-dicarboxylic acid[282]:

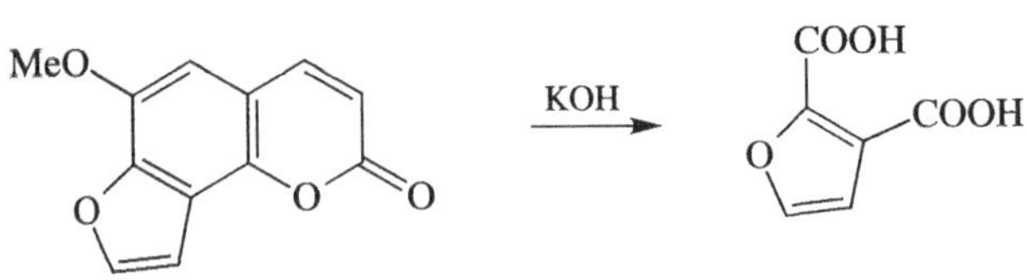

In addition to the various reactions with acid and alkali, and reduction and oxidation reactions, pyrolysis, bromination and esterification of the coumarins have all been reported. Imperatorin is converted to alloimperatorin by heating to 200 °C.[13]:

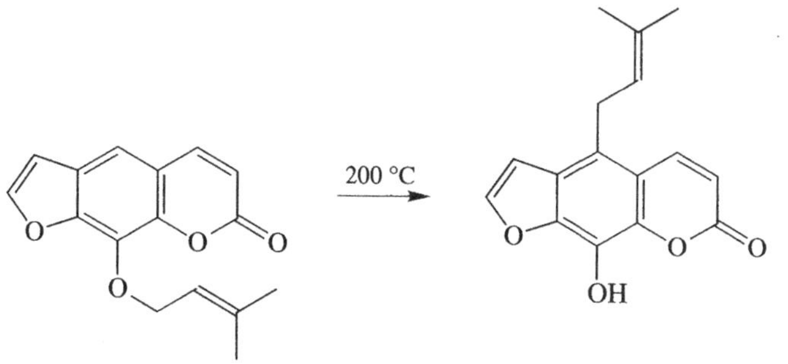

Bruceol has been brominated in dioxan and sodium carbonate to yield 8-bromo-bruceol[283]:

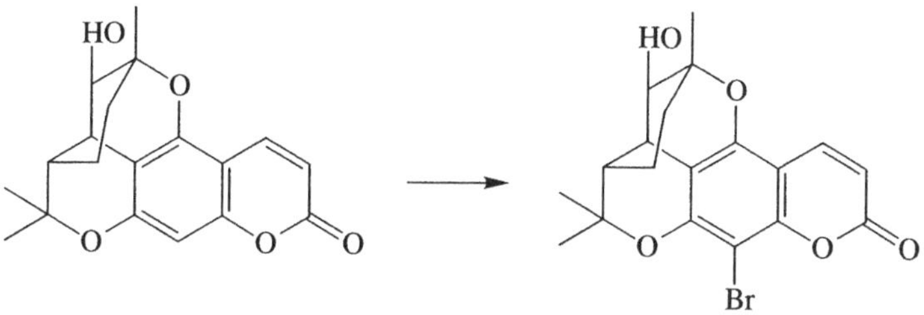

The formation of vaginidin acetate from vaginidin, upon addition of acetic anhydride in pyridine followed by reflux, is an example of an esterification reaction[284]:

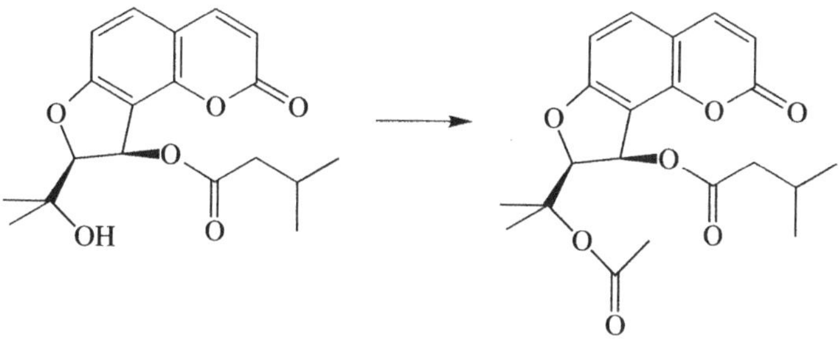

2.7 SUMMARY

It is apparent from the list of almost 300 newly discovered coumarins given in Table 2.1 that much interest still persists in this abundant and useful family of compounds. The coumarins isolated from natural sources since 1988 range from simple members of the group and their glucosides and glycosides to more structurally complex molecules, such as the coumarinolignans and dimeric coumarins, and have been extracted from an ever-growing number and variety of plants.

The chemically synthesised coumarins listed in Table 2.2 have been produced by many different methods, demonstrating the wide diversity of chemical reactions undergone by members of the group.

REFERENCES

1 Thaisrivongs S, Tomich PK, Watenpaugh KD, Chong KT, Howe WJ, Yang CP, Strohbach JW, Turner SR, McGrath JP, Bohanon MJ, Lynn JC, Mulichak AM, Spinelli PA, Hinshaw RR, Pagano PJ, Moon JB, Ruwart MJ, Wilkinson KF, Rush BD, Zipp GL, Dalga RJ, Schwende FJ, Howard GM, Padbury GE, Toth LN, Zhao Z, Koeplinger KA, Kakuk TJ, Cole SL, Zaya RM, Piper RC, Jeffrey P. Structure-based design of HIV protease inhibitors: 4-Hydroxycoumarins and 4-hydroxy-2-pyrones as non-peptide inhibitors. *J Med Chem* 1994; **37**: 3200–4.

2 Park SJ, Lee EJ, Lee DH, Lee SH, Kim SJ. Spectrofluorimetric assay for rapid detection of total and fecal coliforms from surface water. *Appl Environ Microbiol* 1995; **61**: 2027–9.

3 Zimmerman M, Ashe B, Yurewicz EC, Patel G. Sensitive assays for trypsin, elastase and chymotrypsin using new fluorogenic substrates. *Anal Biochem* 1977; **78**: 47–51.

4 de Iongh H, Vles RO, Pelt JG. Milk of mammals fed an aflatoxin-containing diet. *Nature* 1964; **202**: 466.

5 Dalezios JI, Wogan GN, Weinreb SM. Aflatoxin P—new aflatoxin metabolite in monkeys. *Science* 1971; **171**: 584.

6 Hsieh DPH, Dalezios JI, Krieger RI, Masri MS, Haddon WF. Use of monkey liver microsomes in the production of aflatoxin Q. *J Agric Food Chem* 1974; **22**: 515.

7 Salhab AS, Hsieh DPH. Aflatoxicol H_1—major metabolite of aflatoxin B_1, produced by human and rhesus monkey livers. *Res Commun Chem Pathol Pharmacol* 1975; **10**: 419–31.

8 Schabort JC, Steyn M. Substrate and phenobarbital inducible aflatoxin-4-hydroxylation and aflatoxin metabolism by rat liver. *Biochem Pharmacol* 1969; **18**: 2241–52.

9 Menzel D, Kazlauskas R, Reichelt J. Coumarins in the siphonalean green algal family *Dasycladaceae kuetzing* (Chlorophyceae). *Bot Mar* 1983; **26**: 23–9.

10 Nozawa K, Seyea H, Nakajima S, Udagama S, Kawai K. Studies on fungal products. 10. Isolation and structures of novel bicoumarins, desertorin A, desertorin B and desertorin C, from *Emericella desertorum*. *J Chem Soc Perkin Trans 1* 1987; **8**: 1735–8.

11 Suortti T, von Wright A. Isolation of a mutagenic fraction from aqueous extracts of the wild edible mushroom *Lactarius necator*. *J Chromatogr* 1983; **255**: 529–32.

12 Andersen RJ, Faulkner DJ, Cun-heng H, Van Duyne GD, Clardy J. Metabolites of the marine prosobranch mollusk *Lamellaria* sp. *J Am Chem Soc* 1985; **107**: 5492–5.

13 Dean FM. *Naturally Occurring Oxygen Ring Compounds*. London: Butterworth, 1963.

14 Murray RDH, Mendez J, Brown SA. *The Natural Coumarins—Occurrence, Chemistry, and Biochemistry*. Chichester: John Wiley, 1982.

15 Murray RDH. Coumarins. *Natural Product Reports* 1989; 591–624.

16 Chen IS, Lin YC, Tsai IL, Teng CM, Ko FN, Ishikawa T, Ishii H. Coumarins and antiplatelet aggregation constitutents from *Zanthoxylum schinifolium*. *Phytochemistry* 1995; **39**: 1091–9.

17 Appendino G, Ozen HC, Tagliapietra S, Cisero M. Coumarins from *Heptaptera anisoptera*. *Phytochemistry* 1992; **31**: 3211–13.

18 Kashman Y, Gustafson KR, Fuller RW, Cardellina JH, McMahon JB, Currens MJ, Buckheit RW, Hughes SH, Cragg GM, Boyd MR. The calanolides, a novel HIV-inhibitory class of coumarin derivatives from the tropical rainforest tree *Calophyllum lanigerum*. *J Med Chem* 1992; **35**: 2735–43.

19 Kayser O, Kolodziej H. Highly-oxygenated coumarins from *Pelargonium sidoides*. *Phytochemistry* 1995; **39**: 1181–5.

20 Zdero C, Bohlmann F, Niemeyer HM. 5-Methylcoumarin derivatives from *Aphyllocladus denticulatus*. *Phytochemistry* 1988; **27**: 1821–5.

21 Appendino G, Tagliapietra S, Garibaldi P, Nano GM, Picci V. Omega-oxygenated prenylated coumarins from *Ferula communis*. *Phytochemistry* 1988; **27**: 3619–24.

22 Appendino G, Tagliapietra S, Nano GM, Jakupovic J. Sesquiterpene coumarin ethers from *Asafetida*. *Phytochemistry* 1994; **35**: 183–6.

23 Appendino G, Ozen HC, Jakupovic J. Prenylated isocoumarins from *Bunium paucifolium*. *Phytochemistry* 1994; **36**: 531–2.

24 Mata R, Calzada F, Garcia MR. Chemical studies on Mexican plants used in traditional medicine. 6. Additional new 4-phenylcoumarins from *Exostema caribaeum*. *J Nat Prod* 1988; **51**: 851–6.

25 Khan SA, Siddiqui MJ, Dar BA, Shamsuddin KM. Furocoumarins from *Angelica glauca* Edgw. *Indian J Chem B* 1989; **28**: 95–6.

26 Duh CY, Wang SK, Wu YC. Cytotoxic pyranocoumarins from the aerial parts of *Peucedanum japonicum*. *Phytochemistry* 1991; **30**: 2812–14.

27 Ju-Ichi M, Inque M, Kajiura I, Omura M, Ito C, Furukawa H. Constituents of domestic citrus plants. 11. The structure of acrimarins, the first naturally-occurring acridone-coumarin dimers. *Chem Pharm* 1988; **36**: 3202–5.

28 Matsuda H, Kikuchi M. A coumarin glycoside from *Lonicera gracilipes* var *glandulosa*. *Phytochemistry* 1995; **38**: 803–4.

29 Pei YH, Li X, Zhu TR. Studies on the structure of a new isocoumarin glucoside of the root sprouts of *Agrimonia pilosa* Ledeb. *Yao Hsueh Hsueh Pao* 1989; **24**: 837–40.

30 Joshi PC, Mandal S, Das PC, Chaterjee A. Albiflorin I, a coumarin from *Boenninghausen albiflora*. *Phytochemistry* 1991; **30**: 2094–6.

31 Joshi PC, Mandal S, Das PC, Chaterjee A. 2 minor coumarins of *Boenninghausen albiflora*. *Phytochemistry* 1993; **32**: 481–3.

32 Fozdar BI, Khan SA, Shamsuddin T, Shamsuddin KM, Kitzinger JP. Aleuritin, a coumarinolignoid, and a coumarin from *Aleurites fordii*. *Phytochemistry* 1989; **28**: 2459–61.

33 Amaroluis JM, Massanet GM, Pando E, Rodriguez-Luis F, Zubia E. New coumarins from *Pilocarpus goudotianus*. *Planta Med* 1990; 304–6.

34 Liu JH, Xu SX, Yao XS, Kobayashi H. Angelol-type coumarins from *Angelica pubescens* F *biserrata* and their inhibitory effect on platelet aggregation. *Phytochemistry* 1995; **39**: 1099–101.

35 Harmala P, Kaltia S, Vuorela H, Hiltunen R. A furanocoumarin from *Dorstenia contrajerva*. *Phytochemistry* 1991; **30**: 4196–7.

36 Xiao YQ, Lin XH, Sun YF, Baba K, Taniguchi M, Kozawa M. Three new furocoumarins from *Notopterygium incisum* Ting. *Chin Chem Lett* 1994; **5**: 593–6.

37 Ngadjui BT, Ayafor JF, Sondengam BL, Connolly JD. Coumarins from *Clausena anisata*. *Phytochemistry* 1989; **28**: 585–9.

38 Ngadjui BT, Ayafor JF, Sondengam BL, Connolly JD. Prenylated coumarins from the leaves of *Clausena anisata*. *J Nat Prod* 1989; **52**: 243–7.

39 Ngadjui BT, Mouncherou SM, Ayafor JF, Sondengam BL, Tillequin F. Geranyl coumarins from *Clausena anisata*. *Phytochemistry* 1991; **30**: 2809–11.

40 Sati SP, Chankiyal DC, Sati OP, Yamada F, Ono M. 2D-NMR structure elucidation of a new coumarin glycoside from *Xeromphis spinosa*. *J Nat Prod* 1989; **52**: 376–9.

41 Banerji A, Mallick B, Chatterjee A, Budzikiewicz A, Brener M. Assafoetidin and ferocolicin, 2 new sesquiterpenoid coumarins from *Ferula assafoetida* Regel. *Tetrahedron Lett* 1988; **29**: 1557–60.

42 Sarker SD, Armstrong JA, Gray AI, Waterman PG. Pyranocoumarins from *Eriostemon apiculatus*. *Biochem Syst* 1994; **22**: 641–4.

43 Kondo Y, Kato A, Kubota Y, Nozoe S. Bakuchinin, a new simple furanocoumarin from *Psoralea corylifolia* L. *Heterocycle* 1990; **31**: 187–90.

44 Gu ZM, Zhang DX, Yang XW, Hattori M, Namba T. Isolation of two new coumarin glycosides from *Notopterygium forbesii* and evaluation of a Chinese crude drug, qiang-ho, the underground parts of *N. incisum* and *N. forbesii*, by high-performance liquid chromatography. *Chem Pharm Bull* 1990; **38**: 2498–502.

45 Rashid MA, Gray AI, Waterman PG, Armstrong JA. Coumarins from *Phebalium tubercolosum* ssp. *megaphyllum* and *Phebalium filifolium*. *J Nat Prod* 1992; **55**: 851–8.
46 Takemura Y, Nakata T, Ju-Ichi M, Okano M, Fukamiya N, Ito C, Furukawa H. Constituents of domestic citrus plants. 23. Three new biscoumarins from *Citrus hassaku*. *Chem Pharm Bull* 1994; **42**: 1213–15.
47 Ito C, Ju-Ichi M, Inque M, Murayuchi M, Mizuno T, Tanahashi S, Omura M, McPhail DR, McPhail AT, Furukawa H. Constituents of domestic citrus plants. 12. Structure of bisosthenon, a novel dimeric coumarin from citrus plants. *Chem Pharm* 1989; **37**: 1957–9.
48 Ito C, Nakagawa M, Inque M, Takemura Y, Ju-Ichi M, Omura M, Furukawa H. A new biscoumarin from citrus plants. *Chem Pharm* 1993; **41**: 1657–8.
49 Huang SC, Chen MT, Wu TS. Alkaloids and coumarins from stem bark of *Citrus grandis*. *Phytochemistry* 1989; **28**: 3574–6.
50 Wu TS, Huang SC, Lai JS. Stem bark coumarins of *Citrus grandis*. *Phytochemistry* 1994; **30**: 217–19.
51 Gustafson KR, Bokesch HR, Fuller RW, Cardellina JH, Kadushin MR, Soejarto DD, Boyd MR. Calanone, a novel coumarin, from *Calophyllum teysmannii*. *Tetrahedron Lett* 1994; **35**: 5821–4.
52 Garcia EE, Guerreiro E. Sesquiterpene lactones from *Gochnatia polosanto* and coumarins from *G. argentina*. *Phytochemistry* 1988; **27**: 288–90.
53 Ito C, Fujiwara K, Kajita M, Ju-Ichi M, Takemura Y, Suzuki Y, Tanaka K, Omura M, Furukawa H. New coumarins from citrus plants. *Chem Pharm* 1991; **39**: 2509–13.
54 Ito C, Tanahashi S, Omura M, Furukawa H. New coumarins from citrus plants. *Chem Pharm* 1989; **37**: 2217–18.
55 Narantuyaa S, Batsuren D, Rashkes YV, Milgrom EG. Chemical study of plants from mongolian flora—Coumarins of *Stellera chamaejasme*—Structure of Chamaejasme, the new coumarin glucoside. *Khim Prir S* 1994; **2**: 216–19.
56 Ito C, Furukawa H, Ishii H, Ishikawa T, Haginiwa J. The chemical composition of *Murraya paniculata*—The structure of 5 new coumarins and one new alkaloid and the stereochemistry of murrangatin and related coumarins. *J Chem Soc Perkin Trans 1* 1990; **7**: 2047–55.
57 Wu TS. Alkaloids and coumarins of *Citrus grandis*. *Phytochemistry* 1988; **27**: 3717–18.
58 Takemura Y, Nakata Y, Azuma M, Ju-Ichi M, Okano M, Fukamiya N, Omura M, Ito C, Nakagawa K, Furukawa H. 5 new pyranocoumarins from some citrus plants. *Chem Pharm* 1993; **41**: 1530–3.
59 Takemura Y, Nakata T, Uchida H, Ju-Ichi M, Hatano K, Ito C, Furukawa H. Structure of claudimarin A, a novel dimeric coumarin from *Citrus hassaku*. *Chem Pharm* 1993; **41**: 2061–2.
60 Prokska B, Uhrin D, Narantuyaa S, Batsuren D. Cleomiscosin B and cleomiscosin D—new coumarinolignoids from *Salsola laricifolia*. *Pharmazie* 1990; **45**: 804.
61 Kumar S, Ray AB, Konno C, Oshima Y, Hikino H. Cleomiscosin D, a coumarino-lignan from seeds of *Cleome viscosa*. *Phytochemistry* 1988; **27**: 1908–9.
62 Li RZ, He YQ, Chiao M, Zhang QB, Meng JR, Gu Y, Ge LP. Studies of the active constituents of the chinese drug 'duhuo' *Angelica pubescens*. *Yao Hsueh Hsueh Pao* 1989; **24**: 546–51.
63 Fuller RW, Bokesch HR, Gustafson KR, McKee TC, Cardellina JH, McMahon JB, Cragg GM, Soejarto DR, Boyd MR. HIV-inhibitory coumarins from latex of the tropical rain-forest tree *Calophyllum teysmanii* var *inophylloide*. *Bioorg Med* 1994; **4**: 1961–4.
64 Schuster N, Christiansen C, Jakupovic J, Mungai M. An unusual [2 + 2] cycloadduct of terpenoid coumarin from *Ethulia vernonioides*. *Phytochemistry* 1993; **34**: 1179–81.
65 Niwa M, Sugino H, Takashima S, Sakai T, Wu YC, Wu TS, Kuoh CS. A new coumarin glucoside from *Daphne arisanensis*. *Chem Pharm* 1991; **39**: 2422–4.
66 Ahsan M, Gray AI, Leach G, Waterman PG. Novel angular pyranocoumarins from *Boronia lanceolata*. *Phytochemistry* 1994; **36**: 777–80.
67 Rashid MA, Armstrong JA, Gray AI, Waterman PG. Alkaloids, flavonols and coumarins from *Drummondita hasselli* and *D. calida*. *Phytochemistry* 1992; **31**: 1265–9.
68 Kuster RM, Bernardo RR, Parente JP, Mors WB. Furocoumarins from the rhizomes of *Dorstenia brasiliensis*. *Phytochemistry* 1994; **36**: 221–3.

69 Ahluwalia VK, Boyd DR, Jain AK, Khanduri CH, Sharma ND. Furanocoumarin glucosides from the seeds of *Apium graveolens*. *Phytochemistry* 1988; **27**: 1181–5.

70 Reisch J, Achenbach SH. A furanocoumarin glucoside from stembark of *Skimmia japonica*. *Phytochemistry* 1992; **31**: 4376–77.

71 Zdero C, Bohlmann F, Solomon J. Further 5-methylcoumarin derivatives from *Mutisia orbignyana*. *Phytochemistry* 1988; **27**: 891–7.

72 Bevalot F, Armstrong JA, Gray AI, Waterman PG. Coumarins from the leaves of *Phebalium squameum*. *Phytochemistry* 1988; **27**: 1546–7.

73 Gu LH, Li X, Yan SQ, Zhu TR. Studies on antibacterial constituents from *Gerbera anandria* (L.) Sch Bip. IV. *Yao Hsueh Hsueh Pao* 1989; **24**: 744–8.

74 Bilia AR, Cocchini C, Marsili A, Morelli I, Mele S. Coumarins and other constituents of *Prunus prostrata*. *J Nat Prod* 1993; **56**: 2142–8.

75 Vilegas W, Boralle N, Cabrera A, Bernardi AC, Pozetti GL, Arantes SF. Coumarins and a flavonoid from *Pterocaulon alopecuroides*. *Phytochemistry* 1995; **38**: 1017–19.

76 Daily A, Seligman O, Nonnenma G, Fessler B, Wong SM, Wagner H. New chromone, coumarin and coumestan derivatives from *Mutisia acuminata* var. *hirsuta*. *Planta Med* 1988; **3**: 270.

77 Mata R, Calzada F, Garcia MR, Reguero MT. Chemical studies on Mexican plants used in traditional medicine. III. New 4-phenylcoumarins from *Exostema caribaeum*. *J Nat Prod* 1987; **50**: 866–71.

78 Mahmoud ZF, Amer ME, Kader MSA, Abdelsalam NA. A coumestan from *Lotus creticus*. *Phytochemistry* 1990; **29**: 355–6.

79 Baba K, Tabata Y, Taniguti M, Kozawa M. Chemical studies on the constituents of thymelaeceous plants—coumarins from *Edgeworthia chrysantha*. *Phytochemistry* 1989; **28**: 221–5.

80 Mizuno A, Takata M, Okada Y, Okuyama T, Nishino H, Nishino A, Takayasu J, Iwashima A. Structures of new coumarins and antitumour-promoting activity of coumarins from *Angelica edulis*. *Planta Med* 1994; **60**: 333–6.

81 Bittner M, Jakupovic J, Bohlmann F, Grenz M, Silva M. 5-Methylcoumarins and chromones from *Triptilion* species. *Phytochemistry* 1988; **27**: 3263–6.

82 Bhattacharyya P, Jash SS, Dey AK. 6′,7′-Epoxyauraptene—a coumarin from *Aegle marmelos*. *J Ind Chem Soc* 1989; **66**: 424–5.

83 Bittner M, Jakupovic J, Bohlmann F, Silva M. Coumarins and guaianolides from further Chilean representatives of the subtribe *Nassauvinae*. *Phytochemistry* 1990; **28**: 2867–86.

84 Swain LA, Quirke JME, Winkle SA, Downum KR. A furanocoumarin from *Dorstenia contrajerva*. *Phytochemistry* 1991; **30**: 4196–7.

85 El-Turbi JA, Armstrong JA, Gray AI, Waterman PG. Further novel 6,7-dimethoxy-8-prenylated coumarins from the aerial parts of *Phebalium elatius* ssp. *beckleri*. *Z Naturfo C* 1990; **45**: 927–30.

86 Sarker SD, Armstrong JA, Gray AI, Waterman PG. Sesquiterpenyl coumarins and geranyl benzaldehyde derivatives from the aerial parts of *Eriostemon myoporoides*. *Phytochemistry* 1994; **37**: 1287–94.

87 Mizuno M, Tamura KI, Tanaka T, Iinuma M. A novel coumaronochromone from the stems of *Euchrestia japonica* and its antibacterial activity. *Heterocycle* 1988; **27**: 2047–50.

88 Mizuno M, Tanaka T, Tamura K, Iinuma M. Coumaronochromones from the stems of *Euchresta formosana*. *Phytochemistry* 1989; **28**: 2527–9.

89 Agrawal A, Siddiqui IR, Singh J. Coumarins from the roots of *Feronia limonia*. *Phytochemistry* 1989; **28**: 1229–31.

90 Ito C, Matsuoka M, Mizuno T, Sako K, Kimura Y, Motoharu J, Inque M, Kajiura I, Omura M, Furukawa H. Constituents of domestic citrus plants. 9. New coumarins from some citrus plants. *Chem Pharm* 1988; **36**: 3292–5.

91 Kuwajima H, Morita M, Takaishi K, Inque K, Fujita T, He ZD, Yang CR. Secoiridoid, coumarin and secoiridoid-coumarin glucosides from *Fraxinus chinensis*. *Phytochemistry* 1992; **31**: 1277–80.

92 Ju-Ichi M, Motoharu J, Inque M, Ikegami M, Kajiura I, Omura M, Furukawa H. Constituents of domestic citrus plants. 8. New coumarins from *Citrus funadoko*. *Heterocycle* 1988; **27**: 1451–4.

93 Rashid MA, Armstrong JA, Gray AI, Waterman PG, Novel C-geranyl 7-hydroxycoumarins from the aerial parts of *Eriostemon tomentellus*. *Z. Naturfo B* 1992; **47**: 284–7.
94 Masuda T, Muroya Y, Nakatani N. 7-Hydroxycoumarin derivatives from the juice oil of *Citrus hassaku*. *Phytochemistry* 1992; **31**: 1363–6.
95 Jung DJ, Porzel A, Huneck S. Gigasol and other coumarins from *Angelica gigas*. *Phytochemistry* 1991; **30**: 710–12.
96 Takemura Y, Nakata Y, Azuma M, Ju-Ichi M, Okano M, Fukamiya N, Omura M, Ito C, Nakagawa K, Furukawa H. 5 new pyranocoumarins from some citrus plants. *Chem Pharm* 1993; **41**: 1530–3.
97 Ito C, Ono T, Takemura Y, Nakata Y, Ten H, Ju-Ichi M, Okano M, Fukamiya N, Furukawa H. Structure of hassmarin, a novel biscoumarin from citrus plants. *Chem Pharm* 1993; **41**: 1302–4.
98 Das PC, Joshi PC, Mandal S, Das A, Chatterjee A. New coumarinolignoids from *Hemidesmus indicus*. *Ind J Chem B* 1992; **31**: 342–5.
99 Mandal S, Das PC, Joshi PC, Das A, Chatterjee A. Hemidesmine—a new coumarinolignoid from *Hemidesmus indicus*. *Ind J Chem B* 1991; **30**: 712–13.
100 Nkengfack AE, Kouam J, Vouffo TW, Meyer M, Tempesto MS, Fomum ZT. *Erythrina* studies. 25. An isoflavone and a coumestan from *Erythrina sigmoidea*. *Phytochemistry* 1994; **35**: 521–6.
101 Kumar V, Niyaz NMM, Wickramasinghe DB. Coumarins from stem bark of *Paramignya monophylla*. *Phytochemistry* 1995; **38**: 805–6.
102 Rashid MA, Armstrong JA, Gray AI, Waterman PG. Tetracyclic and pentacyclic 6-C-monoterpenyl-5,7-dioxycoumarins from *Eriostemon brucei* and *E. brucei* subspecies *cinerus*. *Phytochemistry* 1992; **31**: 3583–8.
103 Mizuno M, Kojima H, Iinuma M, Tanaka T, Goto K. Coumarin derivatives in *Coptis trifolia*. *Phytochemistry* 1992; **31**: 717–19.
104 Vilegas W, Pozetti GL, Vilegas JHY. Coumarins from *Brosimum gaudichaudi*. *J Nat Prod* 1993; **56**: 416–17.
105 Lemmich J. Monoterpene, chromone and coumarin glucosides of *Diplolophium buchananii*. *Phytochemistry* 1995; **38**: 427–32.
106 Konishi T, Wada S, Kiyosawa S. Constituents of the leaves of *Daphne pseudo-mezereum*. *Yakugaku Zasshi* 1993; **113**: 670–5.
107 Sarker SD, Gray AI, Waterman PG, Armstrong JA. Coumarins from 2 *Asterolasia* species. *J Nat Prod* 1994; **57**: 324–7.
108 Elgamal MHA, Shalaby NMM, Duddeck H, Hiegemann M. Coumarins and coumarin glucosides from the fruits of *Ammi majus*. *Phytochemistry* 1993; **34**: 819–23.
109 Vuorela H, Erdelmeier CA, Nyiredy S, Dallenbach-Tolke K, Anklin C, Hiltunnen R, Sticher O. Isobyakangelicin angelate: a novel furanocoumarin from *Peucedanum palustre*. *Planta Med* 1988; **54**: 538–42.
110 Joshi PC, Mandal S, Das PC. Jayantinin, a dimeric coumarin from *Boenninghausenia albiflora*. *Phytochemistry* 1989; **28**: 1281–3.
111 Ito C, Matsuoka M, Oka T, Ju-Ichi M, Niwa M, Omura M, Furukawa H. New binary coumarins from citrus plants. *Chem Pharm* 1990; **38**: 1230–2.
112 Hatano T, Yasuhara T, Fukuda T, Noro T, Okuda T. Phenolic constituents of licorice. II. Structures of licopyranocoumarin, licoarylcoumarin and glisoflavone, and inhibitory effects of licorice phenolics on xanthine oxidase. *Chem Pharm Bull* 1989; **37**: 3005–9.
113 Binzakaria M, Saito I, Matsuura T. Coumarins of *Merrillia caloxylon*. *Phytochemistry* 1989; **28**: 657–9.
114 Bevalot F, Armstrong JA, Gray AI, Waterman PG. Coumarins from 3 *Phebalium* species. *Biochem Syst* 1988; **16**: 631–3.
115 El-Turbi JA, Armstrong JA, Gray AI, Waterman PG. 6′-Methoxymurranganon derivatives from the aerial parts of *Phebalium elatius* subsp. *beckleri*. *Phytochemistry* 1990; **29**: 3982–3.
116 Rajani P, Sarma PN. A coumestone from the roots of *Tephrosia hamiltonii*. *Phytochemistry* 1988; **27**: 648–9.
117 Maldonado E, Hernandez E, Ortega A. Amides, coumarins and other constituents from *Simsia croquistii*. *Phytochemistry* 1992; **31**: 1413–14.

118 Rahmani M, Taufiqyap YH, Ismail HBM, Sukari A, Waterman PG. New coumarin and dihydrocinnamic acid derivatives from 2 Malaysian populations of *Micromelum minutum*. *Phytochemistry* 1994; **37**: 561–4.

119 Rashid MA, Gray AI, Waterman PG, Armstrong JA. Coumarins from *Eriostemon spicatus*. *J Nat Prod* 1992; **55**: 685–7.

120 Debenedetti SL, Palacios PS, Nadinic EL, Coussio JD, Dekimpe N, Boeykens M, Feneau-Dupont J, Declercq JP. 5-(3-Methyl-2-butenyloxy)-6,7-methylenedioxy-coumarin, a 5,6,7-trioxygenated coumarin from *Pterocaulon virgatum*. *J Nat Prod* 1994; **57**: 1539–42.

121 Sarker SD, Gray AI, Waterman PG, Armstrong JA. Coumarins from *Asterolasia trymalioides*. *J Nat Prod* 1994; **57**: 1549–51.

122 Rahmani M, Hin TYY, Ismail HBM, Sukari MA, Manas AR. Microminutinin—a novel coumarin from *Micromelum minutum*. *Planta Med* 1993; **59**: 93–4.

123 Ingham JL, Tahara S, Dziedzic SZ. Coumestans from the roots of *Pueraria mirifica*. *Z Naturfo C* 1988; **43**: 5–10.

124 Wu TS, Liou MJ, Kuoh CS. Coumarins of the flowers of *Murraya paniculata*. *Phytochemistry* 1989; **28**: 293–4.

125 Wu TS. Coumarins from the leaves of *Murraya paniculata*. *Phytochemistry* 1988; **27**: 2357–8.

126 Bittner M, Jakupovic J, Bohlmann F, Silva M. 5-Methylcoumarins from *Nassauvia* species. *Phytochemistry* 1988; **27**: 3845–7.

127 Takemura Y, Maki S, Ju-Ichi M, Omura M, Ito C, Furukawa H. Studies on the constituents of domestic citrus plants. 15. 2 novel acridone-coumarin dimers, neoacrimarine A and neoacrimarine B, from citrus plants. *Heterocycle* 1993; **36**: 675–80.

128 Takemura Y, Kurozumi T, Ju-Ichi M, Okano M, Fukamiya N, Ito C, Ono T, Furukawa H. Constituents of domestic citrus plants. 17. The structures of neoacrimarine C and neoacrimarine D, 2 new acridone-coumarin dimers from *Citrus hassaku*. *Chem Pharm* 1993; **41**: 1757–9.

129 Ju-Ichi M, Takemura Y, Azuma M, Tanaka K, Okano M, Fukamiya N, Ito C, Furukawa H. Constituents of domestic citrus plants. 10. New coumarins from *Citrus hassaku*. *Chem Pharm* 1991; **39**: 2252–5.

130 Bessonova IA, Batirov EK, Yagudaev MR. Obtusifolin, a new coumarin from roots of *Halophyllum obtusifolium*. *Khim Prir S* 1988; 187–9.

131 Dubois MA, Wierer M, Wagner H. Palustroside—a coumarin glucoside ester from *Ledium palustre*. *Phytochemistry* 1990; **29**: 3369–71.

132 Ito C, Furukawa H. 2 new coumarins from *Murraya* plants. *Chem Pharm* 1989; **37**: 819–20.

133 Takata M, Shibata S, Okuyama T. Studies on coumarins of a chinese drug Qian-hu. 10. Structures of angular pyranocoumarins of Bia-hua Qian-hu, the root of *Peucedanum praeruptorum*. *Planta Med* 1990; **56**: 307–11.

134 Ikeshiro Y, Mase I, Tomita Y. Coumarin glycosides from *Peucedanum japonicum*. *Phytochemistry* 1994; **35**: 1339–41.

135 Ikeshiro Y, Mase I, Tomita Y. Dihydropyranocoumarins from *Peucedanum japonicum*. *Phytochemistry* 1993; **33**: 1543–5.

136 Rasool N, Khan AQ, Ahmad VU, Malik A. A benzoquinone and a coumestan from *Psoralea plicata*. *Phytochemistry* 1991; **30**: 2800–3.

137 Wilzer KA, Fronczek FR, Urbatsch LE, Fischer NH. Coumarins from *Aster praealtus*. *Phytochemistry* 1989; **28**: 1729–35.

138 Takata M, Okuyama T, Shibata S. Studies on coumarins of a chinese drug 'Quian-Hu'. 8. Structures of new coumarin glycosides of 'Bai-Hua Qian-Hu'. *Planta Med* 1988; **4**: 323–7.

139 Cardona L, Garcia B, Pedro JR, Perez J. 6-Prenyloxy-1-methoxycoumarin, a coumarin hemiterpene ether from *Carduus tenuiflorus*. *Phytochemistry* 1992; **31**: 3989–91.

140 Ulubelen A, Mericli AH, Mericli F, Tan N. 2 new coumarins from *Haplophyllum ptilostylum*. *J Nat Prod* 1993; **56**: 1184–6.

141 Ramakrishna KV, Khan RA, Kapil RS. New isoflavone and coumestan from *Pueraria tuberosa*. *Ind J Chem B* 1988; **27**: 285.

142 Ito C, Ono T, Tanaka E, Takemura Y, Nakata T, Ushida H, Ju-Ichi M, Omura M, Furukawa H. Structure of pummeloquinone, a new coumarin naphthaquinone dimer isolated from citrus plants. *Chem Pharm* 1993; **41**: 205–7.
143 Debenedetti SL, Nadinic EL, Gomez MA, Coussio JD, Dekimpe N, Boeykens M. Purpurasol, a highly oxidated coumarin from *Pterocaulon purpurascens*. *Phytochemistry* 1992; **31**: 3284–5.
144 Meng QC, Chen WX. Pygmaeoherin—a novel coumarin from *Pygmaeopremna herbacea*. *Planta Med* 1988; **1**: 48–9.
145 Kong L, Li X, Pei Y, Min Z, Zhu T. Structure elucidation of qianhucoumarin F by 2D NMR. *Bopoxue Zazhi* 1994; **11**: 245–9.
146 Jain N, Alam MS, Kamil M, Ilyas M, Ali M. A coumarin from *Setaria italica*. *Phytochemistry* 1991; **30**: 3826–7.
147 Appendino G, Ozen HC, Nano GM, Cisero M. Sesquiterpene coumarin ethers from the genus *Heptaptera*. *Phytochemistry* 1992; **31**: 4223–6.
148 Ishii H, Kobayashi JI, Ishikawa T. Studies on the chemical constituents of 148 rutaceous plants. 71. Toddalocoumalone, a novel mixed dimer of coumarin and quinolone from *Toddalia asiatica* (L.) Lam (*T. aculeata* Pers). *Tetrahedron Lett* 1991; **32**: 6907–10.
149 Ceccherelli P, Curini M, Marcotullio MC, Madruzza G, Menghini A. Tortuoside—a new natural coumarin glucoside from *Seseli tortuosum*. *J Nat Prod* 1989; **52**: 888–90.
150 Sibanda S, Ndengu G, Multari G, Pompi V, Galeffi C. Research on African medicinal plants. 15. A coumarin glycoside from *Xeromphis obovata*, *Phytochemistry* 1989; **28**: 1550–2.
151 Gestetner B, Conn EE. 2-Hydroxylation of trans-cinnamic acid by chloroplasts from *Melilotus alba* Desr. *Arch Biochem Biophys* 1974; **163**: 617–24.
152 Kleinhofs A, Haskins FA, Gorz HJ. Trans *o*-hydroxycinnamic acid glucosylation in cell free extracts of *Melilotus alba*. *Phytochemistry* 1967; **6**: 1313–18.
153 Haskins FA, Williams LG, Gorz HJ. Light-induced *trans* to *cis* conversion of β-D-glucosyl *o*-hydroxycinnamic acid in *Melilotus alba* leaves. *Plant Physiol* 1964; **39**: 777–81.
154 Edwards KG, Stoker JR. Biosynthesis of herniarin: The isomerisation stage. *Phytochemistry* 1968; **7**: 73–7.
155 Brown SA, El-Dakhakhny M, Steck W. Role of dihydropsoralen in furanocoumarin metabolism. *Can J Biochem* 1970; **48**: 863.
156 Steck W, Brown SA. Biosynthesis of angular furanocoumarins. *Can J Biochem* 1970; **48**: 872.
157 Grisebach H, Barz W. Biogenesis of isoflavones. 7. Biogenesis of coumestrols in alfalfa. *Z Naturfo B* 1963; **18**: 466.
158 Fitzell DL, Hsieh DPH, Yao RC, La Mar GN. Biosynthesis of averufin from acetate by *Aspergillus parasiticus*. *J Agric Food Chem* 1975; **23**: 442–4.
159 Singh R, Hsieh DPH. Aflatoxin biosynthetic pathway—elucidation by using blocked mutants of *Aspergillus parasiticus*. *Arch Biochem Biophys* 1977; **178**: 285–92.
160 Reisch J, Herath HMTB, Kumar NS. Natural product chemistry. 145. Synthesis of rac-acridone-coumarin dimers (acrimarins). *Lieb Ann Ch* 1991; **8**: 839–41.
161 Nicolaides DN, Fylaktakidou KC, Bezergianni-Doubalouktsi C, Litinas KE. Reactions of 4,6-bis(acetyl)resorcinol with alkoxycarbonylalkylidene (triphenyl) phosphoranes—Preparation of coumarin derivatives. *J Hetero Ch* 1994; **31**: 173–6.
162 Iyer RS, Harris TM. Preparation of aflatoxin B_1 8,9-epoxide using m-chloroperbenzoic acid. *Chem Res Toxicol* 1993; **6**: 313–16.
163 Appendino G, Cravotto G, Tagliapietra S, Ferraro S, Nano GM, Palmisano G. The chemistry of coumarin derivatives. 5. Synthesis of 3-alkyl-4-hydroxycoumarins by reductive fragmentation of 3,3′-alkylidene-4,4′-dihydroxybis[coumarins]. *Helv Chim A* 1991; **74**: 1451–8.
164 Rao SSM, Trivedi KN. Studies in the synthesis of furocoumarins. 29. Synthesis of aminomethyl psoralen and angelicin derivatives as potential photochemotherapeutic agents. *Pharmazie* 1991; **46**: 644–6.
165 Rao SSM, Trivedi KN. Studies in the synthesis of furocoumarins. 30. Synthesis of 4-aminomethyl-dihydroangelicin and 4-aminomethyl-dihydropsoralen derivatives. *Ind J Chem B* 1991; **30**: 920–2.

166 Wulff WD, McCallum JS, Kunng FA. 2 regiocomplementary approaches to angular furanocoumarins with chromium carbene complexes—synthesis of sphondin, thiosphondin, heratomin and angelicin. *J Am Chem Soc* 1988; **110**: 7419–34.
167 Ahluwalia VK, Khanduri CH, Sharma ND. Synthesis of 3,4-dimethylpsoralen, angelicin and their derivatives. *Ind J Chem B* 1988; **27**: 748–51.
168 Zubia E, Luis FR, Massanet GM, Collado IG. An efficient synthesis of furanocoumarins. *Tetrahedron* 1992; **48**: 4239–46.
169 Lee YR. A concise new synthesis of angular furanocoumarins—angelicin, oroselone and oroselol. *Tetrahedron* 1995; **51**: 3087–94.
170 Palmer CJ, Josephs JL. Synthesis of the *Calophyllum* coumarins. *Tetrahedron Lett* 1994; **35**: 5363–6.
171 Banerji J, Dhara KP, Das B, Das AK, Chatterjee R. Studies on Rutaceae. 6. Reactions and rearrangements of coumarins. *Ind J Chem B* 1988; **27**: 21–5.
172 Chawla HM, Sharma SK. Control of oxophilicity of cerium (IV) ammonium-nitrate by addition of hydrogen peroxide—A novel preparation of 3-aroyl-2-aryl-4H-1-benzopyran-4-ones. *B S Chim Fr* 1990; **5**: 656–9.
173 Chawla HM, Sharms SK. Novel one-pot photochemical synthesis of 3-C-aroyl-2-aryl-4H-1-benzopyran-4-ones. *Syn Commun* 1990; **20**: 301–6.
174 Rajagopal R, Seshadri S. Synthesis and study of 6-arylazocoumarin derivatives. *Dye Pigment* 1990; **13**: 29–44.
175 Wattansin S. Novel route to 4-aryl coumarins. *Syn Commun* 1988; **18**: 1919–25.
176 Ahluwalia VK, Mann RR, Bala S. Synthesis and anti-bacterial activities of some 4-arylethenyl and 4-arylethyl-6,7-dihydropyranocoumarins. *Ind J Chem B* 1989; **28**: 608–10.
177 Chawla HM, Sharma SK. Regioselective synthesis of 2-arylidene coumarin-3-ones by dye-sensitized photooxygenation of 2-hydroxy phenyl styrylketones in the presence of sodium dodecyl sulphate. *Tetrahedron* 1990; **46**: 1611–24.
178 Ahluwalia VK, Khanduri CH. Synthesis of 6-aryl-5-phenyl-7H-furo[3,2-g]benzopyran-7-ones and 8-aryl-7-phenyl-9H-benzofuro[3,2-*g*][1]benzopyran-9-ones under phase transfer catalysis conditions. *Ind J Chem B* 1989; **28**: 599–601.
179 Soman SS, Trivedi KN. A synthesis of ayapin. *Ind J Chem B* 1993; **32**: 372–3.
180 Cairns N, Harwood LM, Astles DP. Tandem thermal Claisen-Cope rearrangements of coumarate derivatives—total synthesis of the naturally-occurring coumarins suberosin, demethylsuberosin, ostruthin, balsamiferone and gravelliferone. *J Chem Soc Perkin Trans 1* 1994; **21**: 3101–7.
181 Harvey RG, Cortez C, Ananthanarayan TP, Schmolka S. A new coumarin synthesis and its utilization for the synthesis of polycyclic coumarin compounds with anti-carcinogenic properties. *J Org Chem* 1988; **53**: 3936–43.
182 Skripskaya OV, Yagodinets PI, Shevchuk MI. Synthesis and study of phosphorus-containing and nitrogen-containing derivatives of 5,6-benzocoumarin. *Zh Obs Kh* 1992; **62**: 661–7.
183 Ahluwalia VK, Tyagi R, Mehta VD. Synthesis of 3-(benzothiazol-2′-ylthio)-2H-1-benzopyran-2-ones. *Ind J Chem B* 1989; **28**: 349–51.
184 Kocevar M, Polanc S, Tisler M, Vercek B. One-pot synthesis of substituted 3-(benzoylamino)-5-oxo-5,6,7,8-tetrahydrocoumarins. *Syn Commun* 1989; **19**: 1715–19.
185 Kondedeshmukah RS, Paradkar MV. A convenient synthesis of 6-benzylated coumarins. *Ind J Chem B* 1993; **32**: 1159–61.
186 Britto N, Gore VG, Mali RS, Ranade AC. A convenient synthesis of 3-benzyl, 3-benzyl-4-substituted coumarins and their benzo derivatives. *Syn Commun* 1989; **19**: 1899–910.
187 Danheiser RL, Trova MR. Synthesis of linear furocoumarins via a photochemical aromatic annulation strategy—an efficient total synthesis of bergapten. *Syn Lett* 1995; 573–4.
188 Harayama T, Nakatsuka K, Nishioka H, Murakami K, Hayashida N, Ishii H. Convenient synthesis of a simple coumarin from salicaldehyde and Wittig reagent. 2. Synthesis of bromocoumarin and methoxycarbonylcoumarin. *Chem Pharm* 1994; **42**: 2170–3.
189 Borges F, Pinto M. Synthesis of coumarins and derivatives. 1. Biomimetic synthesis of esculetin and halogenated derivatives. *Helv Chim A* 1992; **75**: 1061–8.

190 Clinet JC. Conjugate additions on α-phenyltio-α,β-unsaturated oxazolines—novel synthesis of 3,4-disubstituted coumarins. *Tetrahedron Lett* 1988; **29**: 5901–4.

191 Bonsignore L, Cottiglia F, Maccioni AM, Secci D, Lavagna SM. Synthesis of coumarin 3-*O*-acylisoureas by dicyclohexylcarbodiimide. *J Hetero Ch* 1995; **32**: 573–7.

192 Tkach II, Mikhailova TA, Reznichenko VA, Savvina LP, Lukyanets EA. Synthesis and spectral-luminescent properties of coumarin azomethines. *Khim Getero* 1990; **3**: 321–3.

193 Shirokova EA, Segal GM, Torgov IV. Usage of Meldrum's acid in the synthesis of low molecular weight bioregulators. 3. Synthesis of coumarin-3-carboxylic acids and their derivatives. *Bioorg Khim* 1988; **14**: 236–42.

194 Vishnyakova GM, Tarakanova LA. Synthesis of coumarin-containing phosphonates. *Zh Obs Kh* 1991; **61**: 2123–4.

195 Bulut M, Erk C. The synthesis of novel crown ethers. 1. Coumarin derivatives of crown ethers. *Syn Commun* 1992; **22**: 1259–63.

196 Mahling JA, Schmidt RR. Glycosyl imidates. 70. Synthesis of a coumarin C-glucoside. *Liebigs Ann* 1995; **3**: 467–9.

197 Al-Kindy SMZ, Miller JN. Coumarin-6-sulfonyl chloride—a novel label in fluorimetry and phosphorimetry. 1. Synthesis and luminescence properties. *Analyt Chim* 1989; **227**: 145–53.

198 Pandit SB, Gadre SY. A new synthesis of coumestan—formal synthesis of coumestrol methyl ether. *Syn Commun* 1988; **18**: 157–66.

199 Laschober R, Kappe T. Potential nonsteroidal estrogens and anti-estrogens. 2. A new and efficient synthesis of coumestan and coumestrol. *Synthesis S* 1990; **5**: 387–8.

200 Bhalerao UT, Muralikrishna C, Pandey G. An efficient synthesis of coumestans—a probable biogenetic approach. *Syn Commun* 1989; **19**: 1303–7.

201 Pandey G, Muralikrishna C, Bhalerao UT. Mushroom tyrosinase catalysed synthesis of coumestans, benzofuran derivatives and related heterocyclic compounds. *Tetrahedron* 1989; **45**: 6867–74.

202 Mali RS, Tilve SG. Useful synthesis of coumestans. *Syn Commun* 1990; **20**: 167–73.

203 Litinas KE, Stampelos XN. Reactions of 3-triphenylphosphoranylidene—[b]fused furan-2(3H)-ones with some salicylaldehydes—New synthesis of 3-(2-Hydroxyaryl)-coumarins and coumestans. *J Chem Soc Perkin Trans 1* 1992; **2**: 2981–4.

204 Reisch J, Herath HMTB, Kumar NS. Natural product chemistry. 139. Synthesis of the natural coumarins (E)-suberenol, cyclobisuberodiene and two other related new coumarins. *Lieb Ann Ch* 1990; **9**: 931–3.

205 Reisch J, Bathe A. Natural product chemistry. 118. Synthesis of the coumarins 6-naphthoherniarin, 8-naphthoherniarin, dehydrogeijerin and murraol. *Lieb Ann Chem* 1988; **6**: 543–7.

206 Cairns N, Harwood LM, Astles DP. Syntheses of 6-acylcoumarins via highly regioselective Fries rearrangements—Total syntheses of the linear acylated coumarins geijerin and dehydrogeijerin. *Tetrahedron* 1992; **48**: 7581–90.

207 Murray RDH, Zeghdi S. Synthesis of the natural coumarins murraol (cm-c_2), *trans*-dehydroosthol and swietenocoumarin G. *Phytochemistry* 1989; **28**: 227–30.

208 Tkach II, Mikhailova TA, Savvina LP, Lukyanets EA. Synthesis and spectral luminescence characteristics of 7-diethylamino-3-(2-arylethenyl) coumarins. *Khim Getero* 1992; **9**: 1171–5.

209 Bouyssou P, Chenault J. Phosphonates and phosphine oxides as reagents in a one-pot synthesis of coumarins. *Tetrahedron Lett* 1991; **32**: 5341–4.

210 Ahluwalia VK, Nayal L, Tehim AK. New route to synthesis of some 8-(dihydrocinnamoyl)dihydrocoumarin derivatives. *Ind J Chem B* 1988; **27**: 70–1.

211 Rao SSM, Trivedi KN. Furocoumarins. 34. Synthesis and stereochemistry of dihydrofurocoumarins. *Ind J Chem B* 1992; **31**: 488–91.

212 Nandaiah S, Krupadanam GL. Synthesis of 3,4-dihydro-4-hydroxy pyrano coumarins. *Abs Pap Acs* 1993; **206**: 343.

213 Reisch J, Herath HMTB, Bergenthal D, Kumar NS. Natural product chemistry. 147. Synthesis of some coumarin derivatives. *Lieb Ann Ch* 1991; **11**: 1235.

214 Mizuno T, Nishigucchi I, Hirashima T, Ogawa A, Kambe N, Sonoda N. Facile synthesis of 4-hydroxy-coumarins by sulfur-assisted carbonylation of 2′-hydroxyacetophenones with carbon monoxide. *Synthesis S* 1988; **3**: 257–9.
215 Hormi EOE, Peltonen C, Bergstrom R. A one-pot synthesis of coumarins from dipotassium ortho-methoxy benzylid enemalonates. *J Chem Soc Perkin Trans 1* 1991; 219–21.
216 Reisch J, Wickramasinghe A, Kumar V. Natural product chemistry. 128. Synthesis of dicoumarinylethers with structures possible for oreojasmin-coumarins of Ruta oreojasmine fruits. *J Nat Prod* 1989; **52**: 1379–82.
217 Hernandez Galan R, Massanet GM, Pando E, Rodriguez Luis F, Salva J. Sigmatropic rearrangements in synthesis of C-8 prenylated 3-(1′,1′-dimethylallyl) coumarins: 3-(1′1′-dimethylallyl) seselin. *Heterocycles* 1989; **29**: 297–300.
218 Ahluwalia VK, Mittal B, Mehta VD. Synthesis of 3,4-dimethyldihydropyrano-coumarins. *Ind J Chem B* 1988; **27**: 1148–9.
219 Prashant A, Krupadanam GLD. A single regioselective synthesis of 1,5-dioxano coumarin derivatives. *Abs Pap Acs* 1993; **206**: 109.
220 Ahluwalia VK, Gupta R, Grover M, Mukherjee I, Khanduri CH. Synthesis of 2,3-diphenyl-9-methyl-7H-furo[3,2-*g*]benzopyran-7-ones. *Ind J Chem B* 1988; **27**: 1138–9.
221 Baburaj K, Azam N, Udgaonkar D, Durani S. HOCGO and DMACGO—2 coumarin-derived α-dicarbonyls suitable as pH and polarity sensitive fluorescent reporters for proteins that can be targeted at reactive arginines. *Biochim Biophys Acta* 1994; **1199**: 255–65.
222 Ahluwalia VK, Mehta VD, Khanduri CH. Syntheses of 7-ethoxy-3,4-dimethylcoumarin and 5-methoxy-4-methylcoumarin. *Ind J Chem B* 1988; **27**: 68–9.
223 Patra A, Misra SK. A new synthesis of 4-substituted coumarins. *Ind J Chem B* 1990; **29**: 66–9.
224 Hua DY, Saha S, Roche D, Maeng JC, Iguchi S, Baldwin C. An improved procedure of the Pechman condensation in the synthesis of 8-ethyltrimethoxy-6,4-benzo[d]naphtho[1,2-*b*]pyran-6-ones structurally related to aglycon of gilvocarcins. *J Org Chem* 1992; **57**: 399–403.
225 Reisch J, Wickramasinghe A, Kumar V. Natural product chemistry. 121. Synthesis of dicoumarinyl ethers with the structures proposed for fatagarine and oreojasmine. *Monats Chem* 1988; **119**: 781–91.
226 Bode ML, English RB, Kaye PT. Novel synthesis and x-ray crystal structure of a coumarin derivative. *S Afr J Chem S* 1992; **45**: 25–7.
227 Rao CP, Krupadanam GLD, Srimmanarayana G. Claisen rearrangement of 7-propargyloxycoumarins—formation of pyranocoumarins and furanocoumarins. *Ind J Chem B* 1991; **30**: 666–71.
228 Hager AJ, Evans SP, Otsuki T. Novel furocoumarin derivatives as a model related to DNA-intercalating molecules—Synthesis and photochemical reactivity. *Chem Lett* 1990; **3**: 409–12.
229 Soman SS, Trivedi KN. Synthesis of linear and angular furocoumarins using intramolecular Wittig reaction. 31. *Ind J Chem B* 1992; **31**: 532–4.
230 Atta SMS, Hafez TS, Mahran MR. Organophosphorus chemistry. 25. The utilization of Wittig reagents in lactone ring formation—application to the synthesis of linear furocoumarins and pyranocoumarins. *Phosphor Su* 1993; **80**: 109–16.
231 Rao CP, Prashant A, Krupadanam GL. The Claisen rearrangement of 6-propargyloxycoumarins—formation of pyranocoumarins or furocoumarins and their antibacterial activity. *Ind J Chem B* 1994; **33**: 593–6.
232 Soman SS, Trivedi KN. Studies in synthesis of furocoumarins. 36. Synthesis of pyrano-isocoumestans and furo-coumestans. *Ind J Chem B* 1993; **32**: 477–80.
233 Bhide BH, Akolkar VD, Prabhu VG. Trifluoroacetic acid catalysed Claisen rearrangement—Synthesis of 3-methyl-3,4-dihydro-5-hydroxyisocoumarin and furodihydroisocoumarin. *Ind J Chem B* 1992; **31**: 143–7.
234 Soman SS, Trivedi KN. Studies in the synthesis of furocoumarins. 32. Synthesis of furoisocoumestans. *Ind J Chem B* 1991; **30**: 923–6.
235 Minami T, Matsumoto Y, Nakamura S, Koyanagi S, Yamaguchi M. 3-Vinylcoumarins and 3-vinylchromenes as dienes—application to the synthesis of 3,4-fused coumarins and chromenes. *J Org Chem* 1992; **57**: 167–73.

236 Reisch J, Wickramasinghe A, Wickremarathe DB. Natural product chemistry. 134. Synthesis of the natural coumarins gleinene and gleinadiene. *Lieb Ann Chem* 1990; **9**: 931–3.

237 Reisch J, Zappel J. Chemistry of natural products. 155. A simple synthesis of 4-(herniarin-3-yl)-2-methyl-3-buten-2-ol. *Z Naturfo B* 1992; **47**: 1801–2.

238 Steinfuehrer T, Hantschmann A, Pietsch M, Weissenfels M. Heterocyclic [c]-condensed coumarins from 4-azido-3-coumarin-carbaldehydes. *Lieb Ann Chem* 1992; **1**: 25–8.

239 Harayama T, Katsuno K, Nishioka H, Fujii M, Nishita Y, Ishii H, Kaneko Y. A convenient synthesis of a simple coumarin from salicylaldehyde and Wittig reagent. 1. A synthesis of methoxycoumarin and hydroxycoumarins. *Heterocycle* 1994; **39**: 613–22.

240 Mali RS, Jagtap PG, Patil SR, Pawar PN. Novel $AlCl_3$ catalyzed syntheses of naturally-occurring (±)-8-hydroxy-3-methyl-3,4-dihydroisocoumarins. *J Chem Soc Perkin Trans 1* 1992; **12**: 883–4.

241 Oyman U, Gunaydin K. Condensation of ethyl acetoacetate with naphthalene-diols—the synthesis of some novel coumarins and chromones. *B S Chim Br* 1994; 763–4.

242 Zubkov VA, Kovalenko SN, Chernykh VP, Ivkov SM. New coumarin derivatives—2-(N-R-Imino)-2H-1-benzopyrones. *Khim Getero* 1994; **6**: 760–6.

243 Allison S, Burks SJ, Taylor RT. Total syntheses of microminutin and other coumarins through the key intermediate isomurralonginol. *Tetrahedron* 1991; **47**: 9737–42.

244 Reed MW, Moore HW. Efficient synthesis of furochromone and furocoumarin natural products (khellin, pimpellin, isophellopterin) by thermal rearrangement of 4-furyl-4-hydroxycyclobutenone. *J Org Chem* 1988; **53**: 4166–71.

245 Appendino G, Cravotto G, Nano GM, Palmisano G. A regioselective synthesis of 3-isoprenyl-4-hydroxycoumarins. *Syn Commun* 1992; **22**: 2205–12.

246 Ibrahim TM, Ahmed FSM, Shedid SA. Synthesis and antimicrobial activity of some new 7-methoxy-4-methylcoumarin-6-sulfonylamino acid derivatives. *Proc Ind Natl Sci Acad, Part A* 1994; **60**: 433–9.

247 Soman SS, Trivedi KN. Synthesis of furocoumarins. 33. Synthesis of 6/7-methoxy-8/9-methylcycloalkeno[c]psoralens. *Ind J Chem B* 1992; **31**: 30–4.

248 Biswas GK, Basa K, Barua AK, Bhattacharyya P. Montmorillonite clay as condensing agent in Pechman reaction for the synthesis of coumarin derivatives. *Ind J Chem B* 1992; **31**: 628.

249 Guiotto A, Chilin A, Pastorini G, Palumbo M. Methylfuroquinilones—new furocoumarin isoesters as potential photoreagents towards DNA. *J Hetero Chem* 1989; **26**: 917–22.

250 Ahluwalia VK, Mittal B, Mehta VD. A convenient synthesis of methyl-substituted benzofurocoumarins. *Ind J Chem B* 1988; **27**: 179–81.

251 Rodighiero P, Chilin A, Pastorini G, Guiotto A, Mangini P. Methyltriazolocoumarins—new furocoumarin isosters as potential photochemotherapeutic agents. *J Hetero Chem* 1990; **27**: 1153–8.

252 Mali RS, Manekar AR, Tilve SG. Convenient synthesis of nieshoutin (cycloobliquetin), a furocoumarin isolated from the heartwood of *Ptaeroxylon obliquum*. *Org Prep Pr* 1992; **24**: 52–4.

253 Harayama T, Nakatsuka K, Nishioka H, Murakami K, Ohmori Y, Takenchi Y, Ishii H, Kenmotsu K. Convenient synthesis of a simple coumarin from salicylaldehyde and Wittig reagent. 13. Synthesis of nitrocoumarins. *Heterocycle* 1994; **38**: 2729–38.

254 Matsui M, Shibata K, Muramatsu H, Sawada H, Nakayama M. Synthesis of 3-perfluoroalkylated coumarins. *Syn Lett* 1991; **2**: 113–14.

255 Huang BN, Lin JT, Huang WY. Studies on sulfinatodehalogation. 30. Synthesis of 3-perfluoroalkylated coumarins, thiocoumarins and 2-quinolones by direct perfluoroalkylation with perfluoroalkyl iodides and sodium hydroxymethanesulfinate. *J Chem Soc Perkin Trans 1* 1994; **1**: 101–4.

256 Castellani M, Chinsoli GP, Faguola MI, Solari G. A new palladium catalysed synthesis of 3,4-disubstituted coumarins from 3-alkenoates of ortho-iodophenol, phenylacetylene and carbon monoxide. *Tetrahedron Lett* 1994; **35**: 5923–6.

257 Appendino G, Cravotto G, Toma L, Annunziata R, Palmisano G. The chemistry of coumarin derivatives. 6. Diels–Alder trapping of 3-methylene-2,4-chroman dione—a new entry to substituted pyrano[3,2-*c*]coumarins. *J Org Chem* 1994; **59**: 5556–64.

258 Gaon ID, Zavrsnik D, Lovren F, Sote V, Pesicpecek B. Synthesis, structure and biological activity of 3,4-substituted derivatives of coumarins. *Per Biol* 1991; **93**: 253–6.

259 Ishii H, Kaneko Y, Miyazaki H, Harayama T. A convenient synthesis of simple coumarins. *Chem Pharm* 1991; **39**: 3100–2.

260 Koenig AL, Otsuki T. Synthesis and reactivities of 4′-substituted furocoumarin derivatives. *Abs Pap Acs* 1991; **201**: 345.

261 El-Farargy AF, Hamad MM, Said SA, Haikal A. Synthesis and some reactions of 8-tert-butyl-6-hydroxy-4-methyl coumarin. *J Chem Soc Perkin Trans 1* 1991; **13**: 59–62.

262 Ahluwalia VK, Alauddin M, Khanduri CH, Mehta VD. A convenient one-pot synthesis of 3-thiosubstituted coumarins. *Heterocycle* 1989; **29**: 1729–33.

263 Ishikawa T, Kotake KI, Ishii H. Synthesis of toddacoumaquinone, a coumarin-naphthoquinone dimer, and its antiviral activities. *Chem Pharm* 1995; **43**: 1039–41.

264 Rautenstrauch V, Megand P, Gamper B, Bourdin B, Walther E, Bernardinelli G. 2,2′-bifuryldiene-5,5′-diones, coumarins, 3A,7A-dihydro-1H-inden-1-ones, and 5H-furo[3,2-*b*]pyran-5-ones from propane and carbon monoxide. *Helv Chim A* 1989; **72**: 197–201.

265 Reisch J, Zappel J. Natural product chemistry. 154. Biomimetic synthesis of 3,3′-umbelliferone. *J Hetero Chem* 1992; **29**: 1035–6.

266 Tawada H, Natsugari H, Ishikawa E, Sugiyama Y, Ikeda H, Meguro K. Synthesis of 3-ureido derivatives of coumarin and 2-quinolone as potent acyl-coA-cholesterol acyltransferase inhibitors. *Chem Pharm* 1995; **43**: 616–25.

267 Rodighiero G, Antonello G. Synthesis of more 3-aminocoumarin derivatives and their antibacterial properties. *Boll Cheim Farm* 1958; **97**: 592–601.

268 Morgan GT, Michlethwait FMG, Winfield M. Comparative experiments on the behaviour of 4-bromo-tetrahydro-α-naphthylamine and 4-bromo-α-naphthylamine towards diazo-compounds. *J Biochem Soc* 1904; **85**: 1230–8.

269 Baker MC, Campbell DH, Epstein SI, Singer SJ. Physical-chemical studies of the soluble antigen-antibody complexes. 7. Thermodynamics of the reaction between benzenearsonic acid-azobovine serum albumin and rabbit antibodies to benzenearsonic acid. *J Am Chem Soc* 1956; **78**: 312–16.

270 Satoh H, Hayashi M, Satoh S. Anti-warfarin antibody preparation and its characterisation for radioimmunoassay. *J Pharm Pharmacol* 1982; **34**: 429–33.

271 Finnegan RA, Djerassi C. Naturally-occurring oxygen heterocycles. 4-Phenyl-5,7-dihydroxy-6-iso-valeryl-8-isopentenylcoumarin. *Tetrahedron Lett* 1959; **13**: 11–15.

272 Kariyone T, Hata K. Chemical components of umbelliferone plants. I. Components of the roots of *Angelica glabra. J Pharm Soc Japan* 1956; **76**: 649–51.

273 Dean FM, Parton B, Somvichien N, Taylor DAH. The coumarins of *Ptaeroxylon* obliquum. *Tetrahedron Lett* 1967; 2147.

274 Gonzalez AG, Estevez Reyes R, Jaraiz I. New sources of natural coumarins. 17. Structure of sabandinin and other coumarins isolated from roots of *Ruta pinnata. Phytochemistry* 1971; **10**: 1621.

275 Sharma P, Sharma M, Rangaswami S. Structure of leptophyllidin, leptophyllin, new dihydrofuro-coumarins from seeds of *Apium leptophyllum. Ind J Chem B* 1978; **16B**: 563–6.

276 Yamada M, Matsuhashi M, Torii M. Effect of novobiocin and ethylenediamene-tetraacetate on formation of cell pockets in *Micrococcus luteus. J Gen Appl Microbiol* 1978; **24**: 307–15.

277 Steck W, Wetter LR, Prairie plants. 3. Apterin, an unusual glucoside of *Zizia aptera. Phytochemistry* 1974; **13**: 1925–27.

278 Mitsuhasi H, Itoh T. Constituents of Umbelliferae plants. *Chem Pharm Bull* 1962; **10**: 514.

279 Estevez Reyes R, Gonzalez AG. Structure of pinnarin and furopinnarin, 2 new coumarins from roots of *Ruta pinnata. Phytochemistry* 1970; **9**: 833.

280 Lemmich E, Lemmich J, Nielsen BE. Constituents of umbelliferous plants. 14. Coumarins of *Peucedanum oroselinum* (L) Moench. *Acta Chem Scand* 1970; **24**: 2893.

281 Pozzi H, Sanchez E, Comin J. Studies on argentine plants. 22. Heliettin, a new furocoumarin from *Helietta longifoliata* Britt. *Tetrahedron* 1967; **23**: 1129.

282 Soine TO. Naturally occurring coumarins and related physiological activities. *J Pharm Sci* 1964; **53**: 231.

283 Duffield AM, Jeffries PR, Maslen EN, Rae AIM. The structure of bruceol. *Tetrahedron* 1963; **19**: 593–607.
284 Seshadri TR, Vishwapaul B. Chemical components of roots of *Selinum vaginatum*. 2. Chemistry of vaginidin and vaginidiol. *Ind J Chem* 1971; **9**: 418.

CHAPTER 3

The Metabolism of Coumarin

Olavi Pelkonen, Hannu Raunio, Arja Rautio, Markku Pasanen
Department of Pharmacology and Toxicology, University of Oulu, Oulu, Finland

and

Matti A. Lang
International Agency for Research on Cancer, Lyon, France

3.1 INTRODUCTION

The metabolism of coumarin displays large interspecies differences. In humans, the major metabolic pathway is via 7-hydroxylation, catalyzed by a specific cytochrome P450 enzyme, CYP2A6. In the mouse, 7-hydroxylation is variably active depending on the strain; in addition the ring-opening pathway is also significant. In the rat, hepatic (but not pulmonary or nasal mucosal) 7-hydroxylation is practically absent. These metabolic differences may be of significance for the toxicity and therapeutic effects of coumarin. CYP2A enzymes of human and mouse tissues exhibiting high coumarin 7-hydroxylation activity display special characteristics within the P450 superfamily, in terms of tissue- and sex-specific expression, mode of regulation, as well as substrate and inhibitor selectivities. Perhaps the most striking feature is the inducibility by various structurally unrelated hepatotoxins and viral and parasitic liver damage.

Because of their extremely high expression in the nasal epithelium and catalytic activities towards certain carcinogens, it is important to elucidate the physiological, pharmacological and toxicological significance of coumarin metabolism-related CYP2A enzymes, with respect to olfaction, alkaloid metabolism and activation of certain carcinogens. One interesting way in which these problems can be examined

Coumarins: Biology, Applications and Mode of Action. Edited by R. O'Kennedy and R.D. Thornes.

in human populations is to take advantage of the possible polymorphisms of the *CYP2A6* gene, which seems to lead to impaired enzymatic activity.

3.2 METABOLIC PATHWAYS OF COUMARIN

Elucidation of metabolic fate of coumarin is of interest for two reasons: to assess the possible dependence of coumarin-induced toxicity on metabolism and to make use of the fact that it is metabolized at several sites (Figure 3.1). Thus it serves as a model for the rather complicated metabolism of a relatively simple organic substance[1,2]. Oxidative metabolism at various carbon atoms of the molecule (actually hydroxylation can occur at all available sites) is a prerequisite for phase II conjugative reactions. The various hydroxylations are shown or suggested to be catalyzed primarily by P450 enzymes and conjugations by UDP-glucuronosyl transferase(s) and sulfotransferase(s).

Among the various metabolic steps, 7-hydroxylation has received most attention, principally because it is the major metabolic route in humans, but also partially because of easy analysis: the metabolite, 7-hydroxycoumarin or umbelliferone, a highly fluorescent substance, is easily detectable with specific excitation and emission wavelengths. Therefore, several alkoxycoumarin derivatives have also been used as substrates for studying P450-mediated alkylation reactions in addition to ring hydroxylation. Umbelliferone has also been used as a substrate for measuring glucuronidation. The appearance of 7-hydroxycoumarin in the reaction vessel provides a sensitive and convenient assay for metabolism and several early investigators took advantage of this fact[3]. The fluorescent derivative can be measured directly in the incubation mixture by increasing the pH[4]. The sensitivity can be enhanced by the extraction with organic solvents and this is appropriate for samples of low activity[5]. The 7-hydroxycoumarin (as well as other metabolites) has also been determined by HPLC (see below and Chapter 12).

The 3-hydroxylation pathway, which leads via lactone ring-opening to several distal metabolites, is enzymatically less well-characterized. The nature of the P450 enzyme catalyzing 3-hydroxylation and enzymes catalyzing distal steps of this pathway are not definitely identified; however, 3-hydroxylation seems to be inducible by phenobarbital[6,7]. As a result of the postulated intermediate, 3,4-epoxide (which may be the species bound covalently to proteins), this pathway may be of considerable importance for the hepatotoxicity and carcinogenicity of coumarin in certain animal species[8].

Species differences in the oxidative metabolism of coumarin are striking; some principal features are presented in Table 3.1. In primates, as in humans, coumarin is predominantly hydroxylated at carbon 7. However, recent studies on the metabolic fate of coumarin in human liver microsomes and *in vivo* indicate that in some individuals a considerable proportion of coumarin is metabolized through pathways other than 7-hydroxylation[9–11]. Because coumarin-induced hepatotoxicity is

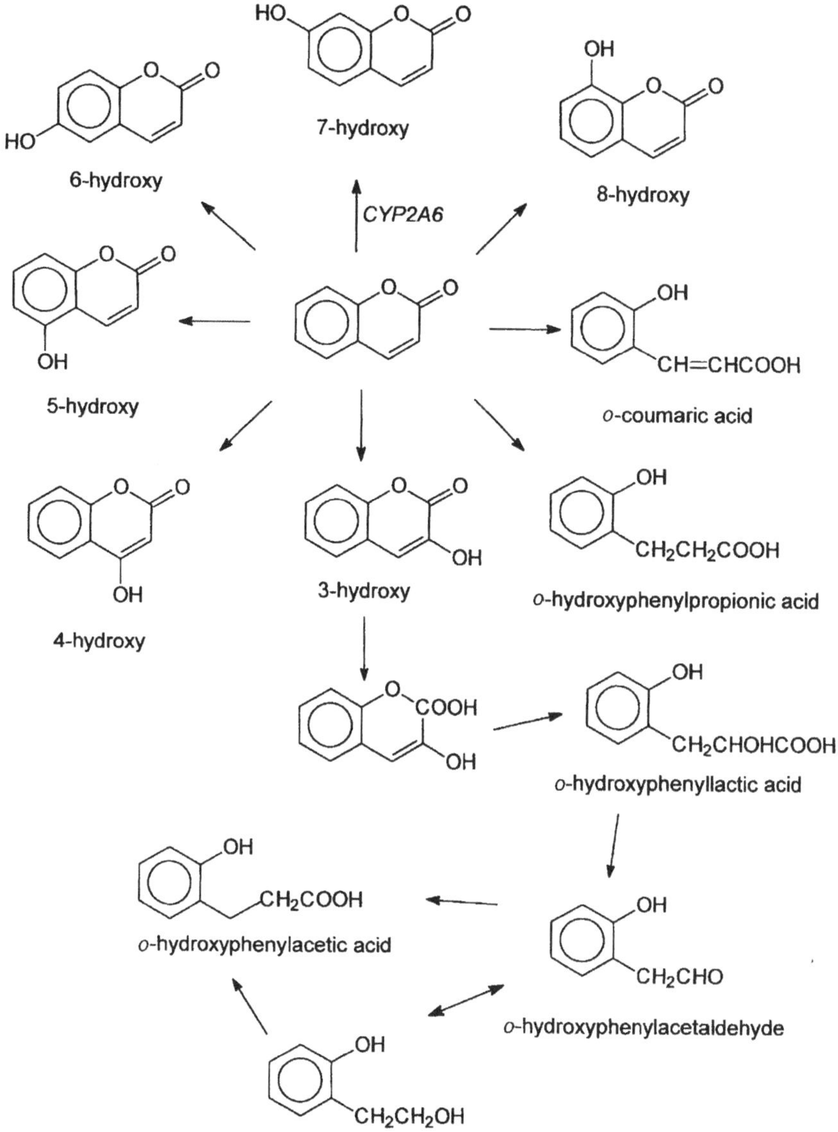

Figure 3.1. Pathways of metabolism of coumarin. For further details concerning 3-hydroxylation and ring-opening pathway, see refs 2, 8, 15. There is indirect evidence that 3,4-epoxide is formed as a covalently bound and toxic metabolite[7,99,100].

Table 3.1. Oxidative metabolic pathways of coumarin in different species.

Species	7-OH	Ring-opening	3-OH	4-OH	5-OH	6-OH	8-OH	DH[a]
Human	79[b]	4	+	(+)	nd	(+)	nd	nd
Baboon	60							
Cynomolgus monkey	(++)	(+)	(+)					
Cat	19							
Dog	3							
Ferret	1		+		+	+	+	
Guinea-pig	1	(+)	+	(+)	+	+	+	
Hamster	5							
Mouse	3 (2–40)	(+)	+	(+)	+	+		
Pig	12							
Rabbit	12	23	22	0.6	0.4	3	2	+
Rat	0.4	20	2	0.3	+	0.3	0.4	+
Squirrel monkey	1	(++)						

The data have been compiled mainly from Cohen[2], Egan *et al.*[15] and Fentem and Fry[8]. Blank space indicates that no studies are available. Parentheses indicate that results have been obtained from studies on liver slices, hepatocytes or liver microsomes[11,13].

[a] 6,7-Dihydroxycoumarin.

[b] An average figure. For variability, see Section 3.6.

presumably associated with the 3-hydroxylation pathway, it is important to elucidate the contribution of this pathway to coumarin metabolism in humans, in particular since coumarin is increasingly used for therapeutic purposes. In a recent *in vitro* study on human liver microsomes[12], the 3-hydroxylation pathway seemed to be quantitatively almost as important as the 7-hydroxylation pathway, and in a single case its activity far exceeded that of 7-hydroxylation. It is probable that 7-hydroxylation is the major pathway at low coumarin concentrations, whereas 3-hydroxylation becomes significant at higher concentrations and in cases in which there is a relative deficiency of the 7-hydroxylation pathway[12].

In the rat, the major metabolic pathway seems to be through 3-hydroxylation, with several distal end-products such as *o*-hydroxyphenylacetic acid and *o*-hydroxyphenyllactic acid as major metabolites and the 3-hydroxyderivative itself and other hydroxylated metabolites as minor metabolites. Interestingly, coumarin seems to be more hepatotoxic to rat than to many other experimental animals and, indeed, the prominent 3-hydroxylation may be the reason for this species difference.

In the rabbit, the major urinary metabolites are *o*-hydroxyphenylacetic acid (18–22%), 3-hydroxycoumarin (18–23%), and 7-hydroxycoumarin (10–16%). Also trace amounts of 4-, 5-, 6- and 8-hydroxy metabolites have been identified. It is not known which P450 enzymes catalyze those minor hydroxylation reactions.

In the mouse, the activity of coumarin 7-hydroxylation is usually higher than in the rat or rabbit; however, there are large interstrain differences (see below). Mouse liver microsomes also catalyze the 3-hydroxylation, although the catalyzing enzyme has not been identified.

In addition there are some sporadic studies on coumarin metabolism, especially on 7-hydroxylation, in a few other animal species (Table 3.1). It seems to be difficult to find any logical evolutionary principle behind the wide interspecies variability of coumarin metabolism.

Recently, novel experimental approaches such as the use of precision-cut liver slices[13] and isolated primary hepatocytes[14] have provided confirmatory data to metabolic pathways of coumarin in different species and have also revealed the existence of some minor potential pathways, such as 4- and 6-hydroxylation in human liver microsomes (see Table 3.1). However, in general, findings agree well with *in vivo* data on metabolic pathways.

3.3 PHARMACOKINETICS OF COUMARIN

3.3.1 Experimental Animals

The pharmacokinetics of coumarin, including the excretion of various metabolites, have been relatively thoroughly elucidated[2,8,15]. In all of the species studied, coumarin is rapidly and practically completely metabolized and only little unchanged compound is excreted. There are, however, qualitative and quantitative interspecies differences in the routes of elimination. Biliary excretion is insignificant in most species, including humans, but in the rat it represents about 40% of the dose[16]. The most thoroughly studied experimental animal is mouse, because the 7-hydroxylation is, at least in some strains, the prominent metabolite. A relatively simple and easy experimental approach to the study of factors affecting coumarin 7-hydroxylation in the mouse is to place the animal into a metabolic cage, collect urine, and measure the excretion of 7-hydroxycoumarin, both free and conjugated. The first such *in vivo* studies comparing different strains of mice were performed by Wood and Conney[17] and Lush and Andrews[18]. They demonstrated that mouse strains could be classified into two or three groups based on the rate of coumarin 7-hydroxylation. The results for some strains for which both *in vitro* and *in vivo* data are available are presented in Table 3.2.

Legrum and Frahseck[19] were able to show, by employing an *in vivo* approach in which the exhalation of carbon dioxide after the administration of radiolabeled 7-methoxycoumarin was followed, that there was a considerable increase in the metabolism of 7-methoxycoumarin after cobalt pretreatment. They assumed that this increase was due to the induction of a specific monooxygenase in the livers of mice. Subsequent studies have shown that this induced enzyme is CYP2A5.

3.3.2 Pharmacokinetics in Humans

Coumarin is rapidly and completely absorbed from the gastrointestinal tract[20]. It is extensively metabolized by the liver in the first pass and less than 4% reaches the

Table 3.2. Coumarin 7-hydroxylase activity in liver preparations and the excretion of 7-hydroxycoumarin in urine in different strains of mice.

Strain	Activity *in vitro*[a] (pmol/mg liver/min)		7-Hydroxycoumarin excreted *in vivo*[b] (nmol/22 h)
	Control	PB-induced	
C57L	34	451	304
DBA/1	37	267	782
DBA/2	39	246	735
129	56	415	499
AKR	16	38	32
CBA/Ca	19	146	416
C3H/He	17	45	93
C57BL/6	14	60	77

Note that *in vitro* results come from US strains and *in vivo* results mainly from UK strains.
[a] Data from ref. 32.
[b] Data from ref. 18, except the value for AKR mice from ref. 17.

systemic circulation after oral administration[20]. The biological half-life of coumarin is about 1 h after p.o. and i.v. administration, and is not dependent on the route of administration. The 7-hydroxymetabolite is excreted as a glucuronide into urine[15,20–22].

3.3.3 Formation of 7-Hydroxycoumarin *in vivo* in Humans

A number of analytical techniques have been used to measure coumarin metabolism. These include paper and gas chromatography[23], spectrophotofluorometry[10,24], thin-layer chromatography[9] and HPLC[10,22,25–27].

Many studies have been performed on the metabolic fate of coumarin in humans and in most of them 7-hydroxycoumarin has been measured from blood and/or urine samples (Table 3.3). The dose of coumarin affects to a certain extent the conversion of coumarin to 7-hydroxycoumarin; about 60–70% of the total dose, on average, is excreted into urine as 7-hydroxycoumarin after 5 mg p.o.[10,28], 10 mg p.o.[29] and 200 mg p.o.[22]. Of a 5 mg dose, 95% is found within 4 h in healthy volunteers[9,10,28], whereas after a dose of 200 mg p.o., 95% of the amount excreted into urine is recovered within 10 h[22]. After oral administration of 1 or 2 g coumarin, the amount excreted into urine as 7-hydroxycoumarin-glucuronide is 20–39% of the dose given[31]. Of the rest, some is 3-hydroxycoumarin and distal metabolites, but most of it has not yet been accounted for.

3.3.4 Method to Determine Coumarin 7-Hydroxylation *in vivo*

Several approaches to design a method to reliably determine the ability of the liver to metabolize coumarin to the 7-hydroxy metabolite have been employed. In most of

Table 3.3. Population studies of 7-hydroxycoumarin excretion in humans.

Population (no. of subjects)	Coumarin dose (mg)	Collection time	Total excretion range (mg)	References
Healthy volunteers (8)	200	48 h	152–204	23
Volunteers (7)	200	fractioned ad 25 h	up to (127)	22
Volunteers (6)	100	fractioned ad 24 h		25
Volunteers	1000–2000			31
Volunteers (64)	2.5	8 h		9
Finnish volunteers (110)	5	fractioned 2, 4 h	1.0–5.0	10
Turkish volunteers (100)	5	fractioned 2, 4, 8 h	0.85–5.0	28
Volunteers (13)	10	fractioned 1,2,3,4,6,8 h		29
Thai patients (> 100)	5	fractioned		30

them coumarin is given orally and urine samples with or without timed fractionation have been collected for analysis (see Table 3.4). The coumarin dose used has varied from 2.5 mg to 15 mg. After a single dose there is great inter-individual variability in the extent and rate of 7-hydroxycoumarin formation in healthy volunteers[9,10,28].

Table 3.4. Comparison of average coumarin 7-hydroxylase activities in liver microsomes or homogenates of various species.

Species	Activity in microsomes (pmol/mg protein/min)
Chick embryo	1200
Rat	< 10
Mouse	
'unknown' strain	80
DBA/2	400
C57BL/6	60
Rabbit	480
Mongolian gerbil	290
Hamster	590
Syrian golden hamster	
female	40
male	70
Guinea-pig	160
Cat	10
Dog	200
Monkey	1780
Human	100–5000

Data compiled from refs 11, 48, 65 and 101.

On the other hand, the intra-individual variability in the percentage of 7-hydroxycoumarin excreted in 2 h and in the rate of the formation of 7-hydroxycoumarin is relatively small[10,22,28]. In two studies the excretion of 7-hydroxycoumarin was significantly higher in females than males[10,28], but considering the wide inter-individual variation, the biological significance of this difference is not clear. This finding points nevertheless to a possibility that sex hormones may be involved in the regulation of the 7-hydroxylation pathway in humans. On the basis of the *in vivo* coumarin test it is difficult to tell whether there is any polymorphism in the *CYP2A6* gene. In both Finnish[10] and Turkish[28] studies several subjects with particularly 'low'

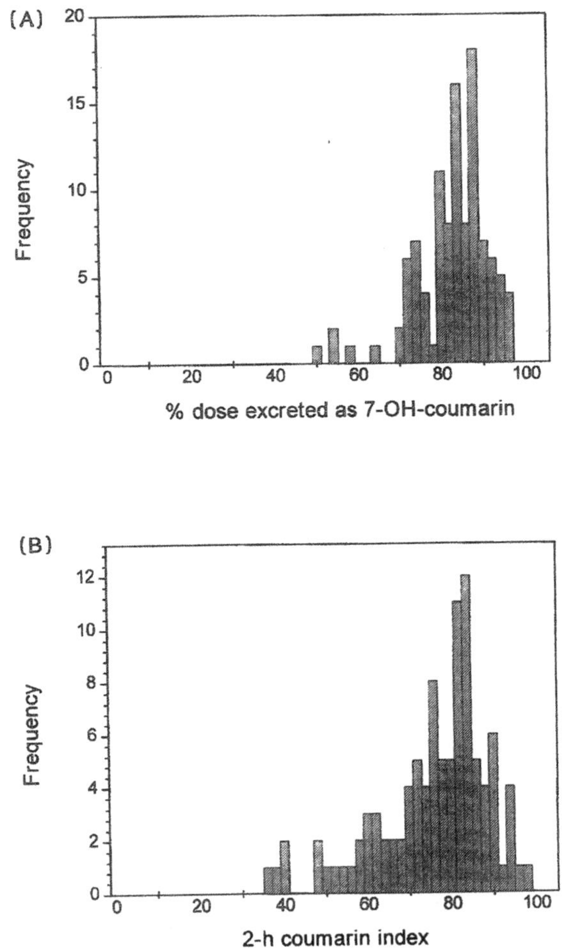

Figure 3.2. Frequency histograms of coumarin test values in two volunteer populations from Finland (A) and Turkey (B). For original data see Rautio *et al.*[10] and Iscan *et al.*[28].

coumarin test values (less than 50%) were found, which may or may not fit the normal distribution pattern (Figure 3.2). The possibility that a pharmacogenetic polymorphism would explain slow coumarin 7-hydroxylation is dealt with later on in detail. In any case, it is conceivable that coumarin is metabolized through other pathways than 7-hydroxylation[15] in these slow metabolizers, but specific pathways have not yet been elucidated.

3.4 COUMARIN-METABOLIZING ENZYMES

3.4.1 CYP2A Enzymes as Coumarin 7-Hydroxylases

The first studies aimed at characterizing the properties and regulation of a P450 enzyme active in coumarin 7-hydroxylation were those of Wood and Conney[17] and Lush and Andrews[18], which demonstrated its inducibility by phenobarbital and large interstrain differences in its activity in mice. Wood[32] and Wood and Taylor[33] extended these studies and characterized the mouse enzyme in different strains and elucidated some aspects of its genetic regulation. It was not until 1984 that Kaipainen *et al.*[34] purified the enzyme from C57BL/6 and DBA/2 mouse liver microsomes and found later on that it is inducible by pyrazole[35], a compound known to be a specific inducer of CYP2E1 in rats and rabbits. They also purified the enzyme from pyrazole-pretreated mouse liver microsomes and made an antibody against it, which was subsequently employed in human studies[36]. Pelkonen *et al.*[37] and Kaipainen *et al.*[38] compared coumarin 7-hydroxylase activities in liver microsomes from several species and found that human liver microsomal activities were by far the highest. Later on, Pelkonen's and Lang's group showed that human and mouse liver microsomes contain apparently identical P450 isoforms in terms of activity, antibody recognition and some other properties[39]. On the other hand, Negishi's group, when studying steroid 15α-hydroxylase in mouse liver and kidney[40,41], found two highly similar genes, one of which (later named *Cyp2a-5*) appeared to be coumarin 7-hydroxylase[42]. At about the same time, Gonzalez's group[43] and Wolf's group[44] finalized the studies to clone and sequence human CYP2A6 and it appeared that CYP2A5 and CYP2A6 are about 85% homologous and that most amino acid residues essential for a high coumarin 7-hydroxylase activity are identical in the two proteins.

3.4.2 Properties of CYP2A5/6 Proteins

The enzyme catalyzing coumarin 7-hydroxylation is a rather typical P450 enzyme located mainly in the endoplasmic reticulum of the liver[33,36]. The microsomal enzyme is actually the most thoroughly studied, although significant coumarin 7-hydroxylase activity can also be found in hepatic mitochondrial fraction and the properties of the enzyme are in most respects similar to those of the microsomal enzyme[45]. It is not known whether the mitochondrial enzyme has any specific function in addition to xenobiotic metabolism.

In humans, coumarin 7-hydroxylation appears to be a relatively specific reaction of CYP2A6. Among 11 heterologously expressed human CYP enzymes, by far the most active isoform was CYP2A6 and the only other P450 enzyme possessing some activity (about 10% of that of CYP2A6) was CYP2B6[46]. This is in striking contrast with another high-affinity substrate of CYP2A6, 7-ethoxycoumarin, which is metabolized by 10 out of 11 human CYP enzymes[46].

3.4.3 Substrate/Inhibitor Specificity of CYP2A5 and CYP2A6

The substrates oxidized by CYP2A5 and CYP2A6 are extremely variable in their K_m and V_{max} values. Coumarin and 7-ethoxycoumarin are high-affinity substrates for both enzymes. With respect to other substrates, it is difficult to categorize them on this basis, because a substrate may have high affinity in one species (aflatoxin B_1 towards CYP2A5), but relatively low affinity in another (aflatoxin B_1 towards CYP2A6). Low-affinity substrates include, for example, certain *N*-nitrosoamines and possibly some other compounds. It should be noted that derivatives of 4-hydroxycoumarin, such as dicumarol and warfarin, which are used as anticoagulants, are not substrates for CYP2A5 or CYP2A6[47,48]. CYP2A6 is one of the P450 enzymes capable of catalyzing C-oxidation of nicotine, but its importance in nicotine metabolism in humans is unclear[49,50].

A large number of substances have been screened with respect to their ability to inhibit coumarin 7-hydroxylation in mouse and human liver microsomes[51]. The mouse enzyme is inhibited by several furanocoumarins, azole antimycotics and metyrapone, although the specificity of inhibition has not been systematically studied[51]. Methoxsalen (8-methoxypsoralen) is the most potent inhibitor, but it is not specific for coumarin 7-hydroxylase, because it inhibits caffeine and theophylline metabolism equally well[52].

Pilocarpine is another relatively potent, and probably more specific, inhibitor of CYP2A5[53,54]. Recently, grapejuice was shown to be an inhibitor of coumarin metabolism in humans *in vivo*[29], suggesting that components of grapejuice like naringenin may interact with CYP2A6.

Coumarin itself has been shown to be a potent inhibitor of activities catalyzed by mouse CYP2A5, human CYP2A6, and related enzymes in hamster liver. However, it seems to interact also with other CYPs. It strongly inhibited NDMA *N*-demethylation—most probably catalyzed by CYP2E1—in hamster liver microsomes[55].

Mouse and human liver coumarin 7-hydroxylases display some interesting differences in their inhibitory properties. Metyrapone is a good inhibitor of the mouse enzyme[33,56], whereas it is a rather poor inhibitor of the human enzyme[57]. Inability to inhibit the human liver enzyme is also characteristic of some furanocoumarins, which are rather potent inhibitors of the mouse enzyme (Figure 3.3). On this basis it seems that active sites of human and mouse enzymes differ in some important characteristics.

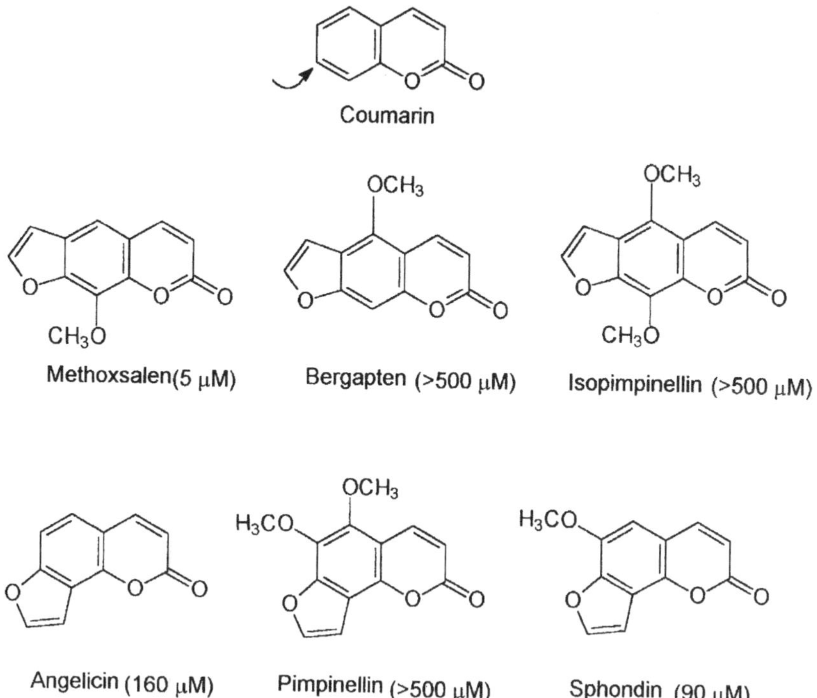

Figure 3.3. Structures of furanocoumarin derivatives active as inhibitors of coumarin 7-hydroxylase in mouse liver microsomes. Mean inhibitory concentrations (IC50 values) are given in parenthesis. For original data see Mäenpää *et al.*[51].

4-Methylpyrazole stimulates CYP2A6-mediated coumarin 7-hydroxylase in human liver microsomes in a manner similar to that in which α-naphthoflavone activates CYP3A isoforms[58].

3.4.4 Interspecies Differences

Coumarin 7-hydroxylase activity displays interspecies variability to an extraordinary degree (see Table 3.4). Of all the species tested so far, humans have the highest specific activity, ranging between 0.1 and 5 nmol/mg microsomal protein/min or 14 and 47 nmol/g liver (although the highest values may represent individuals exposed to inducers; see later). Some primates come close to human activities. In contrast, rat liver is essentially devoid of coumarin 7-hydroxylase activity. This does not mean that the rat does not express the enzyme or that the coumarin 7-hydroxylase gene is deleted, because an enzyme catalyzing coumarin 7-hydroxylase exhibiting properties similar to those of the enzyme in mouse and humans has been found in rat lungs[59] and an enzyme similar to CYP2A5 with a high coumarin 7-hydroxylase activity is expressed in rat nasal mucosa[60].

3.4.5 Multiplicity and Organization of *CYP2A* Genes

In humans, the *CYP2A* locus, which spans over 100 kb, maps into chromosome 19 and contains three complete genes (*CYP2A6*, *CYP2A7* and *CYP2A13*) and two truncated pseudogenes[61]. The product of the *CYP2A6* gene catalyzes coumarin 7-hydroxylase activity. On the other hand, CYP2A7 does not catalyze 7-hydroxylation at all[62]. There is also a third gene, designated *CYP2A13*, but it is unclear whether it is expressed or whether it could catalyze the coumarin 7-hydroxylation[63]. The two pseudogenes seem to be derived from *CYP2A7*. An alternatively spliced version of *CIPY2A7*, missing exon 2, but containing a 10 bp segment of intron and giving rise to a protein of 44 kDa, has been detected[62], but whether the protein has some function is not known.

There seem to be large interspecies differences in the number (and possibly organization) of *CYP2A* genes; for details the reader is referred to a recent review of Fernandez-Salguero and Gonzalez[61].

3.5 REGULATION OF COUMARIN METABOLISM

Because very little is known about the enzymes involved in coumarin metabolism, other than CYP2A-catalyzed 7-hydroxylation, most studies concerning regulation are performed on coumarin 7-hydroxylases.

3.5.1 Constitutive Expression and Genetic Differences

The first studies performed either on liver preparations[17,32,33] or with *in vivo* coumarin administration[18] demonstrated that mouse strains could be classified into two or three groups on the basis of their coumarin 7-hydroxylation (Table 3.2). Genetic studies with these strains performed by Wood and Taylor[33] suggested that the variation in coumarin 7-hydroxylase activity could be due to a structural mutation in the CYP protein specific to the activity. This prediction turned out to be true. Lindberg *et al.*[64] isolated and sequenced full-length CYP2A5 cDNAs from high-activity strains (DBA2/J and 129/J) and low-activity strains (C57BL/6, C3H/HeJ, AKR/J), and found that there is a single nucleotide difference between the cDNAs of the high-activity and low-activity strains. This difference results in a valine to alanine substitution at position 117, which yields a low-activity enzyme. Strains that are homozygous for the high-activity gene (Coh^H) exhibit 10-fold higher V_{max} and fourfold lower K_m values compared with strains homozygotic for the low-activity gene (Coh^L). The genes are inherited in a co-dominant fashion, since F_1 generations between the Coh^H and Coh^L strains express equal amounts of the two genes[64].

3.5.2 Tissue- and Sex-specific Expression

There are a large number of studies addressed to the expression of various CYP2A enzymes in the mouse[65]. The expression pattern of the *Cyp2a-5* gene is known to depend substantially on tissue and sex. In addition, members of the CYP2A subfamily mediating coumarin 7-hydroxylase activity are expressed at least in the liver, kidney, lung and nasal epithelial cells of various rodent species (Table 3.5).

CYP2A5 is expressed at higher levels in female than male mouse liver[11,40,42]. Sex-dependent differences are found also in the induction of CYP2A-related activities in liver and extrahepatic tissues. In female mice of strains DBA/2, 129 and CBA/Ca the induction by phenobarbital is significantly higher than in males[11].

In humans, there appears to be no significant gender difference in the hepatic CYP2A6 as measured at coumarin 7-hydroxylase activity, protein or mRNA levels[3,37,39]. The same is true for cynomolgus monkeys[48]. However, two studies on *in vivo* excretion of 7-hydroxycoumarin indicate slightly higher rates for females than males[10,28].

In addition to the liver, a protein or proteins related to CYP2A have been shown to be present in nasal mucosa[66]. Furthermore, studies on CYP2A6-specific inhibitors indicate that a CYP2A6-like enzyme is present in human kidney microsomes, which may participate in the defluorination of the anesthetic methoxyflurane[67].

3.5.3 Hormone-dependent and Developmental Activation of CYP2A Genes

In mice, it is not known precisely at which time of development coumarin 7-hydroxylase activity appears, but already at four weeks of age the activity, protein and

Table 3.5. Presence of coumarin 7-hydroxylating CYP2A members in various species and tissues.

Species	Organ/tissue	References
Mouse	Liver	42
	Kidney	40, 42
	Lung	59
Rat	Liver	39, 102
	Lung	59, 103
	Nasal mucosa	60
Hamster	Liver	
	Kidney	
	Lung	55
Rabbit	Liver	48
	Nasal mucosa	98
Guinea-pig	Liver	48
Dog	Liver	48
Chick embryo	Liver	101
Monkey	Liver	48, 91, 104
Human	Liver	39, 43, 44, 105
	Nasal epithelia	66

Very low coumarin 7-hydroxylase activity in cat and rat liver.

mRNA are detectable and are significantly higher in female than male liver[35]. In males an activity peak at the age of four weeks has been found. Gonadectomy increases and hormonal supplementation decreases the expression of *CYP2A5* (Kojo *et al.*, unpublished). Furthermore, growth hormone depletion increases the expression of coumarin 7-hydroxylase. Therefore, it is clear that coumarin 7-hydroxylase activity is both hormonally and developmentally regulated.

In humans, *CYP2A6* is not expressed in fetal tissues, at least not during the first half of pregnancy as measured by enzymatic activity, by Western blots or by RT-PCR[68,69]. The excretion of 7-hydroxycoumarin is at the adult level in children of 6–13 years of age (Pasanen *et al.*, unpublished). It seems that the expression is triggered some time between the second half of pregnancy and early infancy. This is consistent with, but does not prove, the hypothesis that CYP2A6 mediates the metabolism of dietary compounds.

3.5.4 Induction by Chemicals

Perhaps the most intriguing aspect of coumarin 7-hydroxylase is its extraordinary inducibility by a wide variety of chemicals and manipulations. A compilation of inducers is presented in Figure 3.4. It has to be stressed that the term 'inducer' is used as a generic term to indicate the increase ('induction') of coumarin 7-hydroxylase activity. Except for few experiments which will be discussed later, the mechanisms of induction have not been elucidated.

'Classical' Inducers

Coumarin 7-hydroxylase is one of the monooxygenase activities elevated by the archetypal inducer phenobarbital[17]. Also some other phenobarbital-type inducers such as TCPOBOP increase the activity[70], although some differences in the induction spectrum between phenobarbital and TCPOBOP have been observed[71].

Pyrazole and Derivatives

Pyrazole, an effective inducer of CYP2E1 (a P450 form metabolizing ethanol and being induced by ethanol) in the rat and rabbit, but not in the mouse, increases the mouse hepatic coumarin 7-hydroxylation several-fold[34]. With the exception of 7-ethoxycoumarin *O*-deethylation, most other P450-associated reactions were decreased by the pyrazole pretreatment. Also several pyrazole derivatives are inducers of coumarin 7-hydroxylation[56].

Metal Ions

In addition to phenobarbital (PB) and pyrazole, some heavy metals, especially cobalt and indium, were found to elevate murine liver coumarin 7-hydroxylase activity[72,73]. These findings were surprising, since cobalt was considered to be a universal P450

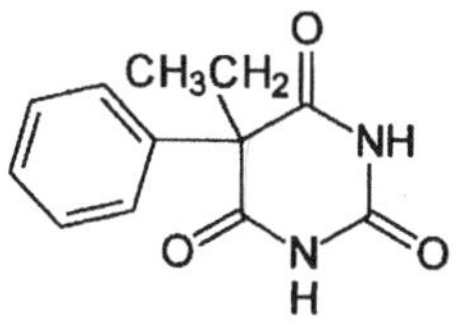

Phenobarbital

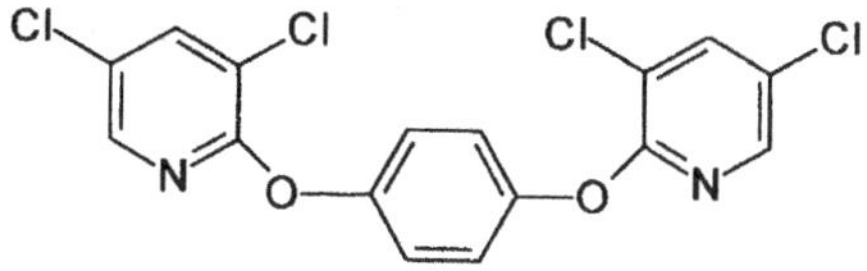

TCPOBOP

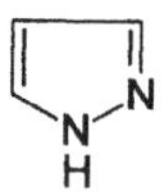

Pyrazole

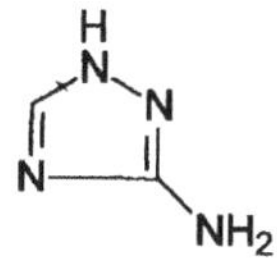

Aminotriazole

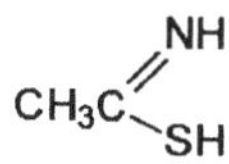

Thioacetamide

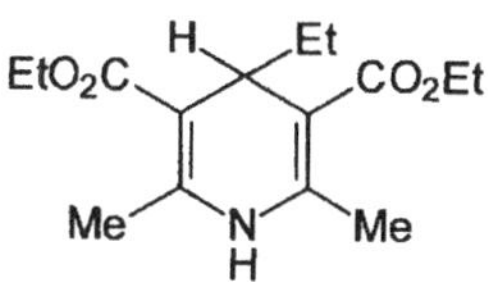

Ethyl-DDC

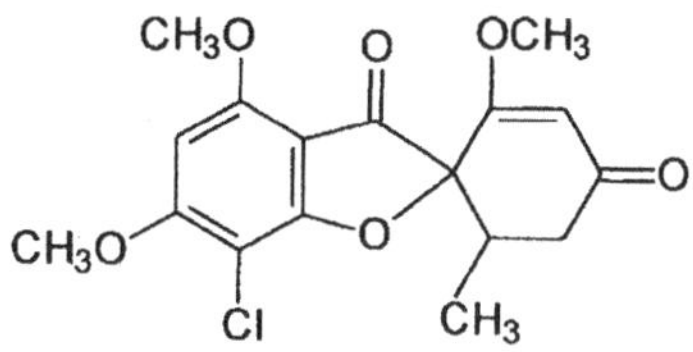

Griseofulvin

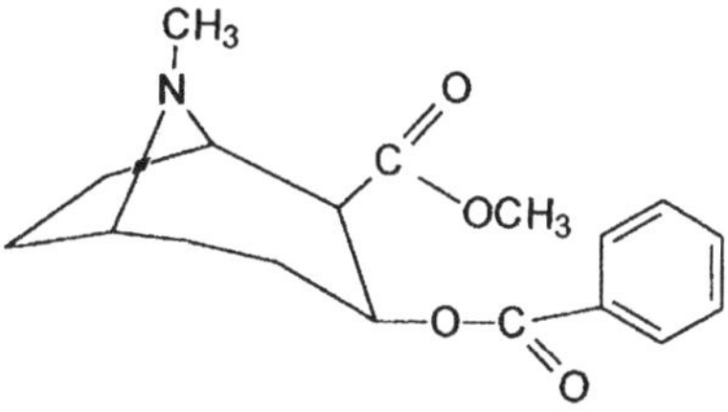

Cocaine

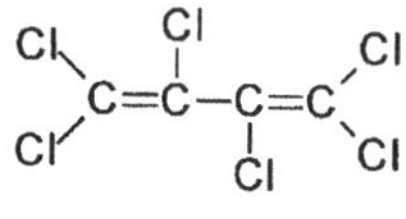

Hexachlorobutadiene

$CoCl_2$ $CeCl_3$ CCl_4

Figure 3.4. Chemical structures of various 'inducers' of coumarin 7-hydroxylation and CYP2A5 in mouse liver. For details see Pelkonen *et al.*[65].

repressor by activating heme oxygenase and, thus, causing the breakdown of cellular heme necessary for the active enzyme[74]. Cobalt was later shown to increase mouse liver coumarin 7-hydroxylase activity by increasing CYP2A5 protein and mRNA levels[75,76].

Cerium, a member of the light lanthanoids, produces profound liver damage and a general decrease in hepatic CYP-mediated oxidative metabolism in rodents (for a review see ref. 77). In mice, cerium-caused hepatotoxicity is strain-dependent, the DBA/2 mice being much more susceptible than the C57BL/6. Administration of toxic doses of cerium to these mice results in an elevation of liver coumarin 7-hydroxylation activity and the amount of CYP2A5 mRNA in the susceptible DBA/2 strain but not in the resistant C57BL/6 strain[77,78]. The increase in *CYP2A5* expression precedes the development of the overt liver damage.

A recent study from Netter's group demonstrates that tin chloride is a potent inducer of coumarin hydroxylase activity and Cyp2a-5 in the DBA/2 mouse liver and kidney (Legrum *et al.*, unpublished) causing as much as 20- to 30-fold increase in the enzyme activity. Like cobalt, tin is known to affect heme metabolism.

Miscellaneous Hepatotoxic Substances

Carbon tetrachloride, hexachlorobutadiene and cocaine increased hepatic coumarin 7-hydroxylase activity several-fold in DBA/2, but not appreciably in C57BL/6 mice, a finding reminiscent of that observed with cerium. On the other hand, allyl alcohol and dimethylnitrosamine did not increase coumarin 7-hydroxylase activity despite their clear hepatotoxicity. Later detailed studies on the hepatotoxicity and CYP2A5 induction by cocaine demonstrated that even a single dose of cocaine causes a progressive deterioration of liver condition. Coumarin 7-hydroxylase activity is very high ('induced') during the first few days and, thereafter, there is a gradual decrease to a practically zero level in a few days[79].

Porphyrinogenic Agents

All CYP enzymes contain heme (Fe-protoporphyrin IX) as a prosthetic group. Several CYP inducers are known to disrupt cellular heme homeostasis by either interfering with the biosynthesis or breakdown of heme[80]. Thus, it is of interest to study the effects of porphyrinogenic agents on the expression of *CYP2A5*. It has been demonstrated that griseofulvin, thioacetamide and aminotriazole produce up to 10-fold increases in coumarin 7-hydroxylase activity and up to 130-fold increases in the amount of CYP2A5 mRNA in DBA/2 mouse liver[81], while some other P450-catalyzed activities were decreased. The increased *CYP2A5* expression could be blocked in some cases by co-administration of exogenous heme to the mice, suggesting that the regulation of *CYP2A5* expression could involve heme-sensitive factors[81].

Mechanisms of Induction

Several mechanisms are involved in the regulation of expression of xenobiotic-metabolizing enzymes. These include gene transcription, pre-mRNA processing, mRNA stabilization, translational efficiency and protein stabilization[82]. As discussed above, coumarin 7-hydroxylase activity is inducible by a multitude of structurally different chemicals (Figure 3.4). Concerning the mechanisms of induction, a process postulating a single receptor, therefore, seems rather difficult to reconcile. It is possible that different agents act through different mechanisms and that different sites of action are involved in the induction of coumarin 7-hydroxylation by different compounds: for example transcription in the case of phenobarbital and mRNA stabilization in the case of pyrazole and cobalt[76,83].

Studies with isolated DBA/2 mouse hepatocytes have revealed new aspects of the induction of CYP2A5[84]. It appears that transcriptional control is essential, since coumarin 7-hydroxylase induction by phenobarbital can be completely blocked by actinomycin D. Superinduction of coumarin 7-hydroxylase activity can be elicited by short-term treatment with cycloheximide, implying that a short-lived repressor protein is suppressing the transcription of CYP2A5 mRNA. Such repressor proteins have been found to regulate phenobarbital-inducible *Bacillus megaterium* genes[85], but the existence and nature of corresponding mammalian repressors are still elusive.

Liver Disease-causing Biological Agents

In a hepatitis B virus (HBV)-transgenic mouse model, over-production of the HBV large envelope protein results in progressive liver cell injury, inflammation, and regenerative hyperplasia. In a recent study of Kirby *et al.*[86], an increased expression and altered distribution of CYP2A5 was demonstrated, by immunohistochemical analysis, to be associated with the development of liver injury in mice. The increase of *CYP2A5* expression was not uniform, but shifted from peracinar localization typical for non-transgenic animals to scattered, intensively stained cells in transgenic animals. It is of interest to notice that the total P450 content decreased as the HBV-related liver injury progressed.

Liver parasitism by *Orpisthorchis viverrini* (liver fluke) causes a periportal inflammation in hamsters. This infestation causes a considerable local increase in the expression of hamster CYP enzymes immunorelated to mouse CYP2A5 immediately adjacent to areas of periportal inflammation[87]. It should be noted that the expression of CYP2A5-related immunoreactivity in normal mice and hamsters is localized around central veins. Another point of importance is that the increase in the *in vitro* microsomal metabolism of coumarin 7-hydroxylation or other related activities or in Western blots is marginal, indicating that localized response could be easily missed by standard approaches of CYP measurements.

The mechanism by which liver fluke infestation or HBV-associated liver injury increase the activity of *CYP2A5*-related enzymes is not known. However, taking into consideration that also several hepatotoxic substances increase *CYP2A5* expression in the mouse liver, it is tempting to speculate that the increase might not be caused by an etiological agent *per se*, but might be a consequence of liver cell injury or the associated inflammatory response. It is possible that an inflammatory factor, i.e. a cytokine or a growth factor, may be regulating the expression of *CYP2A5*, thus altering hepatocyte function in liver disease.

3.6 FACTORS AFFECTING COUMARIN 7-HYDROXYLATION IN HUMANS

3.6.1 Effects of Inducers and Inhibitors

A number of potentially modifying factors of coumarin 7-hydroxylation *in vivo* have been investigated. In a study with 20 young healthy volunteers[88], the excretion rate of 7-hydroxycoumarin did not change after acute alcohol ingestion or after treatment with rifampicin (450 mg daily) or cimetidine (600 mg daily) for six days, which suggests that some known inhibitors (acute alcohol and cimetidine) and inducers (rifampicin) of CYP enzymes appear not to affect CYP2A6 function *in vivo*. On the other hand, 7-hydroxycoumarin excretion was lower in both female and male smokers compared to non-smokers[28], although the effect was not very strong. Methoxsalen (8-methoxypsoralen), an anti-psoriatic agent, has been shown to inhibit the *in vivo* formation of 7-hydroxycoumarin in humans[52] and Merkel *et al.*[29] have shown that grapefruit juice flavonoids inhibit the 7-hydroxycoumarin formation in healthy volunteers. Also wheat germ juice inhibits 7-hydroxycoumarin formation, while a vegan diet has no significant effect[89] (Table 3.6).

Epileptic patients treated with inducing antiepileptic drugs (carbamazepine, phenobarbital) seem to have an increased rate of 7-hydroxycoumarin formation, because they excrete all the measurable 7-hydroxycoumarin during the first 4 h[90]. This finding suggests that *CYP2A6* is inducible by some antiepileptic drugs and is consistent with the induction in isolated primary human hepatocytes by phenobarbital and rifampicin[91].

3.6.2 Effect of Diseases on Coumarin 7-Hydroxylation

Alcohol-induced liver damage seems to reduce coumarin 7-hydroxylation: whereas for healthy volunteers the excretion within 4 h is over 95% of the total excretion, for patients with severe alcoholic liver disease, the excretion in 4 h is only 40%[90]. The 7-hydroxycoumarin formation is also somewhat decreased in moderate and mild alcoholic liver diseases[90]. The 24-h excretion values of 7-hydroxycoumarin are clearly lower in patients with severe liver disease than in healthy volunteers, suggesting that coumarin is metabolized via other pathways or is excreted unchanged in these patients. The latter possibility has not been rigorously checked.

The attenuation of 7-hydroxycoumarin formation by liver disease is apparently in contradiction with mouse studies, which sometimes demonstrate a large increase of

Table 3.6. Factors affecting 7-hydroxycoumarin formation rate in humans.

Subjects/drugs	Treatment	7-Hydroxycoumarin excretion (% of control)	
		0–2 h	Total
Healthy volunteers			
Cimetidine	600 mg p.o. daily for 6 days	101	97
Methoxsalen	45 min p.o. 1 h before coumarin	55	53
Rifampicin	450 mg p.o. daily for 6 days	100	97
Heme arginate	3 mg/kg 1 h before coumarin	96	—
Ethanol	Social dinner (7 p.m.–2 a.m.)	100	97
Cigarette smoking		91	—
Vegans	Uncooked vegan diet	92	90
Wheat-grass juice		86	—
Grapefruit juice flavonoids		74	97
Patients			
Porphyria variegate		112	—
Porphyria variegate	Heme arginate 3 mg/kg 1 h before coumarin	96	—
Alcoholic liver disease		32	70
Kidney disease		49	103
Epilepsy	Carbamazepine, clonazepam, phenobarbital, phenytoin	124	107

Coumarin test is performed by giving 5 mg coumarin p.o., except grapefruit flavonoids study 10 mg coumarin p.o. Data from refs 29, 88–90 and 93.

coumarin 7-hydroxylase after hepatotoxic agents. However, in one study with human liver, an increase of *CYP2A6* mRNA expression, as measured by *in situ* hybridization, was found near cirrhotic areas[92]. It is possible that in alcohol-induced liver damage and liver cirrhosis the stage and time of the disease is of importance with respect to changes in enzyme activities. Furthermore, it is possible that local increases in the amount of CYP2A6 cannot be detected by the *in vivo* test.

In patients with variegate porphyria, the 7-hydroxycoumarin excretion rate was similar to controls and heme arginate treatment did not change it[93]. In contrast, a severely delayed elimination of antipyrine was observed in these same porphyric patients and was restored to normal by heme arginate treatment[94]. This shows that CYP2A6 and the CYP enzymes metabolizing antipyrine are under different regulatory control. The results are also in accordance with the mouse studies demonstrating the inducing effect of porphyrinogenic drugs[81].

Patients with impaired kidney function have delayed 7-hydroxycoumarin excretion, which correlates to serum creatinine levels and creatinine clearance; however, the total recovery of 7-hydroxycoumarin is similar to that of controls[90], suggesting the excretion rather than formation of 7-hydroxycoumarin is compromised in renal impairment.

3.6.3 Polymorphisms of Coumarin 7-Hydroxylation

While the population studies have not been able to convincingly demonstrate any discontinuity in the frequency histograms on the rate or extent of 7-hydroxycoumarin excretion[9,10,28], several individuals who apparently lack the ability to form 7-hydroxycoumarin have been identified in different centers[63]. Thus far, genetic analysis has demonstrated the presence of at least two variant alleles of the *CYP2A6* gene[63]. Yamano *et al.*[43] demonstrated a variant mRNA, leading to only a single amino acid change in the protein (Leu160His). The cDNA, when transfecting into COS cells, produced a protein that did not incorporate heme and appeared to be non-functional. It is possible that this genetic change is responsible for at least some cases of deficient *in vivo* coumarin 7-hydroxylation. The frequency of this allele, termed *CYP2A6v1*, in human populations seems to be variable[63]. Another variant of the *CYP2A6* gene involving several mutations has also been reported[63], but studies linking this allele with coumarin metabolism are not yet available. At the present state of knowledge, the complexity of the *CYP2A* locus in humans prevents any far-reaching conclusions on the relationships between the genotype and the phenotype.

3.7 PHYSIOLOGICAL, PHARMACOLOGICAL AND TOXICOLOGICAL SIGNIFICANCE

3.7.1 Species Differences in Coumarin Metabolism

It is possible that the CYP enzyme catalyzing coumarin 7-hydroxylation protects against coumarin-induced toxicity. Species for which 7-hydroxylation is the major metabolic pathway are less at risk from coumarin-induced hepatotoxicity than those species in which 3-hydroxylation pathway is more prominent. Extensive 7-hydroxylation may be of significance also in the therapeutic applications of coumarin so far as 7-hydroxycoumarin is an active form. Thus the P450 enzyme catalyzing coumarin 7-hydroxylation could, in some cases, be important from both pharmacological and toxicological points of view.

3.7.2 Alkaloid Metabolism and Actions

So far no clinically important pharmaceutical substances have been found to be metabolized by CYP2A6. Most of its substrates and inhibitors are of natural origin, coumarin itself as a foremost example. It may be of interest that many coumarin-related substances have cancer chemopreventive properties. Coumarin also has immunomodulatory properties. It may be speculated that CYP2A may somehow control these actions.

3.7.3 Regulation in Liver Injury

Considering the regulation of other P450 enzymes it is of interest that an increased expression of *CYP2A5* is achieved by various chemical and biological treatments, many of which are associated with liver injury (and typically reduce the expression of a majority of P450 enzymes). Elevated levels of CYP2A5 also exist in chemically induced hepatomas[95,96]. These findings point to the possibility that the expression of the coumarin-hydroxylating enzyme might be regulated by some physiological factors, released or induced in relation to the tissue damage. Speculatively, therefore, the enzyme might have a physiological role.

3.7.4 Olfaction

Coumarin is an odorant, and therefore it is not surprising that CYP2A-like enzymes are found in high levels in rat, rabbit and bovine nasal mucosa[97,98]. The possible function in this case would be to enhance the clearance of coumarin-like substances from the olfactory area and receptors. In support of this view, Bereziat *et al.*[60] have recently shown extremely high coumarin 7-hydroxylase activity in rat nasal epithelium, catalyzed by an enzyme similar to CYP2A5. The highest activity was shown in the olfactory epithelium and the expression of the protein as well as the enzyme activity was increased by coumarin. Another, seemingly unrelated finding suggesting the role of CYP2A in olfaction is the curious finding of cAMP regulation of *CYP2A5* in the liver, because it is known that odorant signal transduction involves cAMP pathways. Naturally, it is also possible that this relation is just a coincidence.

ACKNOWLEDGEMENT

This review contributes to the goals of the Action COST B1.

REFERENCES

1 Williams RT, ed. *Detoxication Mechanisms*. London: Chapman & Hall, 1959.

2 Cohen AJ. Critical review of the toxicology of coumarin with special reference to interspecies differences in metabolism and hepatotoxic response and their significance to man. *Fd Cosmet Toxicol* 1979; **17**: 277–89.

3 Kratz F. Coumarin-7-hydroxylase activity in microsomes from needle biopsies of normal and diseased human liver. *Eur J Clin Pharmacol* 1976; **10**: 133–7.

4 Aitio A. A simple and sensitive assay of 7-ethoxycoumarin deethylation. *Anal Biochem* 1978; **85**: 488–91.

5 Greenlee WF, Poland A. An improved assay of 7-ethoxycoumarin O-deethylase activity: induction of hepatic enzyme activity in C57BL/6J and DBA/2J mice by phenobarbital, 3-methylcholantherene and 2,3,7,8-tetrachloro-p-dioxin. *J Pharmacol Exp Ther* 1978; **205**: 596–605.

6 Peters MMCG, Walters DG, van Ommen B, van Bladeren PJ, Lake BG. Effect of inducers of cytochrome P-450 on the metabolism of [3-^{14}C] coumarin by rat hepatic microsomes. *Xenobiotica* 1991; **21**: 499–514.

7 Fentem JH, Hammond AH, Garle M, Fry JR. Toxicity of coumarin and various methyl derivatives in cultures of rat hepatocytes and V79 cells. *Toxicology In Vitro* 1992; **6**: 21–5.

8 Fentem JH, Fry JR. Species differences in the metabolism and hepatotoxicity of coumarin. *Comp Biochem Physiol* 1993; **104C**: 1–8.

9 Cholerton S, Idle ME, Vas A, Gonzalez FJ, Idle JR. Comparison of a novel thin-layer chromatographic-fluorescence detection method with a spectrofluorometric method for the determination of 7-hydroxycoumarin in human urine. *J Chromatogr* 1992; **575**: 325–30.

10 Rautio A, Kraul H, Kojo A, Salmela E, Pelkonen O. Interindividual variability of coumarin 7-hydroxylation in healthy individuals. *Pharmacogenetics* 1992; **2**: 227–33.

11 van Iersel M, Walters DG, Price RJ, Lovell DP, Lake BG. Sex and strain differences in mouse hepatic microsomal coumarin 7-hydroxylase activity. *Fd Chem Toxicol* 1994; **32**: 387–90.

12 van Iersel MLPS, Henderson CJ, Walters DG, Price RJ, Wolf CR, Lake BG. Metabolism of (3-^{14}C) coumarin by human liver microsomes. *Xenobiotica* 1994; **24**: 795–803.

13 Steensma A, Beamand JA, Walters DG, Price RJ, Lake BG. Metabolism of coumarin and 7-ethoxycoumarin by rat, mouse, guinea pig, cynomolgus monkey and human precision-cut liver slices. *Xenobiotica* 1994; **24**: 893–907.

14 Den Besten C, Körösi SA, Beamand JA, Walters DG, Lake BG. Studies on the mechanism of coumarin-induced hepatotoxicity in rat hepatocytes. *Toxicology In Vitro* 1990; **4**: 518–21.

15 Egan D, O'Kennedy R, Moran E, Cox D, Prosser E, Thornes RD. The pharmacology, metabolism, analysis, and application of coumarins and coumarin-related compounds. *Drug Metab Rev* 1990; **22**: 503–29.

16 Kaighen M, Williams RT. The metabolism of (3-^{14}C)-coumarin. *J Med Pharm Chem* 1961; **3**: 25–43.

17 Wood AW, Conney AH. Genetic variation in coumarin hydroxylase activity in the mouse (*Mus musculus*). *Science* 1974; **185**: 612–14.

18 Lush IE, Andrews KM. Genetic variation between mice in their metabolism of coumarin and its derivatives. *Genet Res Camb* 1978; **31**: 177–86.

19 Legrum W, Frahseck A. Acceleration of 7-(methoxy-^{14}C) coumarin-derived carbon dioxide exhalation by cobalt pretreatment in mice. *J Pharmacol Exp Ther* 1982; **221**: 790–4.

20 Ritschel WA, Brady ME, Tan HSI, Hoffman KA, Yiu IM, Grummich KW. Pharmacokinetics of coumarin and its 7-hydroxymetabolites upon intravenous and peroral administration of coumarin in man. *Eur J Clin Pharmacol* 1977; **12**: 457–61.

21 Ritschel WA, Brady ME, Tan HSI. First-pass effect of coumarin in man. *Int J Clin Pharmacol Biopharmac* 1979; **17**: 99–103.

22 Moran E, O'Kennedy R, Thornes RD. Analysis of coumarin and its urinary metabolites by high-performance liquid chromatography. *J Chromatogr* 1987; **416**: 165–9.

23 Shilling WH, Crampton RF, Longland RC. Metabolism of coumarin in man. *Nature* 1969; **221**: 664–5.

24 Tan HS, Ritschel WA, Sanders PR. Determination of coumarin and umbelliferone mixtures in whole blood by spectrophotofluorimetry. *J Pharm Sci* 1976; **65**: 30–2.

25 Egan DA, O'Kennedy R. Rapid and sensitive determination of coumarin and 7-hydroxycoumarin and its glucuronide conjugate in urine and plasma by high-performance liquid chromatography. *J Chromatogr* 1992; **582**: 137–43.

26 Walters DG, Lake BG, Cottrell RC. High performance liquid chromatography of coumarin and its metabolites. *J Chromatogr* 1980; **196**: 501–5.

27 Lamiable D, Vistelle R, Trenque T, Fay R, Millart H, Choisy H. Sensitive high-performance liquid chromatographic method for the determination of coumarin in plasma. *J Chromatogr* 1993; **620**: 273–7.

28 Iscan M, Rostami H, Guray T, Pelkonen O, Rautio A. A study on the interindividual variability of coumarin 7-hydroxylation in a Turkish population. *Eur J Clin Pharmacol* 1994; **47**: 315–18.

29 Merkel U, Sigusch H, Hoffmann A. Grapefruit juice inhibits 7-hydroxylation of coumarin in healthy volunteers. *Eur J Clin Pharmacol* 1994; **46**: 175–7.

30 Satarug S, Lang MA, Yongvanit P, Sithithaworn P, Mairiang E, Pelkonen P, Bartsch H, Haswell-Elkins MR. Induction of cytochrome P450 2A6 expression in humans by the carcinogenic parasite infection *Orpisthorchis viverrini*. *Cancer Epidemiol Biomark Prevent* (in press).

31 Sharifi S, Lotterer E, Michaelis HC, Bircher J. Pharmacokinetics of coumarin and its metabolites after intravenous and peroral administration of high-dose coumarin in normal human volunteers. *Naunyn-Schmiedeberg Arch Pharmacol* 1991; **344** (Suppl.): 95.

32 Wood AW. Genetic regulation of coumarin hydroxylase activity in mice. Biochemical characterization of the enzyme from two inbred strains and their F1 hybrid. *J Biol Chem* 1979; **254**: 5641–6.

33 Wood AW, Taylor BA. Genetic regulation of coumarin hydroxylase activity in mice. Evidence for single locus control on chromosome 7. *J Biol Chem* 1979; **254**: 5647–51.

34 Kaipainen P, Nebert DW, Lang MA. Purification and characterization of a microsomal P-450 with high activity of coumarin 7-hydroxylase from mouse liver. *Eur J Biochem* 1984; **144**: 425–31.

35 Juvonen R, Kaipainen PK, Lang MA. Selective induction of coumarin 7-hydroxylase by pyrazole in D2 mice. *Eur J Biochem* 1985; **152**: 3–8.

36 Juvonen R, Shkumatov V, Lang MA. Purification and characterization of a liver microsomal P-450 isoenzyme with a high affinity and metabolic capacity for coumarin from pyrazole-treated D2 mice. *Eur J Biochem* 1988; **171**: 205–11.

37 Pelkonen O, Sotaniemi EA, Ahokas JT. Coumarin 7-hydroxylase activity in human liver microsomes. Properties of the enzyme and interspecies comparisons. *Br J Clin Pharmacol* 1985; **19**: 59–66.

38 Kaipainen P, Koivusaari U, Lang M. Catalytic and immunological comparison of coumarin 7-hydroxylation in different species. *Comp Biochem Physiol* 1985; **81C**: 293–6.

39 Raunio H, Syngelmä T, Pasanen M, Juvonen R, Honkakoski P, Kairaluoma MA, Sotaniemi EA, Lang MA, Pelkonen O. Immunochemical and catalytical studies on hepatic coumarin 7-hydroxylase in man, rat, and mouse. *Biochem Pharmacol* 1988; **37**: 3889–95.

40 Squires EJ, Negishi M. Reciprocal regulation of sex-dependent expression of testosterone 15α-hydroxylase (P-45015a) in liver and kidney of male mice by androgen. Evidence for a single gene. *J Biol Chem* 1988; **263**: 4166–71.

41 Lindberg R, Burkhart B, Ichikawa T, Negishi M. The structure and characterization of type I P-45015a gene as major steroid 15α-hydroxylase and its comparison with type II P-45015a gene. *J Biol Chem* 1989; **264**: 6465–71.

42 Negishi M, Lindberg R, Burkhart B, Ichikawa T, Honkakoski P, Lang M. Mouse steroid 15a-hydroxylase gene family. Identification of type 11 P-45015a as coumarin 7-hydroxylase. *Biochemistry* 1989; **28**: 4169–72.

43 Yamano S, Tatsuno J, Gonzalez FJ. The CYP2A3 gene product catalyzes coumarin 7-hydroxylation in human liver microsomes. *Biochemistry* 1990; **29**: 1322–9.

44 Miles JS, McLaren AW, Forrester LM, Glancey MJ, Lang MA, Wolf CR. Identification of the human liver cytochrome P-450 responsible for coumarin 7-hydroxylase activity. *Biochem J* 1990; **267**: 365–71.

45 Honkakoski P, Kojo A, Raunio H, Pasanen M, Juvonen R, Lang MA. Hepatic mitochondrial coumarin 7-hydroxylase. Comparison with the microsomal enzyme. *Arch Biochem Biophys* 1988; **267**: 558–67.

46 Waxman DJ, Lapenson DP, Aoyama T, Gelboin HV, Gonzalez FJ, Korzekwa K. Steroid hormone hydroxylase specificities of eleven cDNA-expressed human cytochrome P450s. *Arch Biochem Biophys* 1991; **290**: 160–6.

47 Honkakoski P, Arvela P, Juvonen R, Lang MA, Kairaluoma M, Pelkonen O. Human and mouse liver coumarin 7-hydroxylases do not metabolize warfarin *in vitro*. *Br J Clin Pharmacol* 1992; **33**: 313–17.

48 Pearce R, Greenway D, Parkinson A. Species differences and interindividual variation in liver microsomal cytochrome P450 2A enzymes: effects on coumarin, dicumarol, and testosterone oxidation. *Arch Biochem Biophys* 1992; **298**: 211–25.

49 Flammang AM, Gelboin HV, Aoyama T, Gonzalez FJ, McCoy GD. Nicotine metabolism by cDNA-expressed human cytochrome P-450s. *Biochem Arch* 1992; **8**: 1–8.

50 McCracken NW, Cholerton S, Idle J. Cotinine formation by cDNA-expressed human cytochromes P450. *Med Sci Res* 1992; **20**: 877–8.

51 Mäenpää J, Sigusch H, Raunio H, Syngelmä T, Vuorela P, Vuorela H, Pelkonen O. Differential inhibition of coumarin 7-hydroxylase activity in mouse and human liver microsomes. *Biochem Pharmacol* 1993; **45**: 1035–42.

52 Mäenpää J, Juvonen R, Raunio H, Rautio A, Pelkonen O. Methoxsalen and coumarin interactions in man and mouse. *Biochem Pharmacol* 1994; **48**: 1363–9.

53 Kimonen T, Juvonen RO, Alhava E, Pasanen M. The inhibition of CYP enzymes in mouse and human liver by pilocarpine. *Br J Pharmacol* 1995; **114**: 832–6.

54 Kimonen T, Pasanen M, Gynther J, Poso A, Järvinen T, Alhava E, Juvonen RO. Competitive inhibition of coumarin 7-hydroxylation by pilocarpine and its interaction with mouse CYP 2A5 and human CYP 2A6. *Br J Pharmacol* 1995; **116**: 2652–30.

55 Pelkonen P, Lang MA, Pasanen M. Tissue and sex-dependent differences in CYP2A activities in hamsters. *Arch Toxicol* 1994; **68**: 416–22.

56 Kojo A, Heiskanen R, Rytkönen A-L, Honkakoski P, Juvonen R, Lang MA. Inducibility of P450Coh by pyrazole and its derivatives. *Biochem Pharmacol* 1991; **42**: 1751–9.

57 Mäenpää J, Syngelmä T, Honkakoski P, Lang MA, Pelkonen O. Comparative studies on coumarin and testosterone metabolism in mouse and human livers. Differential inhibitions by the anti-P450Coh antibody and metyrapone. *Biochem Pharmacol* 1991; **41**: 1229–35.

58 Yamazaki H, Inui Y, Yun C-H, Guengerich FP, Shimada T. Cytochrome P450 2E1 and 2A6 enzymes as major catalysts for metabolic activation of N-nitrosodialkylamines and tobacco-related nitrosamines in human liver microsomes. *Carcinogenesis* 1992; **13**: 1789–92.

59 Honkakoski P, Mäenpää J, Leikola J, Pasanen M, Juvonen R, Lang MA, Pelkonen O, Raunio H. Cytochrome P4502A-mediated coumarin 7-hydroxylation and testosterone hydroxylation in mouse and rat lung. *Pharmacol Toxicol* 1993; **72**: 107–12.

60 Bereziat J-C, Raffalli F, Schmezer P, Frei E, Geneste O, Lang MA. Cytochrome P450 2A of nasal epithelium: regulation and role in carcinogen metabolism. *Mol Carcinogen* 1995; **14**: 130–9.

61 Fernandez-Salguero P, Gonzalez FJ. The CYP2A gene subfamily: species differences, regulation, catalytic activities and role in chemical carcinogenesis. *Pharmacogenetics* 1995; **5**: S123–8.

62 Ding S, Lake BG, Friedberg T, Wolf CR. Expression and alternative splicing of the cytochrome P-450 CYP2A7. *Biochem J* 1995; **306**: 161–6.

63 Fernandez-Salguero P, Hoffman SMG, Cholerton S, Mohrenweiser H, Raunio H, Rautio A, Pelkonen O, Huang J, Evans WE, Idle JR *et al.* A genetic polymorphism in coumarin 7-hydroxylation: sequence of the human CYP2A genes and identification of variant CYP2A6 alleles. *Am J Hum Genet* 1995; **57**: 651–60.

64 Lindberg RLP, Juvonen R, Negishi M. Molecular characterization of the murine Coh locus: an amino acid difference at position 117 confers high and low coumarin 7-hydroxylase activity in P450coh. *Pharmacogenetics* 1992; **2**: 32–7.

65 Pelkonen O, Raunio H, Rautio A, Mäenpää J, Lang MA. Coumarin 7-hydroxylase: characteristics and regulation in mouse and man. *J Irish Coll Phys Surg* 1993; **22**: 24–8.

66 Getchell ML, Chen Y, Sparks DL, Ding X, Getchell TV. Immunohistochemical localization of a cytochrome P-450 isozyme in human nasal mucosa: age-related trends. *Ann Otol Rhinol Laryngol* 1993; **102**: 368–74.

67 Kharasch ED, Hankins DC, Thummel KE. Human kidney methoxyflurane and sevoflurane metabolism. Intrarenal fluoride production as a possible mechanism of methoxyflurane nephrotoxicity. *Anesthesiology* 1995; **82**: 689–99.

68 Mäenpää J, Rane A, Raunio H, Honkakoski P, Pelkonen O. Cytochrome P450 isoforms in human fetal tissues related to phenobarbital-inducible forms in the mouse. *Biochem Pharmacol* 1993; **45**: 899–907.

69 Hakkola J, Pasanen M, Purkunen R, Saarikoski S, Pelkonen O, Mäenpää J, Rane A, Raunio H. Expression of xenobiotic-metabolizing cytochrome P450 forms in human adult and fetal liver. *Biochem Pharmacol* 1994; **48**: 59–64.

70 Raunio H, Kojo A, Juvonen R, Honkakoski P, Järvinen P, Vähäkangas K, Lang MA, Gelboin HV, Park SS, Pelkonen O. Mouse hepatic cytochrome P450 isozyme induction by 1,4-[2-(3,5-dichloropyridyloxy)]benzene, pyrazole, and phenobarbital. *Biochem Pharmacol* 1988; **37**: 4141–7.

71 Smith G, Henderson CJ, Parker MG, White R, Bars RG, Wolf CR. 1,4-Bis[2-(3,5-dichloropyridyloxy)]benzene, an extremely potent modulator of mouse hepatic cytochrome P-450 gene expression. *Biochem J* 1993; **289**: 807–13.

72 Legrum W, Netter KJ. Characteristics of coumarin metabolism by liver microsomes from cobalt-pretreated mice. *Xenobiotica* 1980; **10**: 271–9.

73 Mangoura SA, Strack A, Legrum W, Netter KJ. Indium selectively increases the cytochrome P-450 dependent O-dealkylation of coumarin derivatives in male mice. *Naunyn-Schmiedeberg Arch Pharmacol* 1989; **339**: 596–602.

74 Maines MD, Kappas A. Metals as regulators of heme metabolism. *Science* 1977; **198**: 1215–21.

75 Kozer Z, Raunio H, Pasanen M, Arvela P, Raiskila T, Honkakoski P, Lang MA, Negishi M, Pelkonen O. Comparison between cobalt and pyrazole in the increased expression of coumarin 7-hydroxylase in mouse liver. *Biochem Pharmacol* 1991; **41**: 461–5.

76 Hahnemann B, Salonpää P, Pasanen M, Mäenpää J, Honkakoski P, Juvonen R, Lang MA, Pelkonen O, Raunio H. Effect of pyrazole, cobalt and phenobarbital on mouse liver cytochrome P-450 (CYP2a-4/5) expression. *Biochem J* 1992; **286**: 289–94.

77 Arvela P, Kraul H, Stenbäck F, Pelkonen O. The cerium-induced liver injury and oxidative drug metabolism in DBA/2 and C57BL/6 mice. *Toxicology* 1991; **69**: 1–9.

78 Salonpää P, Iscan M, Pasanen M, Arvela P, Pelkonen O, Raunio H. Cerium-induced strain-dependent increase in Cyp2a-4/5 (cytochrome P4502a-4/5) expression in the liver and kidneys of inbred mice. *Biochem Pharmacol* 1992; **44**: 1269–74.

79 Pellinen P, Stenbäck F, Raunio H, Pelkonen O, Pasanen M. Modification of hepatic cytochrome P450 profile by cocaine-induced hepatotoxicity in DBA/2 mouse. *Eur J Pharmacol Environ Toxicol Pharmacol* 1994; **292**: 57–65.

80 Marks GS, McCluskey SA, Mackie JE, Riddick DS, James CA. Disruption of hepatic heme biosynthesis after interaction of xenobiotics with cytochrome P-450. *FASEB J* 1988; **2**: 2774–83.

81 Salonpää P, Krause K, Pelkonen O, Raunio H. Up-regulation of CYP2A5 expression by porphyrinogenic agents in mouse liver. *Naunyn-Schmiedeberg Arch Pharmacol* 1995; **351**: 446–52.

82 Porter TD, Coon MJ. Cytochrome P-450. Multiplicity of isoforms, substrates, and catalytic and regulatory mechanisms. *J Biol Chem* 1991; **266**: 13469–72.

83 Aida K, Negishi M. Post-transcriptional regulation of coumarin 7-hydroxylase (P450Coh) induction by xenobiotics in mouse liver: mRNA stabilization by pyrazole. *Biochemistry* 1991; **30**: 8041–5.

84 Salonpää P, Pelkonen O, Kojo A, Pasanen M, Negishi M, Raunio H. CYP2A5 expression and inducibility by phenobarbital is modulated by cAMP in mouse primary hepatocytes. *Biochem Biophys Res Commun* 1994; **205**: 631–7.

85 Waxman DJ, Azaroff L. Phenobarbital induction of cytochrome P-450 gene expression. *Biochem J* 1992; **281**: 577–92.

86 Kirby GM, Chemin I, Montesano R, Chisari FV, Lang MA, Wild CP. Induction of specific cytochrome P450s involved in aflatoxin B1 metabolism in hepatitis B virus transgenic mice. *Mol Carcinogen* 1994; **11**: 74–80.

87 Kirby GM, Pelkonen P, Vatanasapt V, Camus A-M, Wild CP, Lang MA. Association of liver fluke (*Orpisthorchis viverrini*) infestation with increased expression of cytochrome P450 and carcinogen metabolism in male hamster liver. *Mol Carcinogen* 1994; **11**: 81–9.

88 Rautio A, Salmela E, Arvela P *et al*. Assessment of CYP2A6 and CYP3A4 activities in vivo in different diseases in man. In: Lechner MC, ed. *Cytochrome P450. Biochemistry, Biophysics and Molecular Biology*. Paris: John Libbey Eurotext; 1994: 519–21.

89 Rauma A-L, Rautio A, Pasanen M, Pelkonen O, Törrönen R, Mykkänen H. Coumarin 7-hydroxylation in long-term adherents of a strict uncooked vegan diet. *Eur J Clin Pharmacol* 1996; **50**: 133–7.

90 Sotaniemi EA, Rautio A, Bäckstrom M, Arvela P, Pelkonen O. Hepatic cytochrome P450 isozyme (CYP2A6 and CYP3A4) activities and fibrotic process in liver. *Br J Clin Pharmacol* 1995; **39**: 71–6.

91 Dalet-Beluche I, Boulenc X, Fabre G, Maurel P, Bonfils C. Purification of two cytochrome P450 isozymes related to CYP2A and CYP3A gene families from monkey (baboon, *Papio papio*) liver microsomes. Cross reactivity with human forms. *Eur J Biochem* 1992; **204**: 641–8.

92 Palmer CNA, Coates PJ, Davies SE, Shephard EA, Phillips IR. Localization of cytochrome P-450 gene expression in normal and diseased human liver by *in situ* hybridization of wax-embedded archival material. *Hepatology* 1992; **16**: 682–7.

93 Mustajoki P, Mustajoki S, Rautio A, Arvela P, Pelkonen O. Effects of heme arginate on cytochrome P450-mediated metabolism of drugs in patients with variegate porphyria and healthy men. *Clin Pharmacol Ther* 1994; **56**: 9–13.

94 Mustajoki P, Himberg JJ, Tokola O, Tenhunen R. Rapid normalization of antipyrine oxidation by heme in variegate porphyria. *Clin Pharmacol Ther* 1992; **51**: 320–4.

95 Lange R, Perin F, Larroque C, Zajdela F. Isolation and partial characterization of a cytochrome P-450 isozyme (cytochrome P-450tu) from mouse liver tumors. *Biochim Biophys Acta* 1990; **1038**: 130–5.

96 Kobliakov V, Kulikova L, Samoilov D, Lang MA. High expression of cytochrome P450 2a-5 (coumarin 7-hydroxylase) in mouse hepatomas. *Mol Carcinogen* 1993; **7**: 276–80.

97 Lazard D, Tal N, Rubinstein M, Khen M, Lancet D, Zupko K. Identification and biochemical analysis of novel olfactory-specific cytochrome P-450IIA and UDP-glucuronosyl transferase. *Biochemistry* 1990; **29**: 7433–40.

98 Peng H-M, Ding X, Coon MJ. Isolation and heterologous expression of cloned cDNAs for two rabbit nasal microsomal proteins, CYP2A10 and CYP2A11, that are related to nasal microsomal cytochrome P450 form a. *J Biol Chem* 1993; **268**: 17253–60.

99 Lake BG, Gray TJB, Evans JG, Lewis DFV, Beamand JA, Hue KL. Studies on the mechanism of coumarin-induced toxicology in rat hepatocytes: comparison with dihydrocoumarin and other coumarin metabolites. *Toxicol Appl Pharmacol* 1989; **97**: 311–23.

100 Huwer T, Altmann HJ, Grunow W, Lenhardt S, Przybylski M, Eisenbrand G. Coumarin mercapturic acid isolated from rat urine indicates metabolic formation of coumarin 3,4-epoxide. *Chem Res Toxicol* 1991; **4**: 586–90.

101 Goeger DE, Anderson KE. Biochemical characterization of coumarin 7-hydroxylase activity in chick embryo liver microsomes. *Biochem Pharmacol* 1992; **43**: 363–9.

102 Yamazaki H, Mimura J, Sugahara C, Shimada T. Catalytic roles of rat and human cytochrome P450 2A enzymes in testosterone 7α- and coumarin 7-hydroxylations. *Biochem Pharmacol* 1994; **48**: 1524–7.

103 Kimura S, Kozak CA, Gonzalez FJ. Identification of a novel P450 expressed in rat lung: cDNA cloning and sequence, chromosome mapping, and induction by 3-methylcholanthrene. *Biochemistry* 1989; **28**: 3798–803.

104 Ohmori S, Horie T, Guengerich FP, Kiuchi M, Kitada M. Purification and characterization of two forms of hepatic microsomal cytochrome P450 from untreated cynomolgus monkeys. *Arch Biochem Biophys* 1993; **305**: 405–13.

105 Crespi CL, Penman BW, Leakey JAE, Arlotto MP, Stark A, Parkinson A, Turner T, Steimel DT, Rudo K, Davies RL *et al*. Human cytochrome P450IIA3: cDNA sequence, role of the enzyme in the metabolic activation of promutagens, comparison to nitrosamine activation by human cytochrome P450IIE1. *Carcinogenesis* 1990; **11**: 1293–300.

CHAPTER 4

The Effects of Coumarin and its Metabolites on Cell Growth and Development

Barbara Seliger
Department of Hematology/Oncology, University of Mainz, Mainz, Germany

4.1 INTRODUCTION

Coumarin and its derivatives are substances with pleiotropic biological activities, particularly in augmenting immune cell reactivity and exerting antitumoral effects[1–4]. In sensitive tumour cells, these drugs cause significant changes in the regulation of immune responses, cell growth and differentiation. Furthermore, responses to coumarin/7-hydroxycoumarin have been reported *in vivo* for malignant melanoma, metastatic renal cell carcinoma and prostate cancer[1,5–7]. Recent developments in the mode of action of coumarins and their related compounds indicate that the observed antiproliferative effects of these substances might be due to various mechanisms, including induction of apoptosis, cell cycle arrest, downregulation of oncogene expression or block of autocrine loops, as well as induction of differentiation. This article reviews the current knowledge on the mode of coumarin/7-hydroxycoumarin action resulting in the inhibition of cell proliferation.

4.2 CELL DEATH IN THE DEVELOPMENT OF CANCER

An intact multicellular organism must equalise the rates of cell generation and cell death to maintain the constant size and is able to remove senescent, damaged and

Coumarins: Biology, Applications and Mode of Action. Edited by R. O'Kennedy and R.D. Thornes.

abnormal cells. Consequently, abnormalities in the regulation of cell death might contribute to the pathogenesis of both degenerative as well as neoplastic diseases. Such physiological cell death has been named apoptosis[8]. Apoptosis has been defined as a genetically encoded cell death programme which is characterised by morphological alterations including compaction of chromatin against the nuclear membrane, cell shrinkage, detachment from surrounding cells and formation of apoptotic bodies.

Apoptosis takes place during embryogenesis and immunity, in the course of normal tissue turnover, and after withdrawal of a trophic hormone from its target tissue. A number of different factors, like growth factors, intracellular mediators of signal transduction and nuclear proteins regulating gene expression, DNA replication and cell cycle are able to induce apoptosis[9–13]. Malignant transformation is often associated with genetic alterations which could result in increased cell division and proliferation. The altered cell number in neoplastic tissues usually reflects the ignorance of normal homeostatic mechanisms which usually represents a tightly controlled balance between cellular proliferation and death pathways in normal cells. Although tumour cells could divide more slowly than normal, non-malignant cells, they would be able to expand because of their prolonged lifespan. Thus, the comparative rates of cell division and cell death determine how fast tumour cells grow[14] and tumour cells can gain a selective growth advantage by blocking the apoptotic pathway[15].

4.3 INDUCTION OF APOPTOSIS BY ANTINEOPLASTIC SUBSTANCES

It has long been believed that antineoplastic therapeutics cure cancer by selectively targeting rapidly dividing cells. Most anticancer drugs, such as topoisomerase inhibitors, alkylating agents, antimetabolites and hormone antagonists, may induce apoptosis in tumour cells[16]. However, they also arrest the cell cycle of normal cells. Apoptosis is well-characterised as a distinct pathological mechanism in tumours responding to anticancer therapies[17]. At the molecular level a striking correlation between an apoptosis gene and cancer involves *p53*: *E1A/ras* transformed murine fibroblasts are sensitive to apoptosis induction by radiation, adriamycin, 5-fluorouracil and etopside if *p53* is present but not if absent[18].

Many carcinogens are able to damage DNA or interfere with enzymes necessary for accurate DNA replication. A cell can respond to such an injury by delay of cell division until repair of damage, by induction of apoptosis or progression through the cell cycle. Apoptosis seemed to be an efficient method of preventing malignant transformation because it could remove cells with genetic lesions. Abnormal apoptosis promotes cancer development either by accumulation of dividing cells or by obstructing removal of genetic variants with enhanced malignant potential. The threshold of apoptosis is changed by apoptosis inducers, such as certain oncoproteins[19]. A dual activity of the *myc* and *E1A* oncogene is the transformation of cells on the one hand and, on the other hand, sensitisation of cells to apoptosis[20]. In contrast, the *bcl2* gene is a general cell death suppressor that regulates apoptosis[21]. Overexpression of *bcl2* could also protect cells from c-*myc*-induced apoptosis[22]. In contrast, the tumour suppressor gene *p53* could delay cell cycle progression at G_0/G_1 and is an important regulator of apoptosis[23,24]. One role of *p53* in regulating tumorigenesis may involve its ability to induce apoptosis in the presence of oncogenic triggers[25]. Thus, an understanding of what determines the apoptosis threshold may be a key factor for cancer therapy.

Recent studies have demonstrated that apoptosis may be involved in the cell death induced by chemotherapeutic reagents such as cisplatin, cytarabine, camphotecin, etoposide, teniposide, amsacrine and quercetin[17,26–31]. There is accumulating evidence that the efficiency of antitumour agents is related to the intrinsic properties of the target tumour cells to respond to these agents by apoptosis. Thus, the knowledge of the mechanisms by which apoptosis in normal and tumour cells could be modulated will provide an important basis for optimising cancer therapy.

4.4 CYTOTOXIC EFFECTS OF COUMARIN AND ITS DERIVATIVES

It has recently been demonstrated that coumarin and some of its derivatives exert cytotoxic activities on a number of human tumour cells. For example, these substances, especially *o*-coumaric acid, had cytotoxic activities against the mastocytoma cell line P815 at a concentration of > 100 mg/ml[32]. This cytotoxic action was neither due to induction of cytotoxic macrophages against P815 nor due to increased release of the cytokines TNF-α and IL-2 which are able to trigger 'immune'-induced apoptosis. Indeed, the observed cytotoxic effects of coumarin and its related compounds were accompanied by morphological changes characteristic for apoptosis: brightly red-fluorescent condensed nuclei observed by fluorescence microscopy of propidium iodide-stained nuclei, blebbing, reduction of cell volume, condensation of nuclear chromatin, nuclear fragmentation and apoptotic bodies[7,33,34]. Similar findings were recorded in my laboratory (unpublished). Agarose gel electrophoresis of coumarin-treated cells demonstrated a ladder-like pattern of DNA fragments of a size of 180–200 base pairs. The apoptosis-induced effects were dose-dependent and only occurred at high concentrations of coumarin/7-hydroxycoumarin (> 100 mg/ml) with a maximum reached at 500 mg/ml.

4.5 CYTOSTATIC EFFECTS OF COUMARIN AND ITS DERIVATIVES

Coumarin and 7-hydroxycoumarin have cytostatic effects on a number of human malignant cells, especially renal cell carcinoma, colon carcinoma, breast carcinoma, glioblastoma, prostatic carcinoma and leukaemic cell lines (A549; ACHN; Caki-2; A-498; 786-0; G-CCM; G-UVW; Dakiki; HS-Sultan; H727; HCT-15; HL60; K562; CCRF-CEM; LnCaP; PC-3; Du 145; Caco-2; Colo-232; MCF-7 and RP-1788). The sensitivity of these tumour cells to both substances varies as determined by different ID_{50} concentrations. Interestingly, at equal concentrations, a comparison of the growth inhibitory activity of 7-hydroxycoumarin and coumarin showed that 7-hydroxycoumarin was more effective than coumarin. In all cases tested, the drug-induced antiproliferative activity was transient, time- and dose-dependent[7,34–38]. In representative tumour cell lines analysed, both substances inhibited the [^{3}H]-thymidine, [^{3}H]uridine and [^{3}H]leucine incorporation[34].

However, coumarin and its derivatives have also been shown to affect the growth rate of normal cells[39,40]. For example, esculetin, a coumarin derivative with lipoxygenase inhibitor activity, inhibits the foetal calf serum (FCS) stimulated proliferation of vascular smooth muscle cells as determined by uptake of tritiated thymidine. The antiproliferative action of esculetin on these cells might be partially due to a decrease of the protein tyrosine kinase enzyme activity which is modulated by inhibition of lipoxygenase. Furthermore, combined treatment of coumarin with other antineoplastic substances, such as suramin, had, at concentrations of up to 400 mg/ml coumarin, synergistic effects when compared to the growth inhibitory action of coumarin alone[38].

4.6 CELL CYCLE AND CANCER

During the last couple of years, knowledge about the connections between malignant transformation and components regulating the cell cycle has increased. The current model of cell cycle control holds that the transitions between different cell cycle states are regulated at checkpoints, like the initiation and completion of DNA replication (S phase) and of cell division[41]. The checkpoints are tightly regulated by cyclin-cyclin-dependent kinases (cyclin-CDK) and their activating components, the cyclins. One of the most important checkpoints in mammalian cells is the restriction point START in the late G_1 phase. In many tumour cells, START is deregulated due to alterations of the cyclin-CDK complexes. As a consequence, cells can bypass the normal restriction on entry into the S phase. This allows replication of damaged DNA which might contribute to the malignant phenotype of tumour cells[42,43].

4.7 COUMARIN AND CELL CYCLE

As already described above, coumarin is able to strongly inhibit cell growth of a number of tumour cells of distinct origin in a concentration- and time-dependent manner. Recently, it has been shown that coumarin exerts its antiproliferative effect by modulation of cell cycle[33]. The authors employed the dexamethasone-sensitive

MMTV*ras* transformed cells as an *in vitro* model of inducible oncogene expression. Dexamethasone treatment of the cells was associated with a time-dependent transient increase of cells in the G_1/G_0 phase and a three-fold induction of cells in S phase[33] when compared to phosphate buffered saline-(PBS-) treated control cells. Tseng and co-workers[44] had shown that the dexamethasone-mediated increase of *ras* oncogene expression could be inhibited by coumarin treatment and was paralleled by a down-regulation of their proliferative capacity[44]. The effect of coumarin on the cell cycle progression of cells was determined by DNA fluorescence histograms. The number of S cells among total cell number decreased during elevation of dosage and prolongation of incubation time with coumarin. Furthermore, Marshall *et al.*[34] also reported that coumarin and 7-hydroxycoumarin treatment induced apoptosis of the leukaemic cell line HL60. These studies showed for the first time the cell cycle inhibitory effect of coumarin and/or 7-hydroxycoumarin. These substances have recently been shown to exert antitumour activities, although the mechanism responsible for this activity has not yet been clarified. The findings described indicate that coumarin/7-hydroxycoumarin may display antitumour activity by inhibiting progression through the cell cycle.

4.8 MECHANISMS OF GROWTH CONTROL BY RECEPTOR/LIGAND SYSTEMS

The control of cell growth involves cellular production and excretion of polypeptide growth factors which exert their effects through binding to specific receptors. Activation of receptors then leads to the interaction with the intracellular components of the signal transduction pathway, transmitting specific signals to the genome and thus controlling its function.

To achieve the level of control necessary for survival, multicellular organisms have developed several different strategies using factors and receptors for the regulation of specific cellular processes within the body. The endocrine regulation involves a distinct partial separation between the source of factor production and the receptor-producing target cell. Paracrine regulation is a short-distance mechanism where the target cell is close to the signalling cell. This markedly limits the target population. When a cell produces and responds to a factor through the co-expression of both factor and corresponding receptor, this is referred to as autocrine growth regulation. Different receptor/ligand systems have been associated with cell growth and proliferation in several *in vitro* systems. These include the epidermal growth factor receptor (EGFR), the platelet growth factor receptor (PDGFR), the fibroblast growth factor receptor (FGFR) and the insulin growth factor receptor (IGFR) systems. Such autocrine loops have been demonstrated in the haemopoietic systems as well as in different tumour types. It has been suggested that cell growth is deregulated in glioblastoma, breast carcinoma and renal cell carcinoma due to anomalies in at least one of those growth factor receptor systems.

4.9 EFFECTS OF COUMARIN/7-HYDROXYCOUMARIN ON AUTOCRINE LOOPS

Growth factors and their corresponding receptors are thought to play important roles in tumour development through paracrine and autocrine growth stimulation. In human malignant gliomas, a class of tumours originating from glial cells, the PDGF, FGF as well as the EFG/TGF-α systems have been postulated to be responsible for or to contribute to the malignant phenotype[45–48].

The PDGF system consists of three ligands, two homodimers, the PDGF-AA and PDGF-BB and one heterodimer, the PDGF-AB. These PDGF isoforms bind with different affinities to two receptor subtypes, the PDGF-α receptor and the PDGF-β receptor. Binding of the different PDGF isoforms causes the formation of different receptor homo- and heterodimers which results in the activation of their tyrosine kinases.

Human glioma cultures and tumours have been shown to express the PDGF-AA, the PDGF-BB as well as the PDGFRs[49], suggesting the possible utilization of this system for glial tumour autocrine growth stimulation. Several authors[36,50] have recently reported that coumarin as well as 7-hydroxycoumarin were able to inhibit the cell growth of a number of glioblastoma. Treatment of these tumour cells with 7-hydroxycoumarin caused a time-dependent selective downregulation of the mRNA levels of the PDGF-AA and PDGF-BB[51]. No changes in the mRNA levels of the PDGF receptors, EGF receptors, EGF and TGF-α were observed. Furthermore ^{125}I-labelled growth factor binding to the respective receptors was not affected by 7-hydroxycoumarin. Upon removal of the substance, PDGF-AA and -BB mRNA expression was fully restored and a gradual recovery of the proliferative capacity of these cells was noted within three days.

The data presented suggest that 7-hydroxycoumarin reversibly inhibits the growth of human malignant glioblastoma cells due to a selective reduction of PDGF-AA and PDGF-BB mRNA levels, implying that 7-hydroxycoumarin is a potent negative regulator of the autocrine pathway. Whether the negative regulation of the PDGF loops is a cause or a consequence of the 7-hydroxycoumarin-mediated growth inhibition is difficult to determine. However, it has been shown for interferons (IFNs) that a reduction in growth factor receptor number is neither a cause nor a consequence of growth inhibition[52,53]. Furthermore, various reports show that PDGF induces proliferation of a number of cell lines. Thus, the block of PDGF expression following 7-hydroxycoumarin treatment may play an important role in the 7-hydroxycoumarin-mediated inhibition of human malignant glioblastoma cells.

4.10 INHIBITORY ACTION OF COUMARIN AND ITS DERIVATIVES ON CELL DIFFERENTIATION

In a number of human malignant glioblastoma cell lines, 7-hydroxycoumarin treatment resulted in a strong growth inhibition which was accompanied by alterations in the differentiation status of these cells[50]. Time kinetic analysis revealed an increase in the expression of the adhesion molecule ICAM as well as the glial-specific marker GFAP (glial fibrillary acidic protein) already after 24 h of 7-hydroxycoumarin treatment. A subsequent increase in GFAP and ICAM expression was demonstrated with a maximum reached 10 days after the onset of 7-hydroxycoumarin treatment. This time-dependent induction of differentiation of glioblastoma cells was accompanied by a decreased cell proliferation.

The promyelocytic leukaemia cell line HL60 expresses high levels of the c-*myc* oncogene due to an approximately 16-fold gene amplification which was not present as episomes[54]. Treatment of HL60 cells with coumarin as well as other inhibitors of the poly(ADP-ribose) polymerase (nicotinamide, benzamide, DMSO, 4-hydroxyquinazoline) resulted in the induction of their differentiation into granulocytes which was preceded by a downregulation of the c-*myc* expression[55]. The observed granulocytic differentiation which was identified by the appearance of tetrazolium-positive cells and mature granulocytes was directly associated with the loss of the c-*myc* expression. In contrast, phorbol-12-myristate-13-acetate or teleocidin, known to promote HL60 differentiation into macrophages, did not affect the *myc* amplification. Coumarin and other poly(ADP-ribose) polymerase inhibitors, such as benzamide, not only eliminated amplified sequences like the c-*myc* gene in HL60 cells, but were also capable of eliminating exogenously introduced oncogenic sequences. This has been demonstrated in murine fibroblasts transformed by the human Ha-*ras*, rat Ki-*ras*, rat c-*raf* and human *ret*II oncogenes[54,56]. However, these results have been controversially discussed: other authors[44,51,57] demonstrated a downregulation of oncogene expression in *in vitro* models of oncogenic transformation by treatment with either coumarin or 7-hydroxycoumarin but no loss of exogenously introduced oncogenes was observed.

Thus, one possible mechanism underlying the growth modulatory activities of coumarin and its derivatives might be the induction of cell differentiation. Similar results have been obtained with cytokines, especially interferons. Interferons were able to induce differentiation of haemopoietic cells. As has been shown for coumarin and related compounds, the differentiation was associated with growth inhibition as well as downregulation of the c-*myc* oncogene expressions[57–59].

4.11 CONCLUSIONS

There could exist different mechanisms whereby coumarin and its derivatives are able to inhibit the proliferative activity of tumour cells. These include effects on the cell cycle, downregulation of oncogene expression, induction of apoptosis, the block of autocrine loops, as well as the induction of differentiation. These findings on the mode and mechanisms of coumarin/7-hydroxycoumarin action may be of importance for the further development of these substances for clinical application.

REFERENCES

1 Thornes RD, Lynch G, Sheehan MV. Cimetidine and coumarin therapy in melanoma. *Lancet* 1989: **2**, 328–33.

2 Marshall ME, Mohler JL. Treatment of renal cell carcinoma with coumarin: A review. *J Irish Coll Phys Surg* 1993: **22**, 6–9.

3 Marshall ME, Conley D, Hollingsworth P, Brown S, Thompson JS. Effects of coumarin (1,2-benzopyrone) on lymphocyte natural killer cell and monocyte functions *in vitro*. *J Biol Response Mod* 1989: **8**, 70–85.

4 Marshall ME. The coumarins: A review of the mechanisms of action and applications to clinical oncology. *Dtsch Zschr Onkol* 1992: **24**, 141–51.

5 Raev LD, Voinova E, Ivanov IC, Popov D. Anti-tumor activity of some coumarin derivatives. *Pharmazie* 1990: **45**, 696.

6 Maucher A, Kager M, von Angerer E. Evaluation of the anti-tumor activity of coumarin in prostate cancer models. *J Cancer Res Clin Oncol* 1993: **119**, 150–4.

7 Marshall ME, Mohler JL, Edmonds K, Williams B, Butler K, Ryles M, Weiss L, Urban D, Beuschen A, Markiewicz M. *et al.* An updated review of the clinical development of coumarin (1,2-benzopyrone) and 7-hydroxycoumarin. *J Cancer Res Clin Oncol* 1994: **120**, 39–42.

8 Kerr J, Wyllie AH, Currie AR. Apoptosis: A basic biological phenomenon with wide ranging implications in tissue kinetics. *Br J Cancer* 1972: **26**, 239–57.

9 Arends MJ, Morris RG, Wyllie AH. Apoptosis: The role of the endonuclease. *Am J Pathol* 1990: **136**, 593–608.

10 Golstein P, Ojcius DM, Young JD. Cell death mechanisms and the immune system. *Immunol Rev* 1991: **121**, 29–65.

11 Brach MA, deVos S, Gruss HJ, Germann F. Prolongation of survival of human polymorphonuclear neutrophils by granulocyte-macrophage colony-stimulating factor is caused in inhibition of programmed cell death. *Blood* 1992: **80**, 2920–4.

12 Cohen JJ, Duke RC, Fadok VA, Sellins KS. Apoptosis and programmed cell immunity. *Annu Rev Immunol* 1992: **10**, 267–93.

13 Cohen JJ. Exponential growth in apoptosis. *Immunol Today* 1995: **16**, 346–8.

14 Carson DA, Ribeiro JM. Apoptosis and disease. *Lancet* 1993: **341**, 1251–4.

15 Reed J. Bcl-2 and the regulation of programmed cell death. *J. Cell Biol* 1994: **124**, 1–6.

16 Hickman JA. Apoptosis induced by anti-cancer drugs. *Cancer Metastasis Rev* 1992: **11**, 121–39.

17 Eastman A. Activation of programmed cell death by anti-cancer agents: Cisplatin as a model system. *Cancer Cells* 1990: **313**, 597–600.

18 Lowe SW, Ruley HE, Jacks T, Housman DE. P53-dependent apoptosis modulates the cytotoxicity of anticancer agents. *Cell* 1993: **74**, 957–67.

19 Oren M. The involvement of oncogenes and tumor suppressor genes in the control of apoptosis. *Cancer Metastasis Rev* 1990: **11**, 141–8.

20 Evan GJ, Wyllie AH, Gilbert CS. Induction of apoptosis in fibroblasts by c-myc proteins. *Cell* 1992: **69**, 119–28.

21 Korsmeyer SJ. Bc12: An antidote to programmed cell death. *Cancer Surveys* 1992: **15**, 105–18.

22 Bissonette RR, Echeverri F, Mahboubi A, Green DR. Apoptotic cell death induced by c-myc is inhibited by bc12. *Nature (Lond)* 1992: **359**, 552–4.

23 Levine AJ, Momand J, Finlay CA. The p53 tumor suppressor gene. *Nature (Lond)* 1991: **351**, 453–6.

24 Yonish-Rouach E, Resnitsky D, Lotem J, Sachs L, Kimchi A, Oren M. Wild type p53 induces apoptosis of myeloid leukaemia cells that is inhibited by interleukin 6. *Nature (Lond)* 1991: **353**, 345–7.

25 Fisher DE. Apoptosis in cancer therapy: crossing the threshold. *Cell* 1994: **78**, 539–42.

26 Barry MA, Behnke CA, Eastman A. Activation of programmed cell death (apoptosis) by cisplatin or other anti-cancer drugs, toxins and hyperthermia. *Biochem Pharmacol* 1990: **40**, 2353–62.

27 Barry MA, Reynolds JE, Eastman A. Etoposide-induced apoptosis in human HL60 cells is associated with intracellular acidification. *Cancer Res* 1993: **53**, 2349–57.

28 Gorczyca W, Gong J, Ardelt B, Traganos F, Darzynkiewicz Z. The cell cycle related difference in susceptibility of HL60 cells to apoptosis induced by various anti-tumor agents. *Cancer Res* 1993: **53**, 3186–92.

29 Ling YH, Priebe W, Perez-Soler R. Apoptosis induced by anthracycline antibiotics in P388 parent and multi-drug resistant cell. *Cancer Res* 1993: **53**, 1845–52.

30 Walton MI, Whysong D, O'Connor PM, Hockenberry D, Korsmeyer SJ, Kohn KW. Constitutive expression of human bcl-2 modulates nitrogen mustard and camphothecin-induced apoptosis. *Cancer Res* 1993: **53**, 1853–61.

31 Wei YQ, Zhao X, Kariya H, Fukata H, Teshigawara K, Uchida A. Induction of apoptosis by quercetin: Involvement of heat shock proteins. *Cancer Res* 1994: **54**, 4952–7.

32 Rosskopf F, Kraus J, Franz G. Immunological and anti-tumor effects of coumarin and some derivatives. *Pharmazie* 1992: **47**, 139–42.

33 Kahn J, Preis P, Waldman F, Tseng A Jr. Coumarin modulates the cell cycle progression of an MTV-EJras cell line. *J Cancer Res Oncol* 1994: **120**, 19–22.

34 Marshall ME, Kervin K, Benefield C, Umerani A, Albainy-Jenei S, Zhao Q, Khazaeli M.B. Growth-inhibitory effects of coumarin (1,2-benzopyrone) and 7-hydroxy-coumarin on human malignant cell lines *in vitro*. *J Cancer Res Clin Oncol* 1991: Suppl. 120, 3–10.

35 Omarbasha B, Fair WR, Heston WD. Effect of coumarin on the normal rat prostate and on the R-3327H prostatic adenocarcinoma. *Cancer Res* 1989: **49**, 3045–9.

36 Moran E, Prosser E, O'Kennedy R, Thornes RD. The effect of coumarin and 7-hydroxycoumarin on the growth of human tumor cell lines. *J Irish Coll Phys Surg* 1993: **22**, 41–8.

37 Siegers CP, Bostelmann HC. Effect of coumarin on cell cycle proliferation in human tumor cell lines. *J Irish Coll Phys Surg* 1993: **22**, 47–50.

38 Myers RB, Parker M, Grizzle WE. The effects of coumarin and suramin on the growth of malignant renal and prostatic cell lines. *J Cancer Res Clin Oncol* 1994: **120**, 11–13.

39 Grainger GJ, Hesketh TR, Weissberg PL, Metcalf JC. Hexamethylenebis-acetamide selectively inhibits the proliferation of human and rat vascular smooth-muscle cells. *Biochem J*, 1992: **283** 403–8.

40 Huang HC, Lai MW, Wang HR, Chung YL, Hsieh LM, Chen CC. Anti-proliferative effect of esculetin on vascular smooth muscle cells: Possible roles of signal transduction pathways. *Eur J Pharmacol.* 1993: **237**, 39–44.

41 Nurse P. Ordering S phase and M phase in the cell cycle. *Cell* 1994: **79** 547–50.

42 Hunter T, Pines J. Cyclins and cancer II: Cyclin D and CDK inhibitors. *Cell* 1994: **79**, 573–82.

43 Sherr CJ. G1 phase progression: Cycling on cue. *Cell* 1994: **79**, 551–5.

44 Tseng A, Lee WMF, Kirsten E, Hkam A, McLick J, Buki K, Kun E. Prevention of tumorigenesis of oncogene-transformed rat fibroblasts with DNA site inhibitors of poly (ADP) ribose polymerase. *Proc Natl Acad Sci USA* 1987: **84**, 1107–11.

45 Libermann TA, Friesel R, Jaye M, Lyall RM, Westermark B, Drohan W, Schmidt A, Maciag T, Schlessinger J. An antigenic growth factor is expressed in human glioma cells. *EMBO J* 1987: **6**, 1627–32.

46 Halden LT, Novak K, Kaye AH, Burgess AW. Selective amplification of the cytoplasmic domain of the epidermal growth factor receptor gene in glioblastoma multiforme. *Cancer Res* 1988; **48**: 2711–14.

47 Hermansson M, Nister M, Betsholtz C, Heldin C, Westermark B, Funa K. Endothelial cell hyperplasia in human glioblastoma: Co-expression of mRNA for platelet derived growth factor (PDGF) B chain and PDGF receptor suggest autocrine growth stimulation. *Proc Natl Acad Sci USA* 1988: **85**, 7748–52.

48 Ekstrand J, James CD, Cavanne W, Seliger B, Pettersson R, Collins PV. The genes for the epidermal growth factor receptor, the transforming growth factor-alpha and the epidermal growth factor and their expression in human gliomas *in vivo*. *Cancer Res* 1991: **51**, 2164–72.

49 Nister M, Libermann TA, Betzholz C, Pettersson M, Claesson-Welsh L, Heldin C, Schlessinger J, Westermark B. Expression of messenger mRNAs for platelet-derived growth factor and transforming

growth factor-alpha and their receptors in human malignant glioma cell lines. *Cancer Res* 1988: **48**, 3910–18.

50 Seliger B. 7-OH-coumarin mediated induction of differentiation of human malignant glioblastoma cells. *J Irish Coll Phys Surg* 1993: **22**, 44–6.

51 Seliger B, Pettersson H. Mechanism of action of 7-OH-coumarin on human malignant glioblastoma cell lines. *Proc Int Soc Coumarin Invest*, Newsletter 1994.

52 Zagari M, Kelper JR, Harris C, Herman B. Inhibition of early platelet-derived growth factor responses in BALB/c-3T3 cells by interferon. *Biochem Biophys Res Commun*. 1980: **36**, 1207–12.

53 Suzuki H, Shibano K, Okano M., Kono I, Matsui Y, Yamane K, Kaskieagi H. Interferon-γ modulates messenger RNA levels of v-*sis* (PDGF-B chain), PDGF-A chain and H-1b genes in human vascular endothelial cells. *Am J Path* 1989: **134**, 35–43.

54 Nagao M, Nakayasu M, Aonuma S, Shima H, Sugimura T. Loss of amplified genes by poly-(ADP-ribose) polymerase inhibitors. *Environ Health Perspect* 1991: **93**, 169–74.

55 Shima H, Nakayasu M, Aonuma, S, Sugimura, T., Nagao, M. Loss of the *MYC* gene amplified in human HL-60 cells after treatment with inhibitor of poly(ADP-ribose) polymerase or with dimethyl sulfoxide. *Proc Natl Acad Sci USA* 1989: **86**, 7442–5.

56 Nakayasu M, Shima H, Aonuma S, Nakagama H, Nagao M, Sugimura T. Deletion of transferred oncogenes from N1H3T3 transformants by inhibitors of poly(ADP-ribose) polymerase. *Proc Natl Acad Sci USA*, 1988: **85**, 9066–70.

57 Zanker K. Coumarin(s) interface between cell growth inhibition, expression of oncogenes, tumor cell locomotion and cell to cell communication. *J Irish Phys Surg* 1993: **22**, 51–5.

58 Pestka S, Langer JA, Zoon K., Samid CE. Interferons and their actions. *Annu Rev Biochem* 1987: **56**, 727–77.

59. Einat M, Resnitzky D, Kimchi A. Close link between reduction of c-*myc* expression by interferon and G_0/G_1 arrest. *Nature* (*Lond.*) 1985: **313**, 597–600.

CHAPTER 5

Coumarin as an Immunomodulator

Gerhard J. Zlabinger
Institute of Immunology, University of Vienna, Vienna, Austria

5.1 INTRODUCTION

The discovery of a substance with an ability to interfere with immune responsiveness, in such a way that particular functions lost or diminished during disease were restored or that aberrant immune reactivity was corrected, would be of major significance. It is apparent, given the complexity of the immune regulatory network, that the requirements for a substance effecting immunomodulation for the benefit of the individual are major. The need for minimal toxicity of the substance means that candidates can be expected to be rather rare. Coumarins, which are constituents of many natural products and have been used to treat infections, do not fulfil all conditions necessary to be classified as immune response modifiers but, nevertheless, they have been shown by classical laboratory techniques as well as in controlled clinical trials, to possess the ability to modulate immune responsiveness, which might contribute to the antitumour effects observed in treated patients.

In the present review *ex vivo* investigations of immune parameters will be correlated with the clinical experience in treated patients and with findings about the influence of coumarins on *in vitro* functions of immune cells covering innate and acquired immunity. Administration of coumarin is associated with very few significant side-effects. This is in contrast with experience using common antitumour drugs. Elucidation of the precise role and potential of coumarins as

Coumarins: Biology, Applications and Mode of Action. Edited by R. O'Kennedy and R.D. Thornes.

immunomodulators still needs considerable investigation and may be dependent on our understanding of the mechanisms of the immune response.

5.2 PRINCIPLES IN IMMUNOMODULATION

5.2.1 Immunological considerations

When attempting to use a regulatory approach for the alteration of immune reactivity, it is necessary to have a thorough understanding of the immune network. Although many aspects of normal immune reactivity have been clarified, the sites of abnormal immune responsiveness involved in the expression of a given disease remains less distinct. Classically, the immune system has been divided into two limbs: humoral and cellular. Originally, they were thought to be discrete, but it is now clear that there is extensive interaction between these systems. Therefore, it may be appropriate to view the immune response as consisting primarily of the thymus-derived limb (T lymphocytes) and the bursa equivalent, or B cell limb, with considerable interaction as well as interdependence between the two.

The induction of each arm follows an orderly progression consisting of initiation of the response (antigen-specific), expansion and differentiation of immunocompetent cells, and finally the effector phase of the immune response. This last state, necessary for elimination of the invading pathogen and restoration of homeostasis may be production of antibodies specific for the triggering antigen by a B cell or its progeny, or induction of a specific T-cell response that performs a regulatory function, such as help or suppression to other types of immunocompetent cells, like B cells, T cells or cells of the monocyte/macrophage lineage. In addition, T-cell responses may represent the endpoint of the immune response as manifested by delayed-type hypersensitivity reactions or cell-mediated cytotoxicity. Another population of lymphoid cells, the natural killer cells, cannot classically be attributed to the aforementioned lineages and may in fact represent a distinct cell population. These cells are thought to represent a first line of defence in immune surveillance against malignant and virus-transformed cells prior to the development of the classical antigen-specific responses. Nevertheless, this cell population has been shown to possess specific recognition structures as well as an ability to discriminate between cells bearing self or non-self/altered self-determinants[1].

Cells of the monocyte/macrophages series clearly play a role in immune reactivity as modifiers of the immune response and as specific effector cells capable of destroying sensitised and non-sensitised target tissues[2]. Macrophages are known to represent a major recognition system for foreign cells, including tumour cells, to play a key role in presentation of antigen to lymphocytes, and to be a major source of a broad range of secretory products controlling inflammation and acting against cancer cells. The use of macrophages as a therapeutic approach against malignant cells has several advantages over the use of other types of leucocytes such as natural killer cells and antitumour-specific T and B lymphocytes:

(1) The spectrum of antitumour activity of macrophages is very wide and is not species-specific[3].
(2) Activated macrophages can react indiscriminately against 'immunogenic', 'weakly immunogenic', and 'non-immunogenic' tumours, and, in view of the fact that most of the spontaneously arising human tumours are considered to belong to the last two categories, this property of macrophages is important in the context of a conceivable use in cancer immunotherapy[4].
(3) Progress of malignancy is often related to a decrease in immune response potential[5–9]. Apparently, the macrophages are the last types of cells affected in situations of decreased immunocompetence caused by tumour presence. However, it should also be mentioned that a decrease in macrophage function(s) in cancer patients has been reported[10].
(4) Non-active resident macrophages can be activated *in vitro* to express antitumour activity. This is usually done by selective interference with the lipoxygenase pathway of arachidonic acid[11,12], and/or by an induced increase in the production and release of macrophage cytokines[11,13–16]. Moreover, human peripheral blood monocytes, which can be easily isolated from cancer patients and reinjected in the same donor patients, can be matured *in vitro* and induced to react against tumour cells[17–20].
(5) The recognition system of macrophages is characterised by overall recognition of foreign cells (including tumour cells) and lack of recognition of normal cells[21].
(6) So far, no cases of induction of resistance of tumour cell lines to macrophages have been reported.

Other cells, such as polymorphonuclear neutrophils, basophils and eosinophils serve as effects or of inflammation elicited by the immune response. The important role of these cells in immunologically mediated diseases with concomitant inflammatory components cannot be overemphasised. Thus, a clear understanding of the immune network and the interrelations among immunocompetent cells is a necessary base for the explanation of the sites of immune aberrancy and for deciding which agents might effect an appropriate change in the aberrant response.

Due to the complexity of the immunological network it is quite probable that potential mechanisms and sites for abnormal reactivity exist that might lead to the appearance of an immunologically mediated disease. For example, systemic lupus erythematosus (SLE) is characterised by autoantibody formation and the appearance of circulating and tissue-bound immune complexes. Conceivably, the production of these antibodies could arise from intrinsic primary hyperreactivity, a loss of suppressor T-cell function or a combination of the two[22]. Considering the findings about reduced expression of type I complement receptors (CR I) on red blood cells of these patients[23] makes exploration of pathogenesis in SLE even more complicated. However, it might offer an extended explanation in so far as reduced elimination of immune complexes as a consequence of lower CR I expression induces chronic inflammation, during which autoantigens are exposed leading to autoreactivity and formation of additional immune complexes which again cannot be cleared properly and the circulus vitiosus is perpetuated. Another example would be innate B-cell hyporeactivity resulting in class-specific immunoglobulin deficiency, which could similarly arise from either deficient T-cell help or excessive suppressor T-cell function. As already indicated above, it becomes even more complicated to explore the pathogenetic traits when several simultaneous *in vitro* or *in vivo* defects have been described for a given disease.

5.2.2 Mechanisms and Problems in Immunomodulation

In principle, regulation of the immune response can be influenced in two ways: the enhancement of the immune response is called immunopotentiation and its down-regulation immunosuppression. The early history of medical immunology is dominated by deliberate attempts to induce resistance to infectious agents by means of eliciting an immune response and by enhancing the immunological defence mechanism. Vaccination or immunisation with an antigen preparation can specifically protect the body against microorganisms like smallpox, rabies or poliomyelitis. More recently, the increasing awareness that several pathological conditions are caused by undesirable reactions of the immune system has led to the attempt to reduce the immune response. Thus, immunosuppression is needed in transplantation to prevent the rejection of a transplanted and histo-incompatible but life-saving organ as well as to induce remission and to prevent relapse in autoaggressive or immune diseases.

Immunopotentiation can be defined as a process which directly enhances one or more specific immune functions, or modifies one or more components of the complex immunoregulatory network to achieve its effects through indirect mechanisms. An enormous array of crude biological extracts, plant and animal substances of varying degrees of purity, and now cloned cytokines and highly purified immunoglobulins have been used as immunopotentiating agents. The subject includes both immunotherapy, directed at pathogenetic microorganisms, malignant cells, immunodeficiency, and the immunoregulatory disturbances of autoimmune disease and the enhancement of immune responses during immunization, or the use of immunological adjuvants. There is substantial overlap between the types of substances used for both these purposes and there may well be a parallel overlap in the mechanisms involved.

It is possible to classify immunopotentiation in a number of ways, one of them would be to discriminate between specific and non-specific mechanisms. Both groups could further be subdivided into active, passive, and adoptive mechanisms. Regarding specific immunopotentiation these strategies would be vaccination, administration of specific antibodies from isologous or xenogenic source, and treatment with specifically sensitised lymphocytes, tumour-infiltrating lymphocytes or lymphocyte extracts, respectively. Active non-specific mechanisms include immunopotentiators of biological origin, thymosins and other hormone-like thymic extracts, cytokines or synthetic chemicals. Passive mechanisms would be immunoglobulin therapy and plasmapheresis and as adoptive strategies cytokine-activated killer cells

(NK cells become LAK cells after activation with interleukin-2) and macrophages activated *in vitro* by cytokines or (OH_2) vitamin D_3 have to be mentioned. There is some overlap between the different categories used and the classification can only be viewed as a framework about which to base discussion. In addition, combinations of specific and non-specific mechanisms are used, such as the addition of adjuvants to specific antigens in many vaccination protocols[24].

The host itself must be seen as being an integrated system of various components acting at the resting (i.e. non-infected) state in a smoothly coordinated manner, with fine (and not yet totally elucidated) regulations in response to environmental stimuli. The introduction of virulent pathogens, or a congenital or acquired disequilibrium of any of these defence components will lead to a clinical situation in which regulatory processes may be different from the original resting situation, defining the infectious state. Pathogens are able to survive in immunocompetent hosts because of their ability to evade or directly inactivate host immune responses, because of their ability to downregulate protective effector mechanisms and also because they acquired the ability to take advantage of the numerous immune defence mechanisms. These evasion mechanisms can be quite sophisticated and represent some of the most interesting and unique aspects of a parasite–host relationship. The simplest form of immune evasion is anatomic seclusion, as exemplified by intracellular pathogens which therefore escape killing by antibodies. Encystment as well as dwelling in immunoprivileged sites would be another example of avoiding contact with immunocompetent cells. Modification of antigenicity leading to adaptation to the immune system, by acquisition of a dense coat of host molecules, is a very effective mechanism resulting in persistence of infection for years. Evasion of the complement system by using decay-accelerating 'factor-like' substances[25], inactivation of antibody molecules through cleavage into Fc and Fab fragments or lysis of lysosomal membranes and subsequent escape into the cytoplasm[26] would be other examples whereby host effector functions could be inactivated. During viral infection direct cytotoxicity on lymphocytes, downmodulation of Class I or Class II MHC antigens, effects on interleukin production, usage of leucocyte surface structures as specific docking receptors, negative selection of peripheral T cells or downmodulation of viral antigens have been documented[27].

Similarly, also, tumour-associated disturbances of immune reactivity can be seen. Some of the earliest observations of inflammatory defects associated with cancer were those of Berg who, in 1959, reported that there was a significant difference in inflammatory infiltrates in breast cancer biopsies in patients who were long-term survivors, compared with those who were not[28]. Later it was demonstrated that the inflammatory responses to implanted cotton thread were impaired in tumours of mice and rats compared with the responses which occurred in non-neoplastic tissue[29]. In correlation with these *in vivo* defects, it was found that chemotactic responses of macrophages from tumour-bearing mice are decreased[30]. One possible explanation of the *in vivo* alteration of the macrophage response at a distance from the tumour site is the presence, in serum of tumour-bearing animals or in cancer patients, of inhibitors. Among other possible candidates, acute-phase proteins such as alpha-1 acid glycoprotein[31], prostaglandins[32], cytokines[32] and P15E-related antigens[33] have been mentioned. The latter bear quite remarkable homologies with a type C retroviral transmembrane protein[34,35] with anti-inflammatory activities[33,36]. One prominent example has emerged to be transforming growth factor β[37], an immunoregulatory protein overexpressed in a variety of malignant cells. As a matter of course these soluble factors can influence the specific arm of the immune system as well. In addition, suppressor cell populations have been identified within the T- and B-cell lineage, which appear to possess specificity for a particular tumour because of their dependency on the presence of malignant cells in the animal[38–40].

Finally, the assumption is generally made that the immune response is protective. To appreciate the concept of immunological mechanisms as pathological processes, the paradigm of the immune system as an only beneficial event has to be shifted completely. Much of the pathology encountered in infectious disease has an aberrant immunologic basis and stems from an excessive or aberrant immune response against the invading pathogens or parts thereof expressed on host cells. Two important features of this immunopathologic response are its sensitivity to regulatory influences and its variable manifestation in different stages of infection and in the context of different host genetic backgrounds. These characteristics suggest that pathology is a dynamic process reflecting the immune system's attempt to neutralise pathogens and their antigens as well as to regulate the potentially harmful consequences of that response. A very prominent example of the ambiguity of mediators released during infection is certainly tumour necrosis factor α, which as far as local infection is concerned can be regarded as a cytokine helping to eliminate the injurious agent but in case of systemic infection would contribute to the perpetuation of septic shock and aggravation of clinical symptoms[41].

Therefore, the ideal approach to modulate the immune system should not only try to stimulate immune responsiveness regardless of what defence mechanism is concerned but should also take account of all aspects of aberrant immune reactivities present during a particular stage of disease. The criteria for an immunomodulator or immune response modifier as formulated by Talmadge *et al.*[42] would therefore have to be: first, augmentation and/or restoration of effector mechanisms or of mediators of host defence; secondly, decreasing the deleterious components of the host's reactions; thirdly, replacement of damaged effector or mediator mechanisms with natural biological or synthetic derivatives, and lastly, augmentation of the host response to vaccines and chemotherapeutic agents.

The most critical and controversial aspect of drug-induced immunomodulation is the clinical relevance of the direct effects of potential immunomodulatory agents on the host defence system. Of these effects, the intracellular uptake and bioactivity of some substances is well-recognised. This property clearly sustains the therapeutic efficacy of drugs that penetrate phagocytes in infections caused by intracellular pathogens. These drugs include macrolides (which in general display the most prominent cellular uptake), antimalarials, quinolones, rifampicin and rifabutin derivatives, isoniazid, clindamycin, vancomycin, teicoplanin and, to a lesser extent, tetracyclins. Whether or not cellular concentration may in turn alter phagocytic function during therapy is a matter of controversy. Non-cell-penetrating agents may also modify specific and/or non-specific immune effector function, which suggests that at least some kind of association to the cell membrane or passage through it with subsequent interaction with intracellular ligands does occur.

Most of the reported effects of immunomodulators on the immune system are obtained *in vitro* with experimentally fixed systems which do not truly reflect the dynamics and multiple interactions occurring *in vivo*. Many controversies arise from the use of non-standardised techniques, cells obtained from various origins (different animal species or humans, as well as different sites) or various cell subsets. For example the antibiotic cefotaxime was shown to stimulate the neutrophil oxidative burst[43] whereas it was reported to inhibit that of monocytes[44]. Similarly, coumarin and its major human metabolites 7-hydroxycoumarin and *o*-hydroxyphenylacetic acid (*o*-HPAA) have been reported to suppress O_2^- production after phorbol myristate acetate stimulation only in monocytes but not in macrophages from healthy volunteers[45]. Also, there are examples that a given function may not be altered by a drug depending on the stimulus and on the technique used. Rifampicin specifically inhibits neutrophil chemotaxis in the case of formyl-methionyl-leucyl-phenylalanine stimulation, but not when activated serum is used as a chemoattractant. This is related to specific competition of rifampicin at the level of the formyl-methionyl-leucyl-phenylalanine receptor[46]. Many other problems related to the analysis of *in vitro* data must also be considered, such as artefactual modifications by interference of the drug with the detection system (e.g. chloramphenicol, ampicillin, and rifampicin)[47], and the still poorly understood phenomenon of interindividual susceptibility to a drug-induced effect[48].

The analysis of drug-induced immunomodulation *ex vivo* is far more difficult to interpret. Aside from the problems related to the *in vitro* techniques, others are specific to *ex vivo* studies themselves, including the choice of a relevant model, chronobiological effect, and removal of drug and/or immune byproducts during cell isolation. Many studies are performed in animals but this may actually not be relevant to the clinical situation. Not only can different animal species or even strains exhibit different susceptibility to drug-induced effects, but also metabolism, chronobiology, pathogen susceptibility and immune functions are fundamentally different in humans and animals. Animal models would be useful only in the case of a drug inducing a very general effect across species, which is unlikely.

Regarding human models, most studies involve human volunteers, which may serve to analyse drug-induced immunosuppression but not enhancement or restoration of immune functions. In this case it would be very important to be able to discriminate between a direct effect on the injurious agent (e.g. pathogen, tumour cells) and the actual immunomodulatory capacity which would contribute to the elimination of the microorganism or noxious cells by this indirect mechanism. Immunocompromised non-infected patients would be useful to assess the effects of some compounds on the restoration of various deficient parameters. However, this may be just a preliminary step in the study of immunomodulators, since these individuals are unlikely to be treated with, for example, antibiotics with modifying capacity in the uninfected state or with immunomodulating antitumour agents when no tumour is present. In the case of infection, *in vitro* antibacterial activity (still the key determinant of antibacterial therapy) has to be considered, and furthermore, as already mentioned above, the infectious disease itself modifies the activity of the immune system, and thus renders the understanding of a drug-induced immunomodulating effect more complex.

Other *ex vivo* problems reflect the extreme inter-individual variability for some immune parameters, and may complicate the study of immunomodulation. Finally, finding some effect in *ex vivo* studies may be considered as a positive effect, but absence of effect should not be interpreted as a true neutral influence of the drug, since the isolation procedures used to study immune activity may remove the drug itself and/or immune byproducts that might be critical in obtaining immune modification.

All the difficulties encountered *in vitro* and *ex vivo* are further amplified when considering the *in vivo* step. Except for a few techniques (e.g. delayed type hypersensitivity, allograft survival, *in vivo* chemotaxis), the only proposed models are those using experimental models of infection. This may be achieved only in animals, which are, as pointed out above, far from the clinical situation. These models are just the first steps in being able to propose a particular drug-related effect when comparators showing similar *in vitro* activity display different therapeutic efficacy, although other parameters such as pharmacokinetics, tissue distribution and tissue concentrations also have to be considered.

Another problem in the use of immunomodulatory agents is that apart from their desired effects, most drugs may interfere with host homeostasis, leading to secondary, more or less deleterious, side-effects. Thus, the various immunological effects of cytokines thought to be important in the destruction of tumour cells are not restricted to tumour tissue. For example, the ability of interferon to enhance the expression of HLA molecules may not only render tumour cells more susceptible to lysis by cytotoxic T lymphocytes but also probably contributes to the predisposition to develop autoimmune thyroid disease known to complicate interferon treatment[49]. The antiproliferative effect of interferon may not only slow tumour growth but may account for leucopenia observed in patients receiving this agent. The activation of natural killer cells by interleukin-2 and their subsequent attachment to the tumour neovascular may be a prelude to tumour infiltration by these cells and may also expose the normal endothelium to toxic concentrations of superoxide and other mediators, resulting in the vascular leak syndrome that is characteristic of interleukin-2-based immunotherapy[50]. The local generation of TNF by tumour-infiltrating lymphocytes may be a critical factor in interleukin-2-induced tumour regression but is also the prime cause of the hypotension that imposes severe restrictions on the use of this cytokine[51].

Taken together, the ideal immunomodulator should be something like a superordinate principle which in every situation knows how to interfere with aberrant reactivities so that homeostasis is restored quite rapidly without causing any inconvenience to the host. In other words one single substance or combinations of immunomodulators should at least be equally effective in surveillance as the extremely highly developed immune system of human beings which took millions of years to be established in its present form. Realistic expectations, therefore, can only involve the discovery of agents with a rather specific spectrum of efficacy which, however, again would require that it is well known what effect would be beneficial in a particular pathological condition. For infectious diseases this approach could be seen to be quite optimistic since most of the important effector mechanisms are well established. However, for malignancies considerable progress may be required before the most effective defence mechanisms are elucidated. Therefore, at the present time only combinations of therapeutic options consisting of surgery, radiation therapy, chemotherapy and immunomodulation can be considered a promising approach which might help improve treatment protocols.

5.3 IMMUNOMODULATION BY COUMARINS

5.3.1 Introductory Remarks

Coumarin has been demonstrated to be of benefit in a small percentage of tumour patients both for the treatment and the prophylaxis of malignancies, in particular of malignant melanoma and renal cell cancer[52–66]. Although the response rate is not very high, there is a significant effect by the drug which cannot be attributed to the spontaneous cure of disease. Chemotherapeutic agents, especially, have produced very poor results for the treatment of advanced disease with low response rates, short durations of response and significant toxicities in these malignancies[67–69]. More promising results have been achieved with the class of substances known as biological response modifiers[68,69]. These have included interferons and interleukin-2 or combinations thereof. In addition, treatment with lymphokine-activated killer cells as well as tumour-infiltrating lymphocytes has been attempted. In these instances response rates very seldom exceeded 25–30% of evaluable patients and toxicities can be quite profound, making withdrawal of treatment necessary[68,69]. Because of the minimal side-effects[70–74] and sometimes comparable efficacy of coumarin, this drug might be a suitable candidate for an alternative strategy to treat tumour patients. Nevertheless, until now the achieved response rates to coumarin are far from being satisfactory. As mentioned above, clinical studies were mainly empirical and little is known about the actual mode of action of coumarin in malignancies.

Based upon results from our own laboratories as well as studies from other investigators, coumarins appear to act either directly on tumour cells, or via modulation of the host's immune system, thereby stimulating immune reactivity which leads to protection against recurrence of a particular tumour or even to activation of host defence mechanisms which also help to eliminate small tumour burdens. Direct (i.e. non-immunologically mediated) antitumour effects have been shown in a number of studies demonstrating a growth inhibitory capacity of coumarin and/or 7-hydroxycoumarin for a number of malignant cell lines *in vitro*. Marshall *et al.*[75,76]

have shown that both drugs are cytostatic for the following human malignant cell lines: K 562 (erythroleukaemia), ACHN and Caki-2 (renal carcinoma), LH-60 (lymphoma), HCT-15 (colon carcinoma), H 727 (malignant carcinoid), LNCaP (prostate carcinoma), and the multiple myeloma lines Dakiki (IgA), HS-Sultan (IgG) and RP-1788 (IgM). At equal concentrations 7-hydroxycoumarin had a greater growth-inhibitory effect than coumarin. These results have been confirmed and extended to other cell lines by many authors[77–79]. For two malignant glioblastoma cell lines it has been demonstrated that 7-hydroxycoumarin acted as a negative regulator of PDGF-A and PDGF-B mRNA, suggesting that the drug blocks one putative autocrine loop of human malignant glioblastoma cells[80,81]. In addition 7-hydroxycoumarin induced differentiation of these cells. Zänker studied the effects of 7-hydroxycoumarin on two cell lines known to overexpress certain oncogenes[82]. The DUKX line overexpresses c-*myc* and the RBA line overexpresses Ha-*ras*. As before, 7-hydroxycoumarin inhibited the growth of both lines in a dose-dependent fashion. Further, this effect was associated with decreased levels of the oncogene mRNA levels in the respective cell lines. In addition, coumarin and its derivatives may act on tumour cells directly by preventing tumorigenesis[83,84] or inhibiting the development of metastasis[85].

Apart from this direct antineoplastic action of coumarins, the drugs have been reported to possess an immunomodulating activity which might also contribute to the antitumour effect observed in treated patients. The purposes of this review are to summarise the present knowledge of the immune response modifying capacity of the parent coumarin compound as well as of its main metabolite in humans, 7-hydroxycoumarin, and to try to develop a consensus as to what the most important remaining questions would be to support this kind of action and what then could contribute to a substantial improvement of current treatment protocols both in the case of tumour treatment as well as in prophylaxis and treatment of infections.

5.3.2 Immunomodulation *in vitro*

The current view of coumarin as an immunomodulator originates from experiments performed more than 20 years ago demonstrating an anti-inflammatory and anti-oedema effect of coumarin[86] which at least, in part, appeared to be due to its action on macrophages, which exhibited an increase in proteolytic capacity after treatment with this drug[87]. The mode of action of coumarin in this indication then was explained by a reduction in the amount of protein present in the connective tissue followed by a reduction of the protein's osmotic effect and clearance of oedema[88,89]. An activity not restricted but nevertheless very important to the immune system is the finding that coumarin, as well as its metabolites 7- and 4-hydroxycoumarin, were able to inhibit the biosynthetic conversion of arachidonic acid to prostaglandin products by a synthetase complex from guinea-pig lung[90]. The inhibition produced by the coumarins was equipotent to that of acetylsalicylic acid. These data further support an anti-inflammatory activity of the drugs but given the multifaceted effects of arachidonic acid intermediates on the immune system it is conceivable that this interference might be of central importance to the mode of action of coumarins. To the best of my knowledge, however, no further studies have been reported on that aspect so far.

Further evidence that cells from the monocyte macrophage lineage might be important targets for coumarin's action on the immune system came from studies about the influence of coumarins on phagocytosis of latex particles[91]. Non-elicited peritoneal macrophages displayed a bell-shaped dose–response curve regarding their phagocytic capacity which was stimulated by increasing amounts of coumarin. The metabolite 7-hydroxycoumarin did not show any effect in this regard. Studying other effector functions of monocytes/macrophages revealed that the generation of reactive oxygen metabolites was influenced in a very particular way. Only superoxide anion but not hydrogen peroxide generation was inhibited after stimulation with phorbol myristate acetate in a dose-dependent manner when peripheral blood monocytes were tested immediately after adherence to plastic but not after 48 h of incubation[45].

These apparently contradictory effects on phagocytosis and respiratory burst might be due to different sources of monocytic cells (mouse versus human) but could also reflect a differential immunomodulatory efficacy of coumarins by reducing a potential harmful effect (respiratory burst) and increasing a mechanism which most probably would result in the elimination of invading pathogens or potentially noxious cells in the case of malignant transformation. The results on the inhibition of superoxide production in addition, however, could be in line also with findings on a peroxyl radical scavenging capacity by a series of coumarins[92] which makes these compounds also attractive candidates for evaluation as protective agents against disorders in which oxidative stress is implicated (e.g. HIV infection and AIDS[93]). Another effect which probably is relevant to monocytes too, is the enhancement of DR and DQ antigen expression by peripheral blood mononuclear cells *in vitro*[94]. Whereas peripheral blood mononuclear cells show a decreased expression of these MHC Class II antigens on continued culture, the presence of coumarin prolongs it. Since expression of DR antigens is regarded as consistent with an activated state these results also might indicate that coumarin is able to act through immune augmentation. There is some limited information on the efficacy of coumarin on other immune cells like NK cells or T cells[45,54] but in this regard, results are not very consistent and the extent of the differences to cultures performed in the absence of coumarins are not that profound so that an influence on these cell populations still has to be viewed as not yet fully established.

Studies from our own group have concentrated on another effector function of monocytes—monokine production—and were designed to get more insights in the actual mode of action of coumarin. Initial experiments indicated that both coumarin and 7-hydroxycoumarin would induce monokine production (especially IL-1) in a dose-dependent manner with concentrations higher than 10 μg/ml resulting in less pronounced levels of mediators[95]. A change in the source of foetal calf serum revealed the dependence of IL-1 induction on that supplement and further indicated that the inducing capacity of coumarin was only operative in the presence of even trace amounts of endotoxin in the serum preparations. Subsequently the foetal calf serum was omitted from cultures and defined lipopolysaccharide (LPS) preparations were added. This demonstrated that even 10 pg/ml LPS, which was ineffective by itself, was able to induce significant amounts of IL-1 when coumarins were present at the same time. With increasing concentrations of LPS there was still a potentiation of IL-1 production, but at optimal LPS concentrations no further effect could be seen. The increase in IL-1 induction was dependent on protein synthesis and could be shown not to be due to an increased release caused by the coumarins since the total amount of IL-1 (cell associated + released form) present in culture after co-stimulation with LPS and coumarin was higher than in cultures not treated with coumarins.

The production of other cytokines was potentiated in a smaller percentage of individuals than production of IL-1. Augmentation of IL-1 induction could be seen in all individuals whereas IL-6 production was increased in about half and TNF-α production in about one-third of the individuals tested.

Attempts to evaluate the mode of action at the molecular level demonstrated that there appeared to be an involvement of protein kinase C and/or lipoxygenase in the potentiating effect of coumarin[96]. This has been demonstrated using a cell culture model where the effect of the two potential agonists LPS and coumarin could be separated by addition at different times. The prerequisite was that when mononuclear cells were pre-incubated with LPS and particular inhibitors were added 3 h later, there was almost no effect on the IL-1-inducing capacity of LPS. Therefore, when coumarin was added 3 h after LPS and together with the respective inhibitors, this setting should indicate the effect of inhibitors on the additional influence of coumarin but not on the one by LPS. When staurosporine or nordiguaretic acid were used in this setting there was a clear inhibitory effect on the additional induction of IL-1 by coumarin, indicating that coumarin might act by interference with signal transduction at the level of protein kinase C or lipoxygenase involvement. In contrast, forskolin, an activator of adenylate cyclase leading to accumulation of cAMP, or indomethacin, an inhibitor of cyclooxygenase, had no influence on coumarin-induced cytokine potentiation. On the one hand, these results demonstrate specificity for particular signalling pathways, and, on the other hand, they show that for potentiation the same pathways are critical as for the primary agonist since, in this instance, only staurosporine and nordiguaretic acid were effective.

With regard to the clinical implications of these findings one might speculate that the synergism between LPS and coumarin could be used to reduce toxic side-effects of LPS, or, another agonist with immunomodulatory potential could also be increased in its stimulatory capacity by coumarin. Several trials have been undertaken to evaluate the efficacy of endotoxin in inducing antitumour activities, but its application in humans has been hampered by the toxicity of the agent[97]. Another important aspect might be that it was mainly patients with lung metastases who were found to respond both to coumarin treatment and also to therapeutic approaches with other immunomodulators[68]. It is arguable, therefore, that the effect of coumarin (and perhaps of other agents) in such patients was also due to a synergism, possibly with endotoxin, which is consistently present in the respiratory tract. This might offer an explanation for the limited response rate if this prerequisite of co-stimulation of endotoxin and coumarin had to be fulfilled.

As has already been mentioned in the introduction, one of the most important aspects in elucidating the mode of action of a substance with potential immune response modifying capacity is to know how this drug interacts with immunocompetent cells. Is it possible to identify a specific membrane receptor or are there intracellular reaction partners which serve as targets for modulating immune responsiveness? Zänker *et al.*[54] have demonstrated the existence of binding sites for coumarin on a variety of tumour cells which appeared to be of heterogeneous as well as differing binding characteristics. Investigations performed in our laboratory revealed that human monocytes express a great number of binding sites for coumarin which show low-affinity binding characteristics[98]. Very interestingly, competition experiments demonstrated that binding of coumarin was inhibited in a clear dose-dependent fashion by the homologous drug but only to a much less extent by 7-hydroxycoumarin. In the case of 7-hydroxycoumarin, no saturating conditions could be found, so that binding of this molecule appears to follow differential binding characteristics with perhaps lower affinity and/or higher numbers of binding sites. Since rather high concentrations of either coumarin or 7-hydroxycoumarin (which never could be reached *in vivo*) had to be used to get saturating conditions, another possibility would be that after the passage through the cell membrane the coumarins interact with their target cells via intracellular ligands in a low-affinity manner such that the functional capacity of the cell is changed. This would be compatible with the hypothesis that coumarin interacts with ubiquitous intracellular receptor proteins able to interact with aromatic hydrocarbons and that might form the basis for enzyme induction and lead to the effects observed *in vitro* and *in vivo*[99].

5.3.3 Immunomodulation *in vivo*

In contrast to the evidence obtained *in vitro*, much less information is available about the interference of coumarin treatment with immunological parameters as determined *in* or *ex vivo*. Patients treated with coumarin and cimetidine because of locally advanced or metastatic renal cell carcinoma presented with a significant decrease in the absolute number of T cells, including both helper and suppressor populations[58]. Results from delayed-type hypersensitivity skin testing using a standardized battery of seven antigens (tuberculin, tetanus and diphtheria toxoid, streptococcus, candida, trichophyton and proteus) showed a decreased reactivity at the beginning of treatment which did not change significantly after two courses of therapy nor was there an apparent correlation to the clinical course of the disease[58]. Another group of investigators did not find any variation in T-cell populations after coumarin treatment in patients with metastatic renal cell carcinoma and malignant melanoma[100]. The discrepancy between these reports might be due to differences in patient populations. It could also be due to the additional use of cimetidine in the first study which has been shown to be capable of influencing the composition of lymphocyte populations[101]. In parallel with the *in vitro* findings mentioned before, coumarin also appeared to increase DR antigen expression by monocytes *in vivo*, indicating the potential of activating this leukocyte population in treated patients as well[100]. Further evidence for immunomodulation by coumarins *in vivo* was obtained from cancer patients receiving coumarin or warfarin without concomitant radiotherapy, showing a significant increase in mitogen-responsiveness[102], which, however, is very difficult to interpret since variations in this parameter are very broad between, as well as within, individuals. These findings demonstrate two problems of *in vivo* monitoring during any potentially immune-mediated disease. First, changes in immune parameters within the peripheral blood do not necessarily reflect processes occurring in a particular tissue, including the tumour site, and, secondly, so far the clinical relevance of fluctuations in immune responsiveness or variations in the composition of leukocyte subpopulations is poorly established unless these variations show very dramatic deviations from normal. Therefore, it would be desirable to have more evidence from *in vivo* studies for a direct immunomodulatory effect.

Since there is no indication that coumarin or 7-hydroxycoumarin possess antibiotic activity, a potential beneficial effect in such a system would have to be caused by interference with immune reactivity irrespective of what arm of the immune system was modulated. As a matter of course subsequent *in vivo*, as well as *in vitro*, experiments then have to elucidate the exact mode of action. This in turn could help to further optimise treatment protocols and could give indications as to what mechanisms were involved in this respect. So far, an animal model for the evaluation of coumarin's efficacy has been set up only with respect to its direct antitumour activity. In these studies—two of them used rats[103,104] which are known to metabolize coumarin in a way which is considerably different from humans[105–107]—some effect, which, however, was inconsistent between both groups, has been observed on prostate-derived tumours with no evidence for interference with immune responsiveness of treated animals. One study in mice demonstrated only a moderate antitumour effect of coumarin against allogenic Sarcoma-180[108]. Therefore, there is a great need for studies in animal models, preferentially in gerbils which have a similar metabolism of coumarin to humans. However, mice, which are much easier to keep, should suffice to provide the key information about a potential influence (immunomodulatory or not) of coumarin in malignancies but also in infectious models.

Another important aspect which can be addressed in such models is the question of the exact nature of the '*in vivo*' active compound. Using various mouse strains with different 7-hydroxylase activity[109] may provide information as to whether it is the parent compound or its metabolite that is the more important substance for *in vivo* immunomodulation.

5.3.4 Implications for Cancer Treatment

Summarising the findings of coumarin's efficacy in immunomodulation, cells from the monocyte/macrophage lineage appear to be a central target population for its modulation of the immune system. There is quite substantial evidence that monocytes/macrophages might be equally important as other lymphocyte populations in maintaining homeostasis or at least play very particular roles in host defence against cancer and infections; they recognise, phagocytose and ultimately dispose of effete cells, cellular debris and foreign invaders. Normal, non-cytotoxic blood monocytes or tissue macrophages can become tumoricidal *in vivo* and *in vitro* if appropriately activated by cytokines or bacterial products[110]. They do so by multiple mechanisms including cell-contact-mediated killing in both antibody-dependent and -independent ways[111], as well as by cytotoxicity that is mediated through soluble and membrane-bound cytotoxic molecules[112,113]. Although tumour cells are heterogeneous with regard to many characteristics, they seem to share susceptibility to destruction by activated macrophages. In fact, activated macrophages can lyse tumour cells that are resistant to other immune effector cells such as T cells or natural killer cells or to chemotherapeutic drugs. Moreover, activated macrophages can discriminate between tumorigenic cells, which they lyse, and non-tumorigenic cells, which they do not, even under co-cultivation conditions[114]. Systemic activation of macrophages can be accomplished by intravenous administration of cytokines like TNF-α[41] but also by synthetic analogues of bacteria-derived immunomodulators entrapped in phospholipid liposomes.

As an example with apparently promising potential, MTP-PE (*N*-acetyl-muramyl-L-alanyl-D-isoglutamyl-L-alanyl-phosphatidylethanolamine), a lipophilic analogue of muramyl dipeptide, is mentioned here briefly for comparative purpose to coumarin. This preparation has been used with success to activate human blood monocytes *in vitro* and mouse macrophages under both *in vitro* and *in vivo* conditions[115–117]. It has also been shown to activate lung macrophages[118] and to eradicate spontaneous metastases in a mouse melanoma system[117] and in dogs with autochthonous osteogenic sarcoma metastases[119]. Results from two parallel phase I trials demonstrated good biologic activity (macrophage activation, production of monokines) and high degree of safety[120,121]. In a phase II study no clear antitumour response could be observed but there was prolongation in time to relapse after treatment of osteosarcoma patients with MTP-PE for six months[122,123]. The most remarkable histological changes were formation of peripheral fibrosis surrounding the tumour lesion and extensive infiltration by host inflammatory cells coupled with neovascularisation[124]. The role of activated macrophages in this respect could be cytokine secretion such as interleukin-1 and tumour necrosis factor, which not only kill tumour cells, including osteosarcoma cells[16,41], but also produce fibrosis and granulomas. Indeed, all of the major monokines have been shown to be induced both *in vitro* and *in vivo* after MTP-PE treatment[125,126], which, however, was followed by significant toxic side-effects[127]. Although macrophage-directed immunotherapy of cancer is still in its infancy, immunostimulation by MTP-PE emerged to be a very encouraging approach. Since it seems to have some common features with coumarins (e.g. the monocyte/macrophage is a target cell that appears central for the antitumour effect), it might serve as a useful model for comparative studies.

Adoptive therapy with differentiated, tumour cytotoxic macrophages which have been generated *in vitro* by, for example, CSF, GM-CSF and (OH)2 vitamin D_3 or interferon γ[20,116,128], may be one modality for bypassing local factors affecting the maturation and suppression of the functional activation of tumour-site-restricted macrophages. Any macrophage-directed immunotherapy, however, might only be successful in early stages of tumour development, or if given to eliminate residual disease, following cytoreductive measures, like surgery and chemo- and radiotherapy.

In conclusion, coumarins certainly possess distinct immunomodulatory activity, which, however, at the present state is not sufficiently established to have the potential of influencing immune responsiveness in a clinically apparent manner. As exemplified by other immune response modifiers the spectrum of activity as well as the target population might be appropriate to interfere at a site of the immune system relevant for antitumour responses. Nevertheless, because of limited response rates in the clinical situations tested so far, more information is needed about the actual mode of action at the cellular and molecular level so that existing treatment schedules can be improved and perhaps also supplemented with synergistically active compounds. With regard to the socioeconomical importance of cancer, this would contribute both to an improvement in the quality of life of patients and to cost reduction because of the relatively small expenditure for the drug.

REFERENCES

1 Malnati MS, Peruzzi M, Parker KC, *et al.* Peptide specificity in the recognition of MHC class I by natural killer cell clones. *Science* 1995; **267**: 1016–18.

2 Adams DO, Hamilton TA. The cell biology of macrophage activation. *Annu Rev Immunol* 1984; **2**: 283–318.

3 Ben-Efraim S, Tak C, Fieren MJWA, Van den Bemd GJCM, Bonta IL. Inflammation amplifies antitumour cytostasis by human peritoneal macrophages. *Med Oncol Tumor Pharmacother* 1991; **8**: 87–94.

4 Fidler IJ. Macrophages and metastasis—a biological approach to cancer therapy. *Cancer Res* 1985; **45**: 4714–26.

5 Stutman O. Immunodepression and malignancy. *Adv Cancer Res* 1975; **22**: 261–422.

6 Kamo I, Friedman H. Immunosuppression and the role of suppressive factors in cancer. *Adv Cancer Res* 1977; **25**: 271–321.

7 Aune TM. Role and function of antigen nonspecific suppressor factors. *Crit Rev Immunol* 1987; **7**: 93–130.

8 Ophir R, Pecht M, Halperin D, *et al.* THF-gamma 2, a thymic hormone, increases immunocompetence and survival in 5-fluorouracil treated mice bearing MOPC-315 plasmacytoma. *Cancer Immunol Immunother* 1989; **30**: 119–25.

9 Sulizeanu D. Immunosuppressive factors in human cancer. *Adv Cancer Res* 1993; **60**: 246–67.

10 Siziopikou KP, Harris JE, Cassey L, Nawas Y, Braun DP. Impaired tumoricidal function of alveolar macrophages from patients with non-small cell lung cancer. *Cancer* 1991; **68**: 1054–6.

11 Ophir R, Ben-Efraim S, Bonta L. Leukotriene D4 and indomethacin enhance additively the macrophage cytostatic activity *in vitro*. *Int J Tissue React* 1987; **9**: 189–94.

12 Bonta IL, Elliott GR, Tak C, Ben-Efraim S. Indomethacin stimulation of macrophage cytostasis against MOPC-315 tumor cells is enhanced by endogenous metabolites of lipoxygenase and counteracted by prostaglandin E2. *Agents Actions* 1989; **26**: 167–9.

13 Chen AR, McKinno KP, Korin HS. Lipopolysaccharide (LPS) stimulates fresh human monocytes to lyse actinomycin D-treated WEHI-164 target cells via increased secretion of a monokine similar to tumor necrosis factor. *J Immunol* 1985; **135**: 3978–87.

14 Feinman R, Henriksen-Destefano M, Tsujimoto M, Vilcek J. Tumor necrosis factor is an important mediator of tumor cell killing by human monocytes. *J Immunol* 1987; **138**: 635–40.

15 Fieren MJWA, Van den Bemd GJCM, Bonta IL. Endotoxin-stimulated peritoneal macrophages obtained from continuous ambulatory peritoneal dialysis patients show an increased capacity to release interleukin-1 *in vitro* during infectious peritonitis. *Eur J Clin Invest* 1990; **20**: 453–7.

16 Onozaki K, Matsushima K, Kleinerman ES, Saito T, Oppenheim JJ. Role of interleukin 1 in promoting human monocyte-mediated tumor cytotoxicity. *J Immunol* 1985; **135**: 314–29.

17 Andreesen R, Brugger W, Gadd S, Lohr GW, Atkins RC. Activation of human monocyte-derived macrophages cultured on response to interferon-gamma during terminal maturation *in vitro*. *Immunobiology* 1988; **177**: 186–98.

18 Dumont S, Hartmann D, Poindron P, Oberling F, Faradji A, Bartholenys J. Control of the antitumoral activity of human macrophages produced in large amounts in view of adoptive transfer. *Eur J Cancer Clin Oncol* 1988; **21**: 1691–8.

19 Andreesen R, Scheibenbogen C, Brugger W, *et al.* Adoptive transfer of tumor cytotoxic macrophages generated *in vitro* from circulating blood monocytes: A new approach to cancer immunotherapy. *Cancer Res* 1990; **50**: 7450–6.

20 Bartholenys J, Lopez M, Andreesen R. Adoptive immunotherapy of solid tumors with activated macrophages: Experimental and clinical results. *Anticancer Res* 1991; **1**: 1201–4.

21 Piessen WF, Hallowell W, Churchill WJ, David JR. Macrophages activated *in vitro* with lymphocyte mediators kill neoplastic but not normal cells. *J Immunol* 1975; **114**: 293–9.

22 Fauci AS. Immunoregulation in autoimmunity. *J Allergy Clin Immunol* 1980; **66**: 5.

23 Walport MJ, Lachmann PJ. Erythrocyte complement receptor 1, immune complexes, and the rheumatic diseases. *Arthritis Rheum* 1988; **31**: 153–8.

24 Gatenby PA. Immunopotentiation. In: Roitt IM, Delves PJ eds. *Encyclopedia of Immunology*. London: Academic Press, 1992: 847–52.

25 Pearce EJ, Hall BF, Sher A. Host-specific evasion of the alternative complement pathway by schistosomes correlates with the presence of a phospholipase C-sensitive surface molecule resembling human decay accelerating factor. *J Immunol* 1990; **144**: 2751–6.

26 Cohen S. Survival of parasites in the immunocompetent host. In: Cohen S, Warren KS eds. *Immunology of Parasitic Infection*. Oxford: Blackwell Scientific, 1982: 138–61.

27 Zinkernagel R. Immunity to viruses. In: Paul WE ed. *Fundamental Immunology*. New York: Raven Press, 1993: 1211–50.

28 Berg JW. Inflammation and prognosis in breast cancer. *Cancer* 1959; **12**: 714.

29 Mahoney MJ, Leighton J. The inflammatory response to a foreign body within transplantable tumors. *Cancer Res* 1961; **22**: 334–8.

30 Cianciolo GJ. Anti-inflammatory effects of neoplasia. In: Gallin JI, Goldstein IM, Snyderman R eds. *Inflammation—Basic Principles and Clinical Correlates*. New York: Raven Press, 1988: 861–74.

31 Samak R, Edelstein R, Israel L. Immunosuppressive effect of acute-phase reactant proteins *in vitro* and its relevance to cancer. *Cancer Immunol Immunother* 1982; **13**: 38–43.

32 Bonta IL, Ben-Efraim S. Involvement of inflammatory mediators in macrophage antitumor activity. *J Leukocyte Biol* 1993; **54**: 613–26.

33 Cianciolo GJ, Matthews TJ, Bolognesi DP, Snyderman R. Macrophage accumulation in mice is inhibited by low molecular weight factors derived from murine leukemia viruses. *J Immunol* 1980; **124**: 2900–05.

34 Cianciolo GJ, Lostrom ME, Tam M.R.S. Murine malignant cells synthesize a 19,000-dalton protein physicochemically and antigenically related to the immunosuppressive retroviral protein P15E. *J Exp Med* 1983; **158**: 885–900.

35 Cianciolo GJ, Kipnis RJ, Snyderman R. Similarity between P15E of murine and feline leukemia viruses and P21 of HTLV. *Nature* 1984; **311**: 515.

36 Cianciolo GJ, Hunter J, Silva J, Haskill JS, Snyderman R. Inhibitors of monocyte responses to chemotaxis are present in human cancerous effusions and react with monoclonal antibodies to the P15E structural protein of retroviruses. *J Clin Invest* 1981; **68**: 831–44.

37 Cianciolo GJ. Inhibition of lymphocyte proliferation by a synthetic peptide corresponding to a region of transforming growth factor beta homologous to CKS-17, an immunosuppressive retroviral-related peptide. *Clin Res* 1990; **38**: 325A.
38 Naor D. Suppressor cells: permitters and promoters of malignancy. *Adv Cancer Res* 1979; **29**: 45–125.
39 Fujimoto S, Greene MI, Sehon AH. Regulation of the immune response to tumor antigen. I. Immunosuppressor cells in tumor-bearing hosts. *J Immunol* 1976; **116**: 791–9.
40 Berendt MJ, North RJ. T-cell-mediated suppression of anti-tumor immunity. An explanation for progressive growth of an immunogenic tumor. *J Exp Med* 1980; **151**: 69–80.
41 Beutler B, Cerami A. Tumour necrosis, cachexia, shock and inflammation: a common mediator. *Annu Rev Biochem* 1988; **57**: 505–18.
42 Talmadge JE, Oldham RK, Fidler IJ. Practical considerations for the establishment of a screening procedure for the assessment of biological response modifiers. *J Biol Resp Modif* 1984; **3**: 88–109.
43 Labro MT, Babin-Chevaye C, Hakim J. Effects of cefotaxime and cefodizime on human granulocyte functions *in vitro*. *J Antimicrob Chemother* 1986; **18**: 233–7.
44 Nielsen H. Antibiotics and human monocyte function II. Phagocytosis and oxidative metabolism. *Acta Pathol Mic Sci* 1989; **97**: 447–51.
45 Marshall ME, Conley D, Hollingsworth P, Brown S, Thompson JS. Effects of coumarin (1,2-benzopyrone) on lymphocyte, natural killer cell and monocyte function *in vitro*. *J Biol Resp Modif* 1989; **8**: 70–85.
46 Gray CG, Smith CW, Hollers JC *et al.* Rifampin affects polymorphonuclear leukocyte interactions with bacterial and synthetic chemotaxins but not interactions with serum-derived chemotaxins. *Antimicrob Agents Chemother* 1983; **24**: 777–83.
47 Briheim G, Dahlgren C. Influence of antibiotics on formyl-methionyl-leucyl-phenylalanine-induced leukocyte chemiluminescence. *Antimicrob Agents Chemother* 1987; **31**: 763–7.
48 Labro MT, El-Benna J, Babin-Chevaye C. Comparison of the *in vitro* effects of several macrolides on the oxidative burst of human neutrophils. *J Antimicrob Chemother* 1989; **24**: 561–72.
49 Fentiman IS, Thomas BS, Balkwill FR, Rubens RD, Hayward JL. Primary hypothyroidism associated with interferon therapy of breast cancer. *Lancet* 1985; **1**: 1166–73.
50 Mier J. Pathogenesis of the interleukin-2-induced vascular leak syndrome. In: Atkins MB, Mier JW eds. *Therapeutic Applications of Interleukin-2*. New York: Marcel Dekker, 1993: 363–79.
51 Gaynor E, Fisher RI. Hemodynamic and cardiovascular effects of interleukin-2. In: Atkins MB, Mier JW eds. *Therapeutic Applications of Interleukin-2*. New York: Marcel Dekker, 1993: 381–8.
52 Thornes RD, Lynch G, Sheehan MV. Cimetidine and coumarin therapy of melanoma. *Lancet* 1982; **2**: 238.
53 Thornes RD, Lynch G. Combination of cimetidine with other drugs for treatment of cancer. *New Engl J Med* 1983; **308**: 591–2.
54 Zänker KS, Blümel G, Lange J, Siewert JR. Coumarin in melanoma patients: an experimental and clinical study. *Drugs Exp Clin Res* 1984; **10**: 767–74.
55 Marshall ME, Mendelsohn L, Butler K *et al.* Treatment of metastatic renal cell carcinoma with coumarin (1,2-benzopyrone) and cimetidine: a pilot study. *J Clin Oncol* 1987; **5**: 862–6.
56 Marshall ME, Butler K, Cantrell J, Wiseman C, Mendelsohn L. Treatment of advanced malignant melanoma with coumarin and cimetidine: a pilot study. *Cancer Chemother Pharmacol* 1989; **24**: 65–6.
57 Thornes RD, Daly L, Lynch G *et al.* Prevention of early recurrence of high risk malignant melanoma by coumarin. *Eur J Surg Oncol* 1989; **15**: 431–5.
58 Dexeus FH, Logothetis CJ, Sella A *et al.* Phase II study of coumarin and cimetidine in patients with metastatic renal cell carcinoma. *J Clin Oncol* 1990; **8**: 325–9.
59 Kokron O, Maca S, Gasser G, Schmidt PR. Cimetidine and coumarin therapy of renal cell carcinoma. *Oncology* 1991; **48**: 102–6.
60 Marshall ME, Butler K, Hermansen D. Treatment of hormonally refractory stage D carcinoma of prostate with coumarin (1,2-benzopyrone) and cimetidine: A pilot study. *Prostate* 1990; **17**: 95–9.

61 Marshall ME, Butler K, Fried A. Phase I evaluation of coumarin (1,2-benzopyrone) and cimetidine in patients with advanced malignancies. *Mol Biother* 1991; **3**: 170–8.

62 Mohler JL, Gomella LG, Crawford ED, *et al.* Phase II evaluation of coumarin (1,2-benzopyrone) in metastatic prostatic carcinoma. *Prostate* 1992; **20**: 123–31.

63 Kokron O, Baumgartner G Theyer G *et al.* Randomised study in metastatic renal cell cancer: Coumarin versus coumarin + cimetidine. *J Irish Coll Phys Surg* 1993; **22**: 10–11.

64 Marshall ME, Stagge S, Moulton J. Successful treatment of malignant carcinoid tumor with 7-hydroxycoumarin. *Proc Am Soc Clin Oncol* 1992; **11**: 206.

65 Mohler J, Williams B, Thompson I, Marshall ME. Coumarin (1,2-benzopyrone) for the treatment of prostatic carcinoma. *J Irish Coll Phys Surg* 1993; **22**: 12–16.

66 Thornes RD. Prevention of recurrence of malignant melanoma by coumarin (1,2-benzopyrone). *J Irish Coll Phys Surg* 1993; **22**: 17–19.

67 Ebert T, Ackermann R. Carcinoma of the kidney. In: Smith PH ed. *Combination Therapy in Urological Malignancy.* London: Springer-Verlag, 1989: 51–81.

68 Stahl M, Wilke HJ, Seeber S, Schmoll HJ. Cytokines and cytotoxic agents in renal cell carcinoma: A review. *Semin Oncol* 1992; **19**: 70–9.

69 Johnson TM, Smith JWI, Nelson BR, Chang A. Current therapy for cutaneous melanoma. *J Am Acad Dermatol* 1995; **32**: 689–707.

70 Cox D, O'Kennedy R, Thornes RD. The rarity of liver toxicity in patients treated with coumarin (1,2-benzopyrone). *Hum Toxicol* 1989; **8**: 501–6.

71 Egan D, O'Kennedy R, Moran E, Cox D, Prosser E, Thornes RD. The pharmacology, metabolism, analysis, and applications of coumarin and coumarin-related compounds. *Drug Metab Rev* 1990; **22**: 503–29.

72 Marshall ME, Butler K, Metcalfe M, Tate M. Coumarin necrosis or coumadin necrosis? *Surgery* 1989; **105**: 237–8.

73 Marshall ME. The coumarins. *Dtsch Zschr Onkol* 1992; **24**: 141–51.

74 Marshall ME, Mohler JL, Edmonds K, *et al.* An updated review of the clinical development of coumarin (1,2-benzopyrone) and 7-hydroxycoumarin. *J Cancer Res Clin* 1994; **120**: 39–42.

75 Marshall ME, Albainy-Jenei S, Umerani A, Zhao Q. Direct cytotoxic effects of coumarin and 7-hydroxycoumarin (7-HC) on human malignant cell lines *in vitro*. *J Irish Coll Phys Surg* 1993; **22** (Suppl. 1): 71–2.

76 Marshall ME, Kervin K, Benefield C *et al.* Growth-inhibitory effects of coumarin (1,2-benzopyrone) and 7-hydroxycoumarin on human malignant cell lines *in vitro*. *J Cancer Res Clin* 1994; **120**: 3–10.

77 Moran E, Prosser E, O'Kennedy R, Thornes RD. The effect of coumarin and 7-hydroxycoumarin on the growth of human tumour cell lines. *J Irish Phys Surg* 1993; **22**: 41–3.

78 Siegers CP, Bostelmann HC. Effect of coumarin on cell proliferation in human tumour cell lines. *J Irish Coll Phys Surg* 1993; **22** (Suppl. 1): 47–8.

79 Prosser E, Cox D, O'Kennedy R, Carroll K, Van der Putten W. Effects of coumarins, haematoporphyrins and acridine orange on the viability and growth of Landshütz ascites tumour cells, in the presence and absence of photoradiation. *Cancer Lett* 1990; **52**: 71–7.

80 Seliger B. 7-Hydroxy-coumarin mediated induction of differentiation of human malignant glioblastoma cells. *J Irish Phys Surg* 1993; **22**: 44–6.

81 Seliger B, Pettersson H. 7-Hydroxycoumarin inhibits oncogene-induced transformation of murine fibroblasts. *J Cancer Res Clin* 1994; **120**: 23–7.

82 Zänker KS. Coumarin(s) interface between cell growth inhibition, expression of oncogenes, tumour cell locomotion and cell-to-cell communication. *J Irish Coll Phys Surg* 1993; **22** (Suppl. 1): 51–5.

83 Wattenberg LW, Lam LKT, Fladmoe AV. Inhibition of chemical carcinogen-induced neoplasia by coumarins and α-angelicalactone. *Cancer Res* 1979; **1979**: 1651–4.

84 Tseng A. Coumarin prevents tumourigenesis in MTV-ras transgenic mice. *J Irish Phys Surg* 1993; **22**: 49–50.

85 Ryan JJ, Ketcham AS, Wexler H. Warfarin treatment of mice bearing autochthonous tumours: effect of metastases. *Science* 1968; **162**: 1493–6.

86 Földi-Börcsök E, Bedall FK, Rahlfs VW. Die antiphlogistische und ödemhemmende Wirkung von Cumarin aus *Melilotus officinalis*. *Arzneimittel-Forsch* 1971; **21**: 2025–30.
87 Bolton T, Casley-Smith JR. An *in vitro* demonstration of proteolysis by macrophages and its increase with coumarin. *Experientia* 1975; **31**: 271–3.
88 Piller NB. An integration of the modes of action of coumarin. *Arzneimittel-Forsch* 1977; **27**: 1135–8.
89 Piller NB. Coumarin and high protein oedemas: relief for some or a cure for all—a story of molecular misunderstanding. *J Irish Phys Surg* 1993; **22**: 63–6.
90 Lee RE, Bykadi G, Ritschel WA. Inhibition of prostaglandin biosynthesis by coumarin, 4-hydroxycoumarin, and 7-hydroxycoumarin. *Arzneimittel-Forsch* 1981; **31**: 640–2.
91 Hardt TJ, Ritschel WA. The effect of coumarin and 7-hydroxycoumarin on *in vitro* macrophage phagocytosis of latex particles. *Meth Find Exp Clin Pharmacol* 1983; **5**: 39–43.
92 Paya M, Halliwell B, Hoult JRS. Peroxyl radical scavenging by a series of coumarins. *Free Radical Res Com* 1992; **17**: 293–8.
93 Greenspan HC, Aruoma OI. Oxidative stress and apoptosis in HIV infection: a role for plant-derived metabolites with synergistic antioxidant activity. *Immunol Today* 1994; **15**: 209–13.
94 Marshall ME, Rhoades JL, Mattingly C, Jennings CD. Coumarin (1,2-benzopyrone) enhances DR and DQ antigen expressions by peripheral blood mononuclear cells *in vitro*. *Mol Biother* 1991; **3**: 204–6.
95 Stuhlmeier K, Theyer G, Baumgartner G, Zlabinger GJ. Synergistic effect of coumarin (1,2-benzopyrone) and endotoxin in the induction of human interleukin-1. *Clin Exp Immunol* 1991; **84**: 317–23.
96 Zlabinger GJ, Böhmig G, Rosenkranz A, Stuhlmeier K. Intracellular signalling in coumarin-enhanced IL-1 production induced by lipopolysaccharide. *J Irish Coll Phys Surg* 1993; **22** (Suppl. 1): 38–40.
97 Ribi E, McLaughlin CA, Cantrell JL *et al.* Immunotherapy for tumors with microbial constituents or their synthetic analogues. A review. In: *Immunotherapy of Human Cancer*. New York: Raven Press, 1978: 131–48.
98 Zlabinger GJ, Nöhammer C, Böhmig GA, Menzel JE. Mode of action of coumarin in immune cells. *J Cancer Res Clin* 1994; **120**: 17–18.
99 Nair RV, Fisher EP, Safe SH, Cortez C, Harvey RG, DiGiovanni J. Novel coumarins as potential anticarcinogenic agents. *Carcinogenesis* 1991; **12**: 65–9.
100 Marshall ME, Riley LK, Rhoades J *et al.* Effects of coumarin (1,2-benzopyrone) and cimetidine on peripheral blood lymphocytes, natural killer cells and monocytes in patients with advanced malignancies. *J Biol Resp Modif* 1989; **8**: 62–9.
101 Brockmeyer NH, Kreuzfelder E, Bluhm C *et al.* Immunomodulation of cimetidine in healthy volunteers. *Klin Wochenschr* 1989; **67**: 26–30.
102 Berkarda B, Bouffard-Eyüboglu H, Derman U. The effect of coumarin derivatives on the immunological system of man. *Agents Actions* 1983; **13**: 50–2.
103 Omarbasha B, Fair WR, Heston WDW. Effect of coumarin on the normal rat prostate and on the R-3327H prostatic adenocarcinoma. *Cancer Res* 1989; **49**: 3045–9.
104 Maucher A, Kager E, von Angerer E. Evaluation of the antitumour activity of coumarin in prostate cancer models. *J Cancer Res Clin* 1992; **119**: 150–4.
105 Fentem JH, Fry JR. Species differences in the metabolism and hepatotoxicity of coumarin. *Comp Biochem Physiol* 1993; **104**: 1–8.
106 Fentem JH, Fry JR. Metabolism of coumarin by rat, gerbil and human liver microsomes. *Xenobiotica* 1992; **22**: 357–67.
107 Lake BG, Gaudin H, Price RJ, Walters DG. Metabolism of coumarin to polar and covalently bound products by hepatic microsomes from the rat, syrian hamster, gerbil and humans. *Food Chem Toxicol* 1992; **30**: 105–15.
108 Rosskopf F, Kraus J, Franz G. Immunological and antitumor effects of coumarin and some derivatives. *Pharmazie* 1992; **47**: 139–42.

109 Wood AW, Conney AH. Genetic variation in coumarin hydroxylase activity in the mouse (*Mus musculus*). *Science* 1974; **185**: 612–14.

110 Witworth P, Pak C, Esgro J, Kleinerman E, Fidler I. Macrophages and cancer. *Cancer Metastasis Rev* 1990; **8**: 319–51.

111 Somers SD, Johnson WJ, Adams DO. Destruction of tumor cells by macrophages: mechanisms of recognition and lysis and their regulation. In: Herberman RB ed. *Cancer Immunology: Innovative Approaches to Therapy.* Boston: M.P. Nijhoff, 1986: 69–122.

112 Männel DN, Moore RN, Mergenhagen SE. Macrophages as a source of tumoricidal activity (tumor-necrotizing factor). *Infect Immun* 1980; **30**: 523–9.

113 Schwamberger G, Flesch I, Ferber E. Characterization and partial purification of high molecular weight tumoricidal activity secreted by murine bone marrow macrophages. *Int Immun* 1992; **4**: 253–64.

114 Fidler IJ, Schroit AJ. Recognition and destruction of neoplastic cells by activated macrophages: discrimination of altered self. *Biochim Biophys Acta* 1988; **948**: 151–73.

115 Kleinerman ES, Schroit AJ, Fogler WE, Fidler IJ. Tumoricidal activity of human monocytes activated *in vitro* by free and liposome-encapsulated human lymphokines. *J Clin Invest* 1983; **72**: 1–12.

116 Fidler IJ, Jessup J, Ojofler W, Staerkel R, Mazumber A. Activation of tumoricidal properties in peripheral blood monocytes of patients with colorectal carcinoma. *Cancer Res* 1986; **46**: 994–8.

117 Fidler IJ, Murray JL, Kleinerman ES. Systemic activation of macrophages by liposomes containing immunomodulators. In: Hellman S, DeVita VTJ, Rosenberg SA eds. *Biologic Therapy of Cancer: Principles and Practice.* Philadelphia: J.B. Lippincott, 1991: 730–42.

118 Alving CR. Delivery of liposome-encapsulated drugs to macrophages. *Pharmacol Ther* 1983; **22**: 407–24.

119 MacEwen EG, Kurzman ID, Rosenthal RC *et al.* Therapy of osteosarcoma in dogs with intravenous injection of liposome-encapsulated muramyl tripeptide. *J Nat Cancer Inst* 1989; **81**: 935–7.

120 Murray JL, Kleinerman ES, Tatom JR *et al.* A pilot phase I trial of liposomal *N*-acetyl-muramyl-L-alanyl-D-isoglutaminyl-L-alanyl-phosphatidylethanolamine (MTP-PE, CGP 19835A) in cancer patients. In: Lopez-Berestein G, Fidler IJ eds. *Liposomes in the Therapy of Infectious Diseases and Cancer*, vol. 89. New York: Alan R. Liss, 1989: 329–42.

121 Urba WJ, Hatmann LC, Longo DL *et al.* Phase I and immunomodulatory study of a muramyl peptide, muramyl tripeptide phosphatidylethanolamine. *Cancer Res* 1990; **50**: 2979–84.

122 Kleinerman ES, Snyder JS, Jaffe N. Influence of chemotherapy administration on monocyte activation by liposomal muramyl tripeptide phosphatidylethanolamine in children with osteosarcoma. *J Clin Oncol* 1991; **9**: 259–67.

123 Kleinerman ES, Gano JB, Johnston DA, Benjamin RS, Jaffe N. Efficacy of liposomal muramyl tripeptide (CGP 19835A) in the treatment of relapsed osteosarcoma. *Am J Clin Oncol* 1995; **18**: 93–99.

124 Kleinerman ES, Raymond AK, Bucana A *et al.* Unique histological changes in lung metastases of osteosarcoma patients following therapy with liposomal MTP-PE. *Cancer Immunol Immunother* 1992; **34**: 211–20.

125 Asano T, McWatters A, An T, Matsushima K, Kleinerman ES. Liposomal muramyl tripeptide up-regulates interleukin-1 alpha, interleukin-1 beta, tumor necrosis factor-alpha, interleukin-6 and interleukin-8 gene expression in human monocytes. *J Pharmacol Exp Ther* 1994; **268**: 1032–9.

126 Kleinerman ES, Jia SF, Griffin J, Seibel NL, Benjamin RS, Jaffe N. Phase II study of liposomal muramyl tripeptide in osteosarcoma: the cytokine cascade and monocyte activation following administration. *J Clin Oncol* 1992; **10**: 1310–16.

127 Fujimaki W, Griffin JR, Kleinerman ES. Effect of ibuprofen on monocyte activation by liposome-encapsulated muramyl tripeptide phosphatidylethanolamine (CGP 19835A): can ibuprofen reduce fever and chills without compromising immune stimulation? *Cancer Immunol Immunother* 1993; **36**: 45–51.

128 Andreesen R, Hennemann B. Adoptive immunotherapy with autologous *ex vivo* generated macrophages—current status and future perspectives. *Pathobiology* 1991; **59**: 259–63.

CHAPTER 6

The Mode of Entry of Coumarin into Cells and its Effects on Cell–Cell Communication and Migration

Kurt S. Zänker
Institute of Immunology, University Witten/Herdecke, Witten, Germany

6.1 INTRODUCTION

The modern era of cancer chemotherapy began in the 1940s with the demonstration that nitrogen mustard showed anticancer activity in malignant lymphoma. At about the same time the coumarin(s) were chemically described[1] as a group of natural compounds, present in a wide variety of vegetables and fruits consumed by humans and animals. (In the following discussion, 'coumarin(s)' refers to the compounds 1,2-benzopyrone and 7-hydroxycoumarin. If not stated otherwise, the experimental results and the mode of actions are almost identical for both substances.) Such chemicals in vegetables, fruits and grains may have significant effects on development and growth of cancer in humans and animals. Research in this area is yielding exciting information on the role of phytochemicals in cancer prevention and treatment, when introduced into adjuvant protocols.

In recent years we have accumulated epidemiological, biological and clinical evidence that a high proportion of apparently early cancers have already produced occult metastatic deposits[2] by the time the primary tumour is diagnosed. Numerous clinical trials are investigating adjuvant systemic therapy[3] in the hope of delaying

Coumarins: Biology, Applications and Mode of Action. Edited by R. O'Kennedy and R.D. Thornes.

clinical recurrence, and, thus, of increasing the duration of survival in patients suspected to have occult spread of cancer.

Available antimitotic and cytotoxic agents are poorly selective and are handicapped by the biological heterogeneity and slow doubling rates characteristic of most solid tumours. As a result of these limitations, we cannot depend on existing cytotoxic agents by themselves to destroy the last remaining ‘clonogenic’ cancer cell in the patient.

In cancer cells it is not just the uncontrolled proliferation but also the capacity of cancer to invade adjacent tissue, release viable cells, lodge those cells at distant sites and establish metastases there, which cause the majority of deaths in cancer. Recent research shows that successful metastases do not result from random survival of cells released from the primary lesion, but are likely to originate in selected clusters of migrating cells from it[4]. Biphasic survival curves have been described for the major types of solid cancer and suggest the existence of at least two biological subsets among patients, one with a highly malignant course and the other with slowly growing disease or delayed metastases[5].

The purpose of adjuvant therapy of cancer is to increase the efficacy of primary surgery, radiation or chemotherapy in order to eradicate or prolong the arrest of hidden metastases of cancer and, hence, to extend disease-free survival. Therapeutic adjuvant approaches have generally focused on the use of cytotoxic, endocrine or biologic agents. In the future, our adjuvant systems may also include drug resistance reversing agents, modulators of inter- and intracellular chemical signal molecules or specific cell migration inhibition compounds. There is an urgent need for agents with minimal side-effects but with significant antitumour activity for particular cancers which might be effective in preventing or delaying the formation of metastases. The oral non-anticoagulative coumarin-type molecules are (among others[6]), promising candidates for adjuvant treatment strategies to arrest or delay metastases formation.

Coumarin and its 7-hydroxy derivative have been studied for their binding properties to tumour cell surfaces/membranes *in vitro*, for their ability to induce gap junction communication and for their ability to inhibit cell proliferation and migration in cells dependent on the overexpression of oncogenes. We report here on the experimental progress made with coumarins focusing on the abovementioned cellular and subcellular parameters.

6.2 CELL SURFACE AND SIGNAL TRANSDUCTION

What happens when the first messenger has crossed the synaptic cleft, the bloodstream or the extracellular fluid, and arrives at its target cell? The interaction of the messenger with the target cell can be highly specific, with even closely related compounds (including stereoisomers) being ineffective, and the messenger can often act at extremely low concentrations, in the nanomolar or even picomolar range. It has always seemed likely that mainly proteins could bind messengers with sufficiently high affinity and specificity to account for processing extracellular signals. A protein which binds a first messenger, and is involved in mediating its biological effects, is known as a receptor. For most first messengers acting as a ligand, the available evidence favours the idea that the receptor domain is located on the outside surface of the cell, that an intramembranous domain spans through, anchors the receptor within the cell membrane and determines the lateral diffusion capacity of the receptor within the cell membrane, whereas an intracytoplasmic part exerts its biological effect.

To start with the hypothesis that coumarin might act as a first messenger on different cell types, such as natural killer cells, T lymphocytes and human tumour cells, we attempted to measure interaction (binding) of radioactively labelled coumarin (ligand) with intact target cells. The binding is taken to reflect a 'receptor interaction' if it exhibits: (a) structural and steric specificity; (b) saturability, indicating a finite and limited number of binding sites; (c) tissue specificity and (d) reversibility. Even when all these criteria are met, the data can be difficult to interpret unless the binding measurements can be evaluated in a system where biological activity can be simultaneously monitored. In this respect we tried to combine the results of several readout systems in order to show biological and clinical relevance for coumarins.

A final concentration of 2×10^6 tumour cells (freshly isolated human melanoma cells; human amelanotic cells; mammary carcinoma and adenocarcinoma of the colon) were equilibrated with increasing amounts of radioactively labelled coumarin. Aliquots were taken from the separated supernatant and pellets to determine radioactivity. Non-specific binding was measured in the presence of an excess of unlabelled coumarin. The binding data were corrected by subtracting the fraction of radioactivity non-specifically bound from the total bound to yield specific binding and plotted according to Scatchard[7]. At low coumarin concentrations the plot was linear and steep, and at high coumarin concentrations it becomes practically horizontal. This indicates that the number of first class binding sites is limited and the affinity high. The capacity of the second class of binding sites is high, but the affinity is low. Metastatic breast carcinoma cells and a metastasis-derived amelanotic cell line showed high-affinity binding properties, whereas an adenocarcinoma raised from a human colorectal primary tumour exhibited low affinity to coumarin[8].

The binding properties of coumarin as a first messenger to target cells are enigmatic. Some cells have high affinity ($K_a = 1.5 \times 10^{-10}\ \mathrm{M}^{-1}$) whereas other tumour cells reveal low affinity but high capacity. From these experiments it can be concluded that coumarin binds as a first messenger to cell surface structures, qualified as receptors by their high coumarin affinities and, secondly, by alteration of cellular effects (proliferation, migration). The low affinity and high capacity of coumarin shown in binding experiments is not necessarily surprising: it is well known that the oestrogen receptor also exhibits two types of binding sites, with type I sites representing the classical oestrogen receptor site which binds oestrogen with high affinity and low capacity, and type II sites binding oestradiol with low affinity but increasing capacity[9].

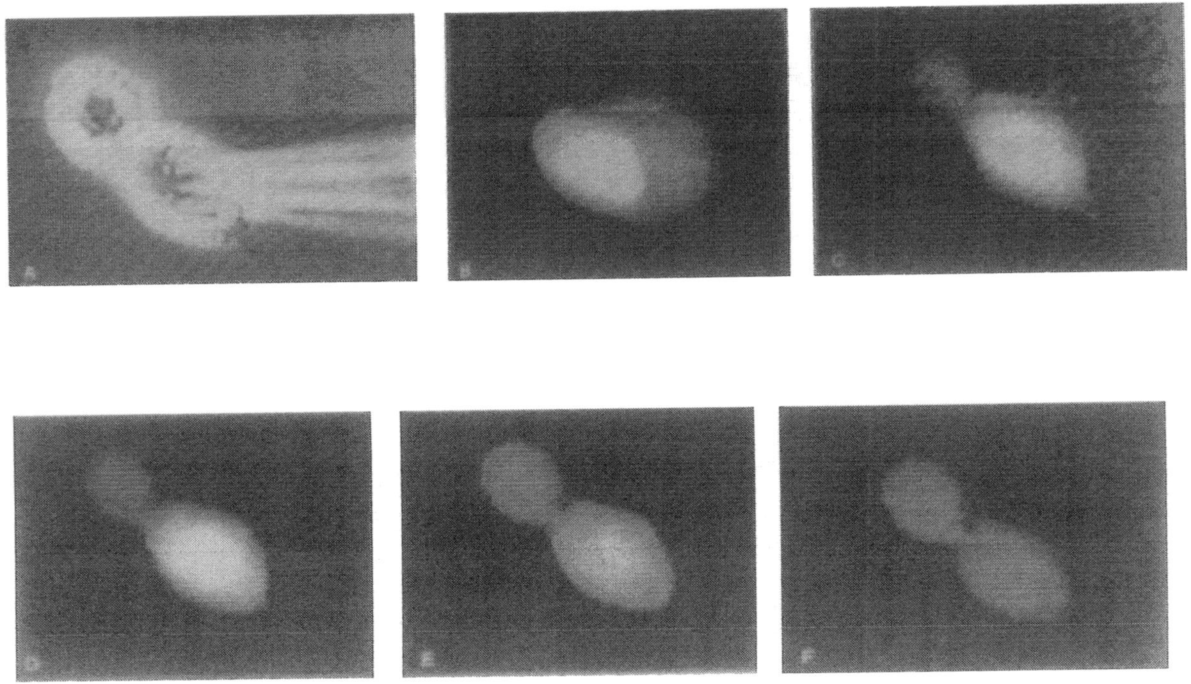

Figure 6.1. Serial documentation of lateral homotypic dye spread to a neighbouring tumour cell (RBA cell line). The dye transfer is depicted in a time frame of about 15 s. (A) The fluorescent dye Lucifer Yellow is microinjected into one cell. (B) The microinjection needle is retracted and there is only faint nonspecific fluorescence around the injected cell. (C) First appearance of pale fluorescence in the adjacent cell next to brightly fluorescent, adherent tumour cell. (D, E) The fluorescent dye outlines the confine of the adjacent cell, which was still invisible in (B). (F) The fluorescent area is equally distributed within the two tumour cells and the fluorescence is significantly weaker in comparison to the single cell injected (B–D).

The uptake of coumarin by target cells seems to be mediated on the one hand by surface membrane structures that show a high affinity for and are saturable by coumarin, and, on the other hand, in certain tumour cells, by the phospholipid composition of the plasma membrane. It is well established that lipophilic compounds, like steroids, steroid-structured molecules and retinoic acid, can rapidly enter the cell and accumulate in the nucleus. Considering the molecular structure of coumarin(s), it is reasonable that the coumarin(s) should not only bind to cell surface structures but also pass the plasma membrane by lipophilic mechanisms due to the qualitative and quantitative phospholipid–protein composition for a given cell membrane. Endocytosis of membrane-associated coumarin molecules is likely to be one way of entry into mammalian cells.

The different modes of actions on and entrance into a cell, as a first messenger, might explain different signal transduction processing mediated by coumarin(s). Basically, there are three mechanisms for signal transduction by cell surface receptors: (a) the messenger causes opening of ion channels; (b) the messenger is coupled to a second messenger generating enzyme activity via a G protein; and (c) the receptor has a messenger-activated enzyme as its cytoplasmic domain. Research in our laboratory indicates that a rat breast adenocarcinoma cell line that overexpresses the *ras* oncogene can be slowed down in its proliferation capacity when incubated with the 7-hydroxycoumarin. Using Western blotting we have demonstrated with these cells that when coumarin(s) are present permanently in the cell culture medium the *ras* p21 protein is decreased and, concomitantly, has almost lost its GTPase activity, which shifts and arrests the *ras* G protein signal cycle in an off state[10]. On a molecular transcription level we have described a diminished 5 kb *ras* transcript and an intensified 1.8 kb *ras* transcription product signal in the presence of 7-hydroxycoumarin[11].

The decreased protein p21 level and altered *ras* mRNA expression found might explain the impaired GTPase function of p21 GTP-binding protein, and raises the intriguing possibility that further intracellular growth signal processing is almost abandoned, leading to inhibition of cell proliferation and migration.

6.3 CELL–CELL COMMUNICATION AND COUMARIN

The ordered growth and functioning of animal cells requires the transfer of information from one cell to another, i.e. cell–cell communication. In multicellular organisms the labour is divided between differentiated cells which are specialised for particular purposes and close cooperation is required between the various cell types in order for the organisms to function efficiently. The information comes from a bewildering variety of sources, both within and outside of the cell. Hormones, neurotransmitters, growth factors and extracellular metabolites interact with specific receptors on the plasma membrane or in the cytoplasm. Inside the cell, information is relayed by intracellular messengers including cyclic nucleotides, inositol

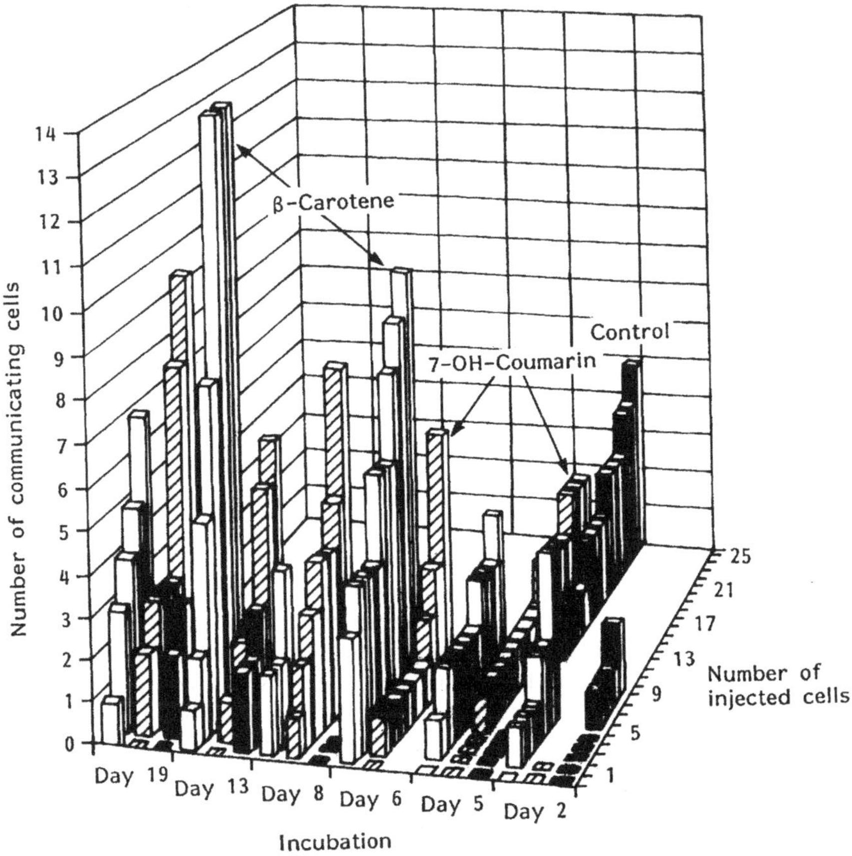

Figure 6.2. Augmentation of gap junction communication within a human tumour bladder carcinoma cell line (T24) raised in the presence of *β*-carotene and 7-hydroxycoumarin. The black bars represent the control experiments, omitting *β*-carotene or 7-hydroxycoumarin. The shaded bars indicate the 7-hydroxycoumarin and the white bars the *β*-carotene supplementation in the medium. When RBA cells were grown for 19 days in the presence of either *β*-carotene or 7-hydroxycoumarin, the cell lines increased the gap junctional communication within the incubation period. The number of injected cells denotes the experiments simultaneously carried out within one cell culture, the number of communicating cells defines those cells which have taken up the dye from one single injected cell.

polyphosphates, diacylglycerol and calcium, GTP-binding proteins, protein kinases, and DNA-binding proteins. Finely balanced interactions between these different signalling pathways ensure that the required response occurs, and abnormalities in any of the components of the system can lead to a variety of diseases, including cancer.

Signalling between cells of multicellular organisms appears to occur by one of three general mechanisms. First, cells that are in direct contact with each other can

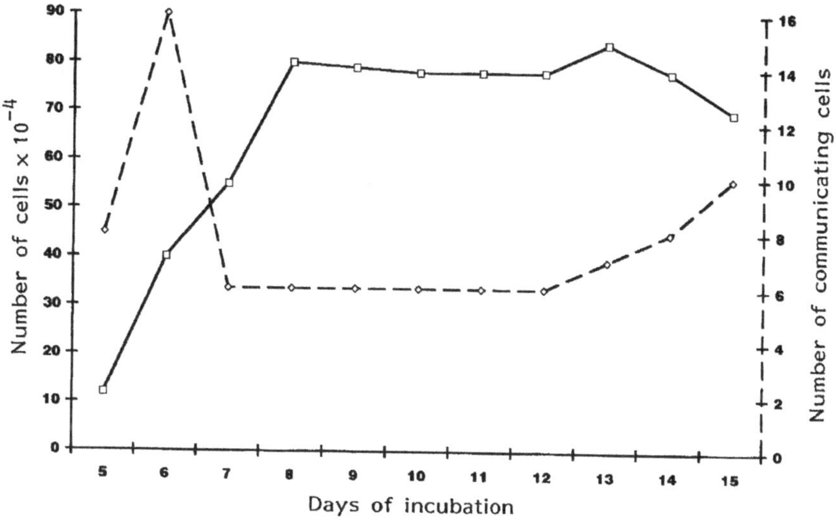

Figure 6.3. An example, obtained with human fibroblasts, showing the correlation between cell proliferation and gap junctional communication. In log-phase growing cultures (solid line) the communication behaviour (dashed line) of the cells rapidly increases within two days (days 5 and 6) and in the plateau phase (days 7–12) cell communication decreases and stays at this level until the next medium change is carried out. The medium was changed on days 5 and 12. Conditioned medium on day 12 induces a slight augmentation of gap junctional communication.

communicate via cytoplasmic bridges, which allow intracellular signal molecules to pass from one cell to the next without being secreted into the extracellular fluid. In higher animals the bridges are known as gap junctions. Secondly, cells can communicate via extracellular signal molecules, which are secreted by one cell and can cause response to another cell. The mode of action involved may be endocrine, paracrine or autocrine. A third possibility—whereby integral surface molecules on one cell could signal, perhaps via a change in conformation, to receptor molecules on a neighbouring cell—is called juxtacrine communication.

Gap junctions were so named because when seen in cross-sections, they appeared as areas where the plasma membranes of the neighbouring cells were closely apposed, but with a very narrow (2–4 nm) gap of low electron density remaining between them. Use of marker dyes with different molecular characteristics showed that this passage of molecules was relatively non-specific, but only occurred with molecules up to about 1000 Da.

In general, it is said that in tumour cells metastases formation is related to gap junction communication[12]. Tumour cells that are candidates for shedding and spreading out as metastatic cells from the primary neoplasm downregulate intercellular communication via gap junctions. Several factors have been shown to decrease the permeability of gap junctions, including a lowering in pH, and a rise in cytosolic Ca^{2+} ion concentration. Connexins, the subunits of the connexon (gap junction), can be covalently modified by phosphorylation, for example by cyclic AMP-dependent protein kinase, and thereby regulate the opening and closing of the gap junction channel. These are biochemical conditions that can be considered to exist in one or the other form within a growing tumour mass. There exist several lines of evidence that morphogens (retinoic acid) and functionally related compounds (carotenes, carotenoids) can re-establish gap junction communication in otherwise non-communicating (tumour) cells[13].

Having shown that coumarin binds in part in a saturable mode to plasma membranes, we further investigated the possibility that coumarin(s) can enhance gap junction formation and might, therefore, be promising candidates for re-differentiation.

Selective communication between a single tumour cell and surrounding tumour cells was measured by microinjecting a fluorescent dye into a single tumour cell and watching the dye transfer to neighbouring tumour cells under a fluorescence microscope. A tumour cell near the border of a focus was microinjected with a 10% (w/v) solution of Lucifer Yellow CH in a 0.33 M lithium chloride solution by means of a microinjector apparatus. The injection pressure was 40 hPa and the injection time 0.1 ms. Transfer of the dye to the surrounding tumour cells was monitored in the presence or absence of 7-hydroxycoumarin under a Leitz fluorescence phase-contrast microscope and recorded photographically. As an internal standard β-carotene

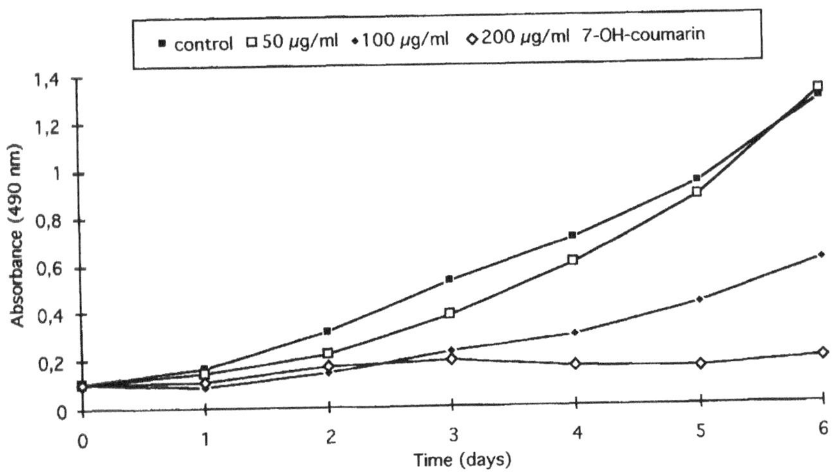

Figure 6.4. Growth inhibition of RBA cells in the presence of increasing concentration of 7-hydroxycoumarin (50–200 µg/ml) *in vitro*. The compound is likely to be more cytostatic than cytotoxic, because at these pharmacologically used dosages the cell cultures are arrested in a non-proliferative state. With increasing incubation time, however, the cells show pigmentation, build up large intracytoplasmic vacuoles, and the trypan blue dye is rapidly taken up, indicating a process of cell death.

was used, because it is well established that this compound is highly selective to induce gap junction communication[14].

A series of experiments were performed with two human cell lines that have been established in our laboratory[11] and with two oncogenic *ras*-driven cell lines (RBA; T24). The number of communicating tumour cells increased from day 2 to day 19 when 7-hydroxycoumarin (100 μg/ml) was freshly replaced every second day during the incubation period. Control cells were similarly treated in respect to medium change with the exception that 7-hydroxycoumarin was not supplemented. However, no significant change in gap junction communication could be observed. Cells growing in the presence of *β*-carotene showed the most pronounced effect in respect to the induction of gap junction communication. Despite the long incubation time in culture (19 days) the dye transfer was generally limited to tumour cells that were in direct contact with neighbouring cells. Cytoplasmic dye transfer between the tumour cells was inhibited by the addition of 3.5 mM of 1-heptanol into the culture medium. This classic inhibitor of gap junction communication impeded transfer of more than 80%. This finding strongly suggests that dye transfer between the tumour cells was achieved by gap junctional channels that could be induced by *β*-carotene, and to a lesser extent by 7-hydroxycoumarin, and was limited to molecules smaller than 1000 Da.

6.4 CELL PROLIFERATION AND COUMARIN

As described in other chapters, there are numerous reports that coumarin(s) cause inhibition of cell growth within an array of tumour cells *in vitro* and *in vivo*. However, if an agent shows pleonastic modes of action its precise molecular target needs to be identified. Therefore, the potential of coumarin(s) to inhibit the proliferation of oncogene-driven cells was investigated.

Cancer can be initiated by agents that cause mutation of DNA, such as mutagenic chemicals, ionizing radiation, ultraviolet light and cigarette smoke. Since a tumour is normally of clonal origin (i.e. it arises from multiple division from a single cell) the idea arose that cancer was due to a mutation in that progenitor cell. Mutations do occur in germ cells and can be passed on to the next generation. Mutations also occur in non-germ cells (somatic cells), but in contrast to mutated germ cells, a harmful mutation in a somatic cell is usually not noticed, because that cell merely dies and neighbouring cells take over its functions. However, things would be different if the mutation caused the cell to shut down the apoptosis programme and to divide continuously. It is now becoming clear that cancer can be caused by mutation or altered expression of key genes, the products of which are involved in regulation of cell growth. These mutated, cancer-causing genes are known as oncogenes.

Using different cell proliferation assays (dye exclusion, MTT colorimetric assay [^{3}H]thymidine incorporation) it was shown that coumarin and its 7-hydroxy derivative inhibit the growth of a rat mammary adenocarcinoma line (RBA, ATCC, CRL

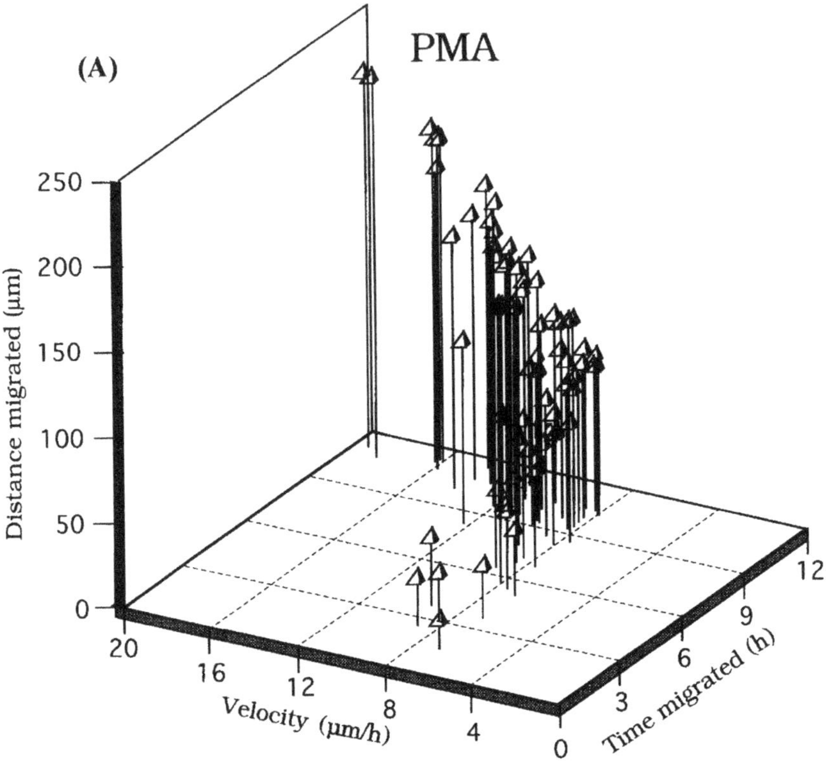

Figure 6.5. Activation (A) or inhibition (C) of the motile machinery of a fraction of cells by PMA and 7-hydroxycoumarin of RBA cells as compared to controls (B). In an unsynchronised culture most of the cells can be modulated in their migration behaviour in respect to the parameters 'time migrated', 'distance migrated' and 'velocity'. The 3-D graphs show the results of three independently performed experiments. Each bar represents the path of one single cell: at least 120 cells were followed up for the indicated parameters. PMA clearly activates and 7-hydroxycoumarin inhibits the motile machinery of RBA cells. (For B and C; see following pages.)

1747) and a human bladder carcinoma cell line (T 24, ATCC, HTB 4) in a concentration- and time-dependent fashion. Both cell lines are characterised by their growth dependency on the overexpression of the H-*ras* oncogene.

In H-*ras* driven tumour cells the mechanism of action can be explained in more detail. Coumarin(s) may act not only as simple cell surface binding compounds, but they may induce changes in signal transduction, via GTP-binding proteins, thus de-amplifying growth signals, which finally leads to growth arrest in the cell culture.

There is some evidence that potentiation of interleukin-1 (IL-1) release from human mononuclear cells by coumarin involves signal transduction pathways similar to the primary stimulus of lipopolysaccharide (LPS)[15]. LPS is a classical example of

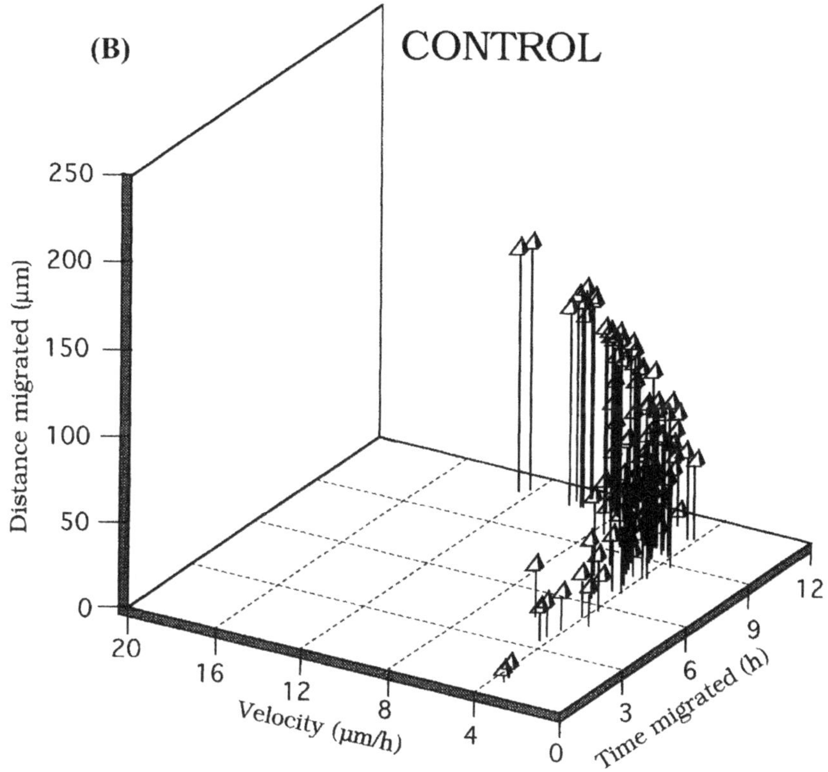

Figure 6.5. (*continued*)

a first messenger. After binding to cell surface proteins, it activates phospholipase C, followed by the generation of diacylglycerol and inositol-1,4,5-triphosphate, released from phosphatidylinositol-4,5-biphosphate. The second messengers induce translocation of protein kinase C from the cytosol to the plasma membrane or release of Ca^{2+} from the endoplasmic reticulum. One enzyme involved in the phosphoinositide breakdown, phosphoinositidase C, appears to be coupled to G proteins. Coumarin closes a circulatory and operational mode, starting from cell surface binding and modifying downstream signals for growth regulation (e.g. cell growth inhibition). The low specificity distinguishes them from very specific ligands, like peptide domains of proteins. This may also reflect the relatively high coumarin(s) concentrations needed to obtain a biological response *in vitro*. Arbitrary diffusion coefficients are determined by the phospholipid composition of the plasma membrane diffusion barrier for coumarin(s); a diffusion gradient between the outside and the inside of the cell is established as one driving force of coumarin entry into the cell. Coumarin(s) are rapidly metabolised (e.g. glucuronidated) and/or find intracellular

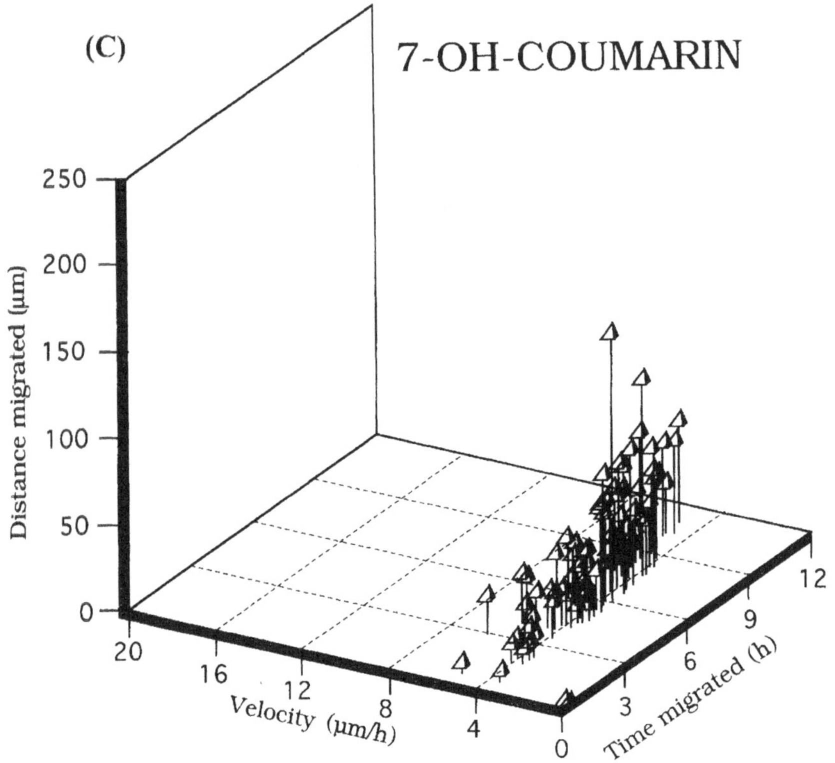

Figure 6.5. (*continued*)

targets. Therefore, a steep outside/inside directed gradient is formed as long as coumarin(s) are present in excess in the medium.

6.5 CELL MIGRATION AND COUMARIN

The process which a tumour cell must undergo to form a metastasis is complex, involving escape from the primary tumour, movement to a new location and establishment of growth at the new site. The various steps associated with this process are called the metastatic cascade and one mandatory prerequisite is cell movement; without cell movement, no distant metastasis formation is possible. One of the largest remaining hurdles of experimental and clinical cancer research is our quest to understand why and how tumour cells develop a greater autonomy by movement. Obviously there is an extensive interconnection between cytoplasmic signalling pathways and the motile machinery: a membrane receptor activates a G protein that activates several different targets and is itself triggered by a number of other membrane receptors. Different protein kinases act on cytoskeletal proteins, phosphorylating them sometimes singly, at other times in combinations.

Many signals travel from the cell surface to the motile machinery along many parallel pathways. Such a parallel distributed network generates different commands for movement, which allow the cells to discriminate between different environmental influences and to respond to each influence by generating a distinct type of motile behaviour; perhaps the molecules of cell signalling and cell motility are 'trained' by evolution and repeated clonal expansion. The form-shaping episodes of embryogenesis depend on coordinated migrations and are controlled ultimately by genetic mechanisms, by the executive expression of genes in response to spatial or temporal cues. Immune surveillance also depends on the highly coordinated migration and communication of immune competent cells throughout the body in order to fight successfully against foreign invaders and to guarantee survival of the host.

Embryogenesis and immune competence are examples of physiologically controlled cell movements for the sake of an organism. Metastasis, however, is a deteriorating cell movement directed towards the destruction of an organism. It is no surprise that people try to understand the motile machinery in a physiological state in molecular terms in order to learn about inputs which convert the motile machinery into a pathological state.

Our laboratory is involved in studies to elucidate cellular and subcellular mechanisms of motility in immune competent cells[16,17] (e.g. cell movement of T lymphocytes and antigen-presenting cells). Coumarin markedly decreases tumour incidence of an MTV-EJ-*ras* fibroblast cell line, when injected subcutaneously in syngeneic Fisher rats[18]. Furthermore, coumarin prevents tumorigenesis in MTV-H-*ras* transgenic mice[19]. We investigated this hypothesis by means of *ras*-driven oncogenic cells, the RBA and the T24 tumour cell lines. The method followed published work[20] with the exception that these tumour cell lines, instead of lymphocytes, were used. Tumour cells were monitored for cell displacement in the presence and absence of coumarin(s) in the medium. A positive control was carried out by incubating the cells with the tumour promoter phorbol myristoyl acetate (PMA). PMA is a product of the Euphorbiaceae family of higher plants and mimics the effect of diacylglycerols on the activation of protein kinase C.

Both cell lines show a very low activity of GTPase, observed even in the presence of GTPase activator protein. This means that the *ras* p21 protein remains permanently in the active *ras* GTP complex and constitutively activates the production of growth signals, which also involves the cycle of protein kinase C phosphorylation and dephosphorylation, which in turn spawns a plethora of secondary responses of tyrosine and serine/threonine kinases.

The motile machinery of a cell is triggered by a complex network of activated and de-activated protein kinases, acting on cytoskeletal proteins. PMA, activating protein kinase C, can be expected to act as a promoter of cell motility and, therefore, serves as a positive control compound. Indeed, it was clearly demonstrated that PMA activated cell locomotion of the *ras*-driven oncogenic cell lines RBA and T24.

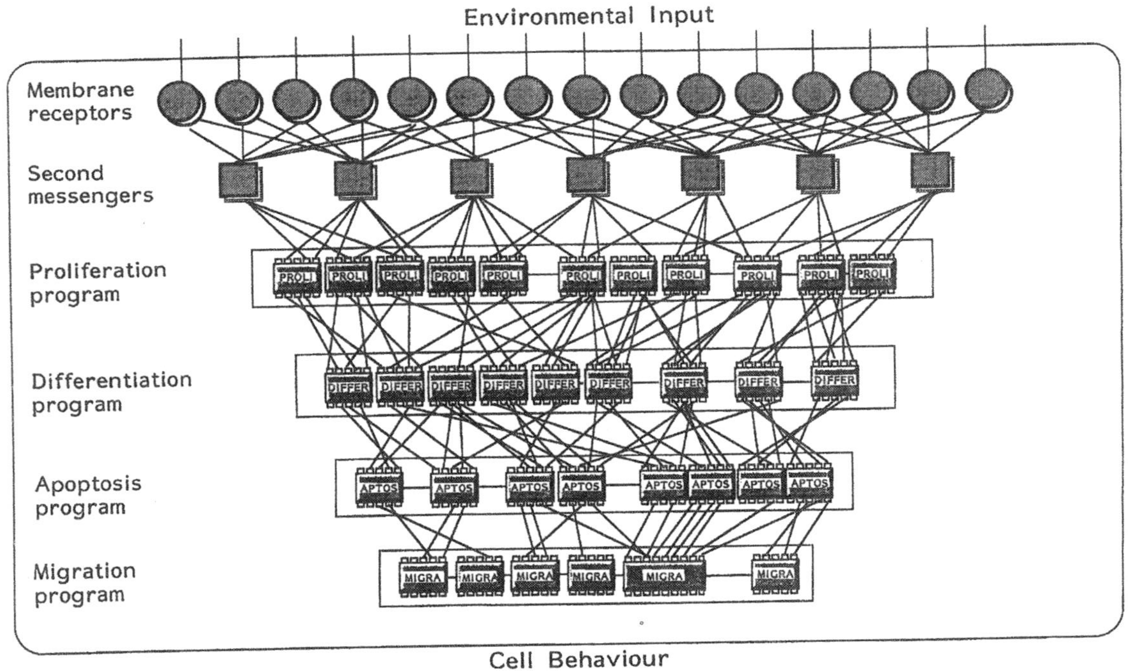

Figure 6.6. The physiology of a eukaryotic cell seen as a parallel distributed network. Signals received on the outside of the cell by specific membrane receptors or metabolic changes within the cell membrane induced by incoming drugs cause changes in the concentration of intracellular messengers, including Ca^{2+}, phosphatidylinositol biphosphate (PIP_2), cAMP, G proteins, protein kinases and intracellular proteases. These messengers then regulate the proliferation, differentiation, apoptosis and migration programs. These different programs are interconnected by molecular activities through allosteric interactions, by post-translational modifications or by degradation of constitutive proteins. Finally the subcellular assemblies act together to generate a particular sequence of growth inhibition (proliferation program), gap junction induction (differentiation program) and migration inhibition (migration program) as shown for coumarin(s). As cross-talks exist at every program level it is not such a striking feature that coumarin(s) induce(s) various cell responses depending on which read-out system the investigator is looking at.

Coumarin(s) clearly suppressed a major portion of these tumour cells in their locomotory behaviour. These results are in agreement with work on a coumarin-related molecule, warfarin, in which it was seen that the locomotory behaviour of V2 cancer cells in rabbits is blocked[21].

6.6 CONCLUSION

For a couple of years, we mainly investigated the biological responses of oncogenic *ras*-driven tumour cells in respect to proliferation, cell–cell communication and locomotory behaviour after exposure to coumarin and 7-hydroxycoumarin, the main metabolite in humans of 1,2-benzopyrone. We could demonstrate that both substances, almost at equimolar concentrations, inhibit cell proliferation. The effect is rather more a cytostatic than a cytotoxic one, because in long-term experiments coumarin has to be freshly supplemented to the medium in order to maintain arrest of tumour cell growth.

Coumarin(s) can also induce cell–cell communication in otherwise impaired intercellular communication in tumour cell cultures. Oncogenes can activate parallel pathways of intercellular communication. Transfection of an established culture of fibroblasts, such as NIH-3T3 fibroblasts, with a single gene of the *ras* family, leads to dramatic changes in cell growth, in gap junction numbers and cell structure. Cells shut down intercellular communication, continue to proliferate in the absence of mitogens and show altered morphology. The synthesis of relatively large amounts of the oncogenic *ras* protein and the actions of this mutant protein dominate the functions conducted by normal *ras* proteins. We have shown that coumarin(s) decrease the amount of *ras* proteins in *ras* overexpressing tumour cells, increase cell–cell communication and inhibit cell growth. Although these results are in line with those reported from *ras*-transfected NIH-3T3 cells, future research on intercellular communication has to reveal new information on steps making up individual pathways from an external stimulus and to define intracellular structures, including the nature of all the genes that influence gap junction protein expression and regulation. Coumarin(s) seem to play a major role because of their activity on *ras* protein expression.

A further interesting effect of coumarin(s) is their ability to inhibit migration. Active cell locomotion is a mandatory prerequisite for tumour cells that are candidates for metastasis formation. The *ras* gene family is also integrated in the signalling pathway for activating the motile machinery. As already mentioned, coumarin(s) modulate the amount of p21 *ras* protein and the functional state of the GTPase; it is therefore no surprise that locomotory behaviour is inhibited.

A number of different protein kinases act on many different proteins, which command cell proliferation, apoptosis or migrations programs. It is not unlikely that coumarin(s) are highly conserved plant-derived non-peptide-type binding domains for cellular targets. The net outcome in cellular response can be proliferation inhibition, enhanced gap junction communication or decrease in cell motility. It is well established that for the opening or closing of gap junction communication the actual proliferation status (proliferation/plateau phase) of a cell within the cell culture determines its susceptibility to a given drug. This differs in no way from the mode of action of most cytotoxic drugs in clinical use, which show activity either in G_1/S or in M phase. Because of their molecular activities on nuclear targets[18,19] and on oncogene products[10,11], coumarin(s) can be regarded as future drugs for somatic gene therapy.

ACKNOWLEDGEMENT

The author is supported by Matthias Lackas-foundation, Germany.

REFERENCES

1 Murray RDH, Mendex J. Brown SA. *The Natural Coumarins: Occurrence, Chemistry and Biochemistry*. New York: John Wiley, 1983.

2 Wheelock EF, Brodovsky HS. Dormant cancer. In: Stoll BA ed. *Prolonged Arrest of Cancer. New Horizons in Oncology*, vol. 1, Chichester: John Wiley, 1982: 87–101.

3 Salmon SE. Adjuvant therapy of cancer. In: Salmon SE ed. *Proceedings of the Seventh International Conference on the Adjuvant Therapy of Cancer*. Philadelphia: J.B. Lippincott, 1993.

4 Friedl P, Noble PB, Walton PA, Laird DW, Chauvin PJ, Tabah RJ, Black M, Zänker KS. Migration of coordinated cell clusters in mesenchymal and epithelial cancer explants *in vitro*. *Cancer Res* 1995; **55**: 4557–60.

5 Blackwood JM, Seelig RF, Hutter RVP, Rush BF. Survival distribution in breast cancer. *Surgery* 1977; **82**: 443–8.

6 Niederprüm HJ, Schweikert HU, Thüroff JW, Zänker KS. Inhibition of steroid 5-alpha-reductase activity by aliphatic fatty acids. Candidates for chemoprevention of prostate cancer. *Ann NY Acad Sci* 1995; **768**: 227–30.

7 Scatchard G. The attractions of proteins for small molecules and ions. *Ann NY Acad Sci* 1949; **51**: 660–4.

8 Zänker KS, Blümel G, Lange J, Siewert JR. Coumarin in melanoma patients: An experimental and clinical study. *Future Trends Chemother* 1985; 473–80.

9 Zänker KS, Schmitz U, Schwarte A. Computer-assisted molecular design as a guide for determination of structure-activity relationships for estrogen receptor substrates by means of a model receptor (human-alpha$_1$-antitrypsin). In: Zeller WJ, D'Incalci M, Newell DR eds. *Novel Approaches in Anticancer Design. Molecular Modelling—New Treatment Strategies. Contrib. Oncol.* vol. 49 Basel: Karger, 1995: 50–9.

10 Lü HQu, Zänker KS. Decreased cell growth and expression of p^{21} H-ras in RBA cell treated by 7-OH coumarin. Abstr. Commun. Intracellular Signalling from RAS to Genes. Iowa State University, 9th MDCB/ISU Symposium, 1994: 35.

11 Zänker KS. Coumarin(s) interface between cell growth inhibition, expression of oncogenes, tumour cell locomotion and cell-to-cell communication. *Irish Coll Physic Surg* 1993; **22**: 51–5.

12 El-Sabban ME, Pauli BU. Cytoplasmic dye transfer between metastatic tumour cells and vascular endothelium. *J Cell Biol* 1991; **115**: 1375–82.

13 Bertram JS. Structure activity relationship among various retinoids and their ability to inhibit neoplastic transformation and to increase cell adhesion in the C3H/10T1½ CL8 cell line. *Cancer Res* 1980; **40**: 3141–6.

14 Matthews-Roth MM. Anti-tumour activity of β-carotene, canthaxanthin and phytoene. *Oncology* 1982; **39**: 33–7.

15 Zablinger G, Böhmig G, Rosenkranz A, Stuhlmeier K. Intracellular signalling in coumarin-enhanced IL-1 production induced by lipopolysacharide. *Irish Coll Physic Surg* 1993; **22**: 38–43.

16 Friedl P, Noble PB, Shields ES, Zänker KS. Locomotor phenotypes of unstimulated CD45RA^{high} and CD45RO^{high} CD4^{+} and CD8^{+} lymphocytes in three-dimensional collagen lattices. *Immunology* 1994; **82**: 617–24.

17 Friedl P, Noble PB, Zänker KS. T lymphocyte locomotion in a three-dimensional collagen matrix. Expression and function of cell adhesion molecules. *J Immunol* 1995; **154**: 4973–85.

18 Tseng A, Lee WMF, Kirsten E, Hakam A, McLick J, Buke K, Kun E. Prevention of tumorigenesis of oncogene-transformed rat fibroblasts with DNA site inhibitors of poly(ADP ribose) polymerase. *Proc Natl Acad Sci USA* 1987; **84**: 1107–11.

19 Tseng A. Coumarin prevents tumorigenesis in MTV-H-ras transgenic mice. *Irish Coll Physic Surg* 1993; **22**: 49–50.

20 Friedl P, Noble PB, Zänker KS. Lymphocyte locomotion in three-dimensional collagen gels. Comparison of three quantitative methods for analysing cell trajectories. *J Immunol Methods* 1993; **165**: 157–65.

21 Thornes RD, Edlow DN, Wood S Jr. The inhibition of locomotion of cancer cells *in vivo* by anticoagulation therapy: Effects of sodium warfarin on V2 cancer cells, granulocytes, lymphocytes and macrophages in rabbits. *Johns Hopkins Med J* 1968; **123**: 305–16.

CHAPTER 7

Coumarin in the Treatment of Lymphoedema and Other High-Protein Oedemas

John R. Casley-Smith and Judith R. Casley-Smith
The Henry Thomas Laboratory, University of Adelaide, Adelaide, Australia

7.1 INTRODUCTION

This chapter describes the effects of coumarin on high-protein oedemas, in particular on lymphoedema and acute inflammation, concentrating on human trials and applications. There have been many animal experiments involving some 50, quite various, acute and chronic high-protein oedemas[1–4], but these are not reviewed in detail here.

7.2 THE NATURE OF HIGH-PROTEIN OEDEMAS

7.2.1 The Definition of High-Protein Oedema

The concept of 'high-' versus 'low-'protein oedema is becoming increasingly important. The terms 'high' and 'low' refer to the concentrations of the plasma proteins in the tissue fluid in the two major groups of oedemas. They are not related to the concentrations in normal tissues, which may be higher than the concentrations

Coumarins: Biology, Applications and Mode of Action. Edited by R. O'Kennedy and R.D. Thornes.

found in many typical high-protein oedemas. It is the concentration in the oedematous state that is important.

An arbitrary line is usually drawn at protein concentrations of 1 g/dl (1 g%, 10 g/l) to separate the two kinds. In typical low-protein oedemas (early venous, cardiac, renal and nutritional) the concentrations are usually less than this; in typical high-protein oedemas the concentrations of proteins are above it. Many diseases have an oedema associated with them, usually a high-protein oedema.

7.2.2 The Importance of Tissue Protein Concentration

The division of oedemas into high- and low-protein types has several benefits:

(1) It divides the oedemas of very different pathogenesis.
(2) The colloidal osmotic pressure of protein increases greatly, in a non-linear manner, as its concentration increases. At 1 g/dl, or higher, this pressure exerts a considerable effect.
(3) Excess protein has many deleterious effects (see Section 7.3) in addition to those common to all oedemas.
(4) Above 1 g/dl, the benzopyrones can remove enough of the excess protein to materially lessen the extent of all its sequelae, such as oedema, chronic inflammation, excess fibrosis, secondary infection and pain.

7.3 DELETERIOUS EFFECTS OF OEDEMAS

7.3.1 All Oedemas

Oedemas (both low- and high-protein types) have many, often seldom realised, deleterious effects on the body[1]:

(1) They all cause swelling. Much of this is trivial, since it rapidly resolves, but if chronic it can cause much disability and may lead to psychological effects, even to suicide.
(2) If oedemas increase rapidly, they cause pain. This may be extreme.
(3) Oedemas all cause poor oxygenation of the tissues. This is mainly due to the severance of the many cell-to-cell contacts which are essential to facilitate the rapid passage of oxygen since it is only poorly soluble in water.
(4) Hypoxia reduces cellular functioning, including wound healing.
(5) If an oedema is massive, the gross functioning of a limb can also be greatly impaired and it becomes prone to injury. Postural deformities and problems often follow.
(6) Severe oedema collapses the lymphatics, thus worsening the condition. Indeed, oedema reduces the effectiveness of many of the safety factors against oedema and hence tends to be self-perpetuating[1,6]. There are often many vicious circles

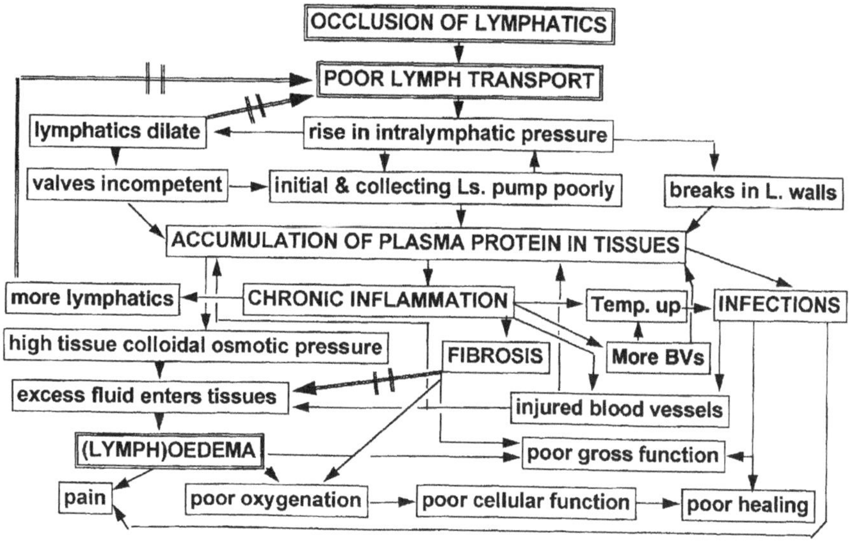

Figure 7.1. Some of the sequences of events that may occur in lymphoedema. Not all necessarily happen, but often they do. (Lines across an arrow mean that this effect tends to reduce the oedema.)

in even the most apparently 'simple' oedema. This is especially the case for high-protein oedemas. Figure 7.1 illustrates an example (lymphoedema) of some of these complexities and interactions.

However, a vicious circle has one advantage: if one can break it, the benefits are much greater than one might expect. This is why the action of the benzopyrones in lysing excess protein (which might seem rather minor) can have such far-reaching consequences—not only on the oedema itself, but on its sequelae.

7.3.2 Chronic High-Protein Oedemas

High-protein oedemas have additional and still more deleterious effects than low-protein ones. If they last more than a few weeks, the excess protein itself acts as a mediator of chronic inflammation[1,6]. This produces dilation and damage of the small blood vessels, and causes new, leaky vessels to form. Thus, it exacerbates the oedema and causes whole blood to enter the tissues. Fibrin from this blood blocks tissue channels and lymphatics. The chronic inflammation also produces excess fibrosis. This impedes the passage of fluid and protein through the tissues. Eventually it contracts and blocks the lymphatics. The excess protein also acts as an excellent incubation medium for bacteria and fungi. This is aided by the warmth from the dilated blood vessels. Hence secondary infections may be facilitated, thus exacerbating the lymphoedema. While macrophages at first accumulate and help to remove much of the excess protein and debris, eventually they become inactive and filled with lipid. Occasionally, chronic high-protein oedemas progress to angiosarcoma (Stewart–Treves syndrome).

7.4 MODE OF ACTION OF COUMARIN IN HIGH-PROTEIN OEDEMAS

Benzopyrones have many actions in the body. At least 50 have been used experimentally[1–4]. However, all the 25 (or so) benzopyrones which have been tested on it have one action in common: they all reduce all high-protein oedemas by increasing normal proteolysis by the macrophages (some 50 different models of such oedemas have been tested)[1–4]. They increase both the numbers of these cells at the site of high-protein oedema and their proteolytic activity. Thus, they provide an alternative pathway to the lymphatic system for the removal of excess plasma proteins from the tissues. In normal tissues the effect of proteolysis is probably minimal, even in the presence of benzopyrones. In high-protein oedemas it is a very important factor in controlling the amount of oedema (Table 7.1)[5].

7.4.1 Evidence that Coumarin Increases Proteolysis in the Tissues

Coumarin increased the clearance of labelled protein, both absolutely and relative to a radiolabelled non-metabolisable similarly sized molecule. This was in normal,

Table 7.1. Results from a mathematical model: alterations from trauma, lymphoedema and benzopyrones.

State of the tissue	Tissue fluid volume (ml)	Total tissue protein (g)	Tissue protein concentration (g/dl)(%)
Normal	15.1	0.38	2.5
Normal, without proteolysis	15.2	0.39	2.6
Trauma	114.0	4.8	4.2
Trauma, without proteolysis	118.0	5.1	4.3
Trauma, plus benzopyrones	68.0	2.6	3.8
Lymphoedema	51.0	2.2	4.3
Lymphoedema, without proteolysis	143.0	10.0	7.0
Lymphoedema, plus benzopyrones	24.0	0.79	3.3

This is just a model (although it returns results very close to those of many experiments)[6]. The normal, steady state is given for 100 g of 'tissue'. This is then altered by simulating various pathological conditions and new steady states occur. It is interesting that preventing proteolysis makes almost no difference in the normal condition, little in trauma, but a great difference in lymphoedema. On the other hand, increasing proteolysis with benzopyrones makes a considerable difference in both trauma and lymphoedema. This is because the various factors limiting oedema have very varying effects under different conditions.

burnt, lymphoedematous, and lymphoedematous-and-burnt rat legs[7]. Since the labelled protein left the tissue, this experiment excludes simple phagocytosis as the mechanism of protein removal; if the protein were merely phagocytosed, it would not be seen to leave the tissues. In addition, the increase in the ratio of the clearance of the metabolisable to that of the non-metabolisable tracer shows that such removal depends on lysis of the protein. The ratio of radiolabelled protein fragments to labelled protein fragments also increased in the legs of animals treated with coumarin, and indeed in their whole bodies[8]. Coumarins (and other benzopyrones) have been shown to increase the levels of proteases in high-protein oedema fluid, in whole tissues, and in lymph[9,10]. The effectiveness of various benzopyrones in reducing high-protein oedemas correlates well with their abilities to induce excess production of proteolytic enzymes[11].

In experimental ascites, the amount of protein in the peritoneal cavity is very greatly reduced, by coumarin, yet this cannot be accounted for either by an increased lymph flow or by a reduced output from the blood. Increased proteolysis was thought to be responsible[12]. Similarly, in acute thrombophlebitic oedema, increased proteolysis was the only likely explanation for the large decrease in the total protein content of the oedema fluid (and of the total amount of this fluid) by coumarin, while the concentration of protein in the fluid and its removal by the lymph system remained constant[13].

It is of interest that, at least in acute lymphoedema, once the dose of coumarin has reached a certain level (somewhat above those used in humans), additional amounts of the drug do not seem to have additional effects[14]. However, the situation in various acute inflammations is quite different[1] (e.g. in burns), with even 'window-effects' being apparent (see Chapter 8).

7.4.2 Evidence Supporting the Role of Macrophages in Increased Proteolysis

Macrophages are the most likely cells to be responsible for the increased proteolysis, for a number of reasons. They are normally very numerous in the tissues, but increase greatly in numbers and in activity in high-protein oedemas[1]. Around the macrophages, in a tissue containing much protein, one can frequently see a pale zone, as if lysis was occurring in their vicinity[15,16]. When plasma was injected into the skin of immunologically tolerant litter-mates every four days for 64 days, taking care to avoid known causes of inflammation, a chronic inflammatory response developed[6]. There was an increase in the numbers of macrophages from 4×10^6 (superficial to 1 cm^2 of deep fascia) in control animals injected with saline, to 160×10^6 in the animals injected with plasma. These increased to 560×10^6 when the animals injected with plasma were treated with coumarin. Such coumarin-treated animals had greatly reduced protein content of their tissues, oedema and excess fibrosis (and other signs of chronic inflammation). Similar results have also been observed using this model with topically applied coumarin in rats and horses[17], topically applied

7-hydroxycoumarin[18], and unguentum lymphaticum (a cream which contains compounds acting similarly to benzopyrones)[19].

Increases in macrophage numbers in acute high-protein oedema has been demonstrated in many models using coumarin and other benzopyrones and drugs with similar actions. The amount of protein in the tissues was greatly reduced, and the numbers of macrophages greatly increased, by coumarin[20–24], 7-hydroxycoumarin[18] and sodium cromoglycate[25]. Non-benzopyrones (e.g. calcium dobesilate[26] and unguentum lymphaticum[27]) had similar actions.

In the subcutaneous tissue of normal rats coumarin, or a mixture of coumarin and troxerutin, greatly increased the numbers of macrophages attaching to implanted coverslips, and the percentage showing stimulation[28]. Such increases and stimulation were also found in chronic lymphoedema (experimental and congenital, in dogs)[29,30]. Without coumarin the macrophages (while still numerous, but less than with coumarin) were usually 'dormant' with few pseudopodia and many fat

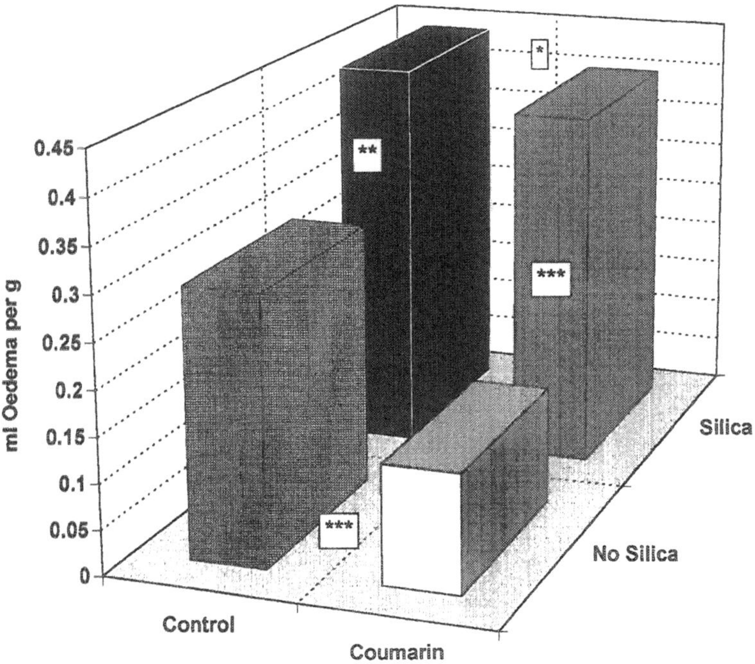

Figure 7.2. The effect of selectively killing the macrophages with tridymite silica is shown (left columns) by the increase in the amount of oedema (ml/g) in acute lymphoedema of the ratlet after four days. In the absence of silica (front columns), coumarin greatly reduces this. However, when silica has been given (rear columns) this reduction, while still slightly evident, is very much reduced (Casley-Smith, unpublished 1990).
Note that statistical significances used here and in all figures and tables are as follows: *** $p < 0.001$; ** $0.001 < p < 0.01$; * $0.01 < p < 0.05$; NSD, $0.05 < p$.

vacuoles. Coumarin stimulated them, greatly increasing the numbers of pseudopodia and reducing the numbers of vacuoles. This has also been found with other benzopyrones[1,31].

Coumarin and other benzopyrones increase not only macrophage numbers but also their level of activity. There is *in vitro* evidence of increased proteolysis per cell. Coumarin (at a dose of 25 μg/ml) increased macrophage proteolysis from 3.7 to 8.1 ($\times 10^{-10}$ mmol of glycine/cell/24 h)[32]. Similar *in vitro* proteolysis occurred with 6-F1 coumarin when the drug was administered to rats before harvesting the peritoneal macrophages and incubating them with albumin for 24 h[33].

All this evidence is supported by the effects observed following macrophage killing using a specific poison. For example, with the use of tridymite silica in a model high-protein oedema, the amount of oedema increases very considerably and coumarin is rendered ineffective. This has also been found using a number of benzopyrones, and similar drugs, in pulmonary contusion[23], acute lymphoedema of a number of tissues[24–27,34] (Figure 7.2) and burns[26,35]. This demonstrates the very considerable importance of proteolysis by macrophages in the limitation of such high-protein oedemas (Figure 7.2), which is also demonstrated by a mathematical model (Table 7.1)[5].

The immediate importance of these experiments is that when the macrophages were selectively destroyed coumarin, oxerutins (HR) and other drugs with similar actions, lost almost all of their efficacy[24–27,34,35] (Figure 7.2). The lack of lysis of proteins was demonstrated by quantitative electron microscopy in most of these experiments. Sometimes not all of the effect is lost (presumably due to the drugs' effects on the lymphatics—see below), but most of it disappeared. This shows how vital macrophages are for the reduction of high-protein oedemas by the benzopyrones.

Additional suggestive evidence of the effect of benzopyrones on macrophages is given by the biphasic effect of oxerutins (HR) on fibroblasts[31]. This is the same as the biphasic effect of the macrophages on the fibroblasts[36].

7.4.3 Coumarin and Phagocytosis

It is frequently considered, quite erroneously, that phagocytosis is a necessary preliminary for proteolysis. While indeed they often go together, proteolysis often occurs without phagocytosis. Many cells release proteolytic enzymes into the tissues[1]. There is considerable evidence that this is true for macrophages[1,9,10]. The increased clearance of radio-protein from the tissues, but not that of labelled non-metabolizable macromolecules of similar size[7], also shows that the action of coumarin is not confined to the mere stimulation of phagocytosis.

While coumarin (and other benzopyrones) may certainly increase phagocytosis[28,37–40], the effect of coumarin *in vitro* varies with the dose[40], and sometimes does not occur[41]. The degree of stimulation of phagocytosis certainly varies with different benzopyrones; for example, it is lacking with 7-hydroxycoumarin[40]. However, to argue that a lack of phagocytosis implies a lack of proteolysis, as some have done[41], shows no appreciation of the essential separation between the two processes.

7.4.4 Other Effects of Coumarin on Oedemas (High- and Low-Protein)

Benzopyrones may have other beneficial effects on high-protein oedemas. Coumarin and other benzopyrones increase the pumping by the collecting lymphatics[42]. However, this action, while undoubtedly most useful, is minor compared with its effect on proteolysis by macrophages. This has been demonstrated by carefully obstructing all the lymphatics of a rat's leg, which would normally cause the animal to die within 14 days (probably from protein loss), and applying topical coumarin. This not only considerably reduces the lymphoedema, but also prevents death, despite the complete lack of lymphatics[43]. Curiously, coumarin usually slightly increases blood capillary permeability[1]. If, however, there is increased capillary fragility (as in P-, C- and some B-avitaminoses) many benzopyrones (including coumarin) improve this, hence removing the cause of the high-protein oedemas in these deficient states[1]. Benzopyrones also improve the strength of weakened interstitial tissue[1] and reduce high-protein oedemas from this cause (e.g. in the lax tissues of ankles of the elderly).

7.4.5 The Effects of Increased Proteolysis on High-Protein Oedemas

Once the excess proteins of high-protein oedema are lysed, their colloidal osmotic activity is removed. The oedema fluid, plus the protein fragments, rapidly return to the blood[1]. Excess proteins in the tissues cannot readily pass back into the blood because of their size and because their concentration gradient is directed towards the tissues. Neither of these handicaps apply to their fragments.

The removal of the oedema improves the oxygenation of the tissues and the functioning of their cells. Importantly, this improves wound-healing. Lessened oedema also improves gross functioning. Mobility is improved and the tendency to injury from the lack of control is lessened. The chronic accumulation of excess protein is—by itself—a mediator of the chronic inflammatory reaction, with all its associated sequelae, including excess fibrosis[1,6]. Removing this stimulus to chronic inflammation removes its consequences: excess fibrosis, blood and lymph vascular neogenesis (with leaky new vessels), dilatation of the blood vessels with an elevated temperature in the tissues and, possibly, a removal of the tendency to the formation of angiosarcoma. Finally, the excess protein provides an ideal culture medium for bacteria and greatly increases the risk of secondary infection. This can cause quantum leaps in the diseased state of the tissues in chronic high-protein oedema (e.g. lymphoedema) and can complicate an acute one. Removal of this excess protein clearly aids the healing process.

Despite their effects on proteins in oedema, benzopyrones have no noticeable effect on the plasma protein levels in the blood. This is because the numbers of macrophages at sites of high-protein oedema are so very much increased[5], as is their proteolysis. This increased proteolysis far outweighs that of the macrophages elsewhere in the body. The fragments are not wasted, but are made into new plasma proteins.

One of the particular benefits of coumarin is that it absorbs to the plasma protein[1]. Hence it chiefly goes to where it is most needed, to the site of a high-protein oedema.

7.5 DISEASES ASSOCIATED WITH HIGH-PROTEIN OEDEMAS

There are more reports about lymphoedema and elephantiasis than about any other oedema. This is undoubtedly the most serious of all high-protein oedemas. Lymphoedema is chronic and gets progressively worse, eventually causing gross deformity, disability and death[44]. No drug treatment exists for it, and those who suffer from it are surprisingly numerous.

One woman in 100 suffers lymphoedema following mastectomy[1]. At least 10% of these are serious and 50% will suffer some disability. Post-mastectomy lymphoedema accounts for about one third of all lymphoedemas in non-filaritic areas, giving a total of about 1.5% of the population who will suffer from lymphoedema during their lifetimes. Worldwide, lymphoedema and elephantiasis affected some 150 million in 1982. Hence, many workers have been concerned to use coumarin in lymphoedema, at least in the first instance.

Other high-protein oedemas, however, are also remarkably common. While many are acute and may resolve quickly, all should benefit to some extent by treatment with benzopyrones. Of the one million people in South Australia in 1982, one in three received medical treatment for a disease which had an associated high-protein oedema[1]. Many of these were relatively trivial (e.g. dental extractions), but the overall mortality rate was still 0.8% and morbidity was considerable. Examples of diseases with associated high-protein oedema are:

(a) Acute inflammation of any kind: infection, accidental and surgical traumata of all kinds (including work and sports injuries), allergy, chemical injuries, burns, ischaemic conditions (e.g. of the heart or brain).
(b) Chronic inflammation of any kind: chronic venous insufficiency—including around ulcers, chronic haemorrhoids, chronic infections (e.g. tuberculosis, chronic hepatitis, pancreatitis, brucellosis, syphilis), repeated traumata (e.g. tenosynovitis and bursitis), ischaemia, around gangrene, post-ischaemic oedema, most ascites and peritonititis, most pleurisies and pleural effusions.
(c) Metabolic and degenerative diseases: diabetes, glaucoma, senile cystic macular degeneration, and many others.

The high-protein oedema associated with these conditions is usually considered to be the least of a patient's problems. However, such oedemas still have many deleterious effects, particularly reduced healing, the importance of which is often not realised. Unlike corticosteroids, coumarin is safe to use in infections and does not encourage the growth of bacteria[45]. It reduces the oedematous component of the inflammatory reaction and improves phagocytosis.

7.5.1 Trials Using a Mixture of Coumarin and Troxerutin

It is an accident of history that many of the trials involving coumarin have been performed with a mixture of two benzopyrones: coumarin and troxerutin (or sodium rutin sulphate). Assessing results from a mixture of drugs is not simple. On the other hand, all benzopyrones have similar effects on high-protein oedema[1], with an efficacy apparently relating only to the benzopyrone part of their molecules. This is contradicted to some extent by the work of Pethö[46] on episiotomy wounds treated with troxerutin, but coincides with animal studies[1] and with trials on human lymphoedema, for example those reported below using coumarin and a similar one using oxerutins (HR)[47]. At the doses used in humans, the efficacies of the benzopyrones in high-protein oedemas appear approximately constant on a molar basis.

7.6 USE OF COUMARIN IN LYMPHOEDEMA

Without adequate treatment, lymphoedema always worsens. The amount of oedema was compared with its duration in the first consecutive 231 post-mastectomy arm, 74 primary leg and 103 secondary leg lymphoedemas seen at some 25 clinics[44]. These were taken as a sample of the general, untreated lymphoedema population. The amount of oedema increased with time (Figure 7.3), as did the grades. Arms increased more rapidly than secondary or primary legs, which did not differ from each other (1.8, 0.63 and 0.68% of the oedema/year, respectively). Grades of primary lymphoedemas increased more slowly than those of either arms or secondary legs, which did not differ from each other. The increases with duration were not caused simply by the patient ageing. While both duration and the age at the onset of lymphoedema were significant for the arms, only the former was for the legs.

Compression garments were not used in any of the following trials, except where coumarin was combined with other treatments. Such garments are an essential part of physical therapies. However, they are impractical in hot, wet or dirty conditions, and are often much disliked by patients. If not maintained properly, they lose effectiveness and the benefits of the rapid reductions are lost since the newly emptied spaces rapidly refill with more protein and fluid. The slower reductions induced by coumarin allow the slowly remodelling fibrous tissue to hold the tissues together and prevent protein and fluid re-entering the tissues. Hence its slower, steady reductions are actually an advantage.

The most important complication of lymphoedema is secondary acute inflammation (SAI). This includes lymphangitis, lymphadenitis, fungal and bacterial infections (often mistermed: 'cellulitis' or 'erysipelas')[48]. Such infections can be very

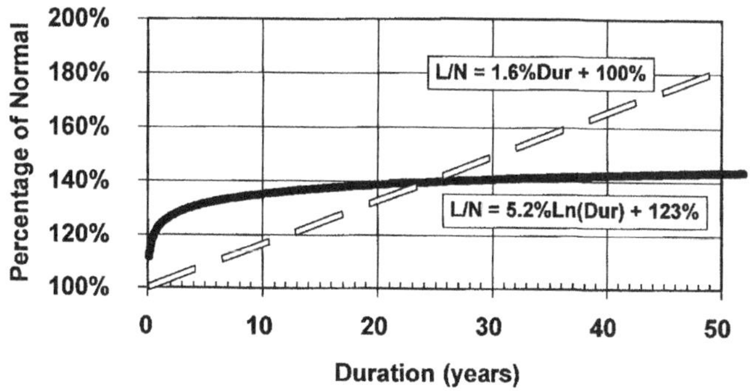

Figure 7.3. The increase with time of untreated (or poorly treated) lymphoedema in the leg[44]. The dashed line shows the linear regression, but lymphoedema increases more rapidly initially, as shown by the solid line (which is the logarithmic regression line). Arms (not shown) increase three times faster than legs. Increases are seldom uniform. Quantum jumps occur with infection, and other stimuli (e.g. stings, trauma, aircraft flights, etc.).

serious in the excellent 'incubation medium' formed by the concentrated protein of lymphoedema, with its elevated temperature from the many (and dilated) blood vessels produced by the chronic inflammatory process[1,5]. SAI always needs adequate treatment, often in hospital, since it is extremely painful, frequently causes a permanent quantum increase in the amount of lymphoedema, and may result in death. Some consider that, even if coumarin were to do nothing else, its effect in reducing SAI alone would more than justify its use in lymphoedema. Other symptoms are also reduced.

7.6.1 Coumarin Toxicity

Our own trials involved 1106 patients, receiving the oral drug for a mean of 15 months. Mild nausea or diarrhoea were reduced by an enteric-coating of the tablets (from 6% to 0.7%). Mild dizziness or drowsiness, rashes or itchiness, general aches and pains amounted to 4%. Our only serious side-effects were two possible cases of idiosyncratic hepatitis (0.2%) in open trials[49]. This idiosyncratic hepatitis usually occurs only after some months, has a complete recovery on withdrawal of the drug, is probably of immune origin, and was found in four (possibly in eight) of 2173 long-term patients (0.2% or 0.4%)[50]. No side-effects have ever been reported for topical coumarin preparations.

7.6.2 Lymphoedema in Western Communities

Coumarin has been shown to produce significant effects in lymphoedema. Oral coumarin (400 mg/day) was used in a randomised, double-blind, crossover (six-month periods), placebo-controlled trial (Figure 7.4)[51]. There were 31 post-mastectomy lymphoedemas and 21 primary and secondary lymphoedemas of legs, all grade 2[48]. There was a 43% reduction in the volume of oedema (reducing from 46% above normal to 26% in the arms and from 25% above normal to 17% in the legs). There were similar improvements in the circumferences and tonometry (a measure of the softness of the tissue). The mildly elevated skin temperatures were also reduced to normal. This confirmed the presence of chronic inflammation. There were improvements in many of the symptoms (Table 7.2). Some others were not significantly improved, but these were often the less frequent ones. SAI was less frequent and less severe. All these alterations were much the same as those obtained in a similar trial using oxerutins (HR), although the significances were less in that trial because of smaller numbers[47].

A randomised, double-blind, placebo-controlled, matched-group (39 post-mastectomy lymphoedemas per group) trial for nine months showed that the oedema reduced by 15% with 90 mg/day of coumarin[52]. It increased by 8% in the controls. Symptoms, isotope clearances, lymph speed and the half-life of isotopes in the tissues were improved by the active drug. An open, multicentre (54 centres) study of 90 mg/day was conducted for 12 months on 591 post-mastectomy patients (reducing to 481), 60% of whom also had 'some physiotherapy'[53]. The circumferences were very significantly reduced, as were the volumes of oedema (by 21%). These results were used for a cost-effectiveness evaluation[54]. An uncontrolled open trial of patients with grade 2 post-surgical head and neck lymphoedema given 80 mg/day coumarin intravenously (plus some 'massage') for four weeks gave similar results to those described above[55], as did another on primary and secondary leg lymphoedema[56]. Other trials are noted in Section 7.6.4.

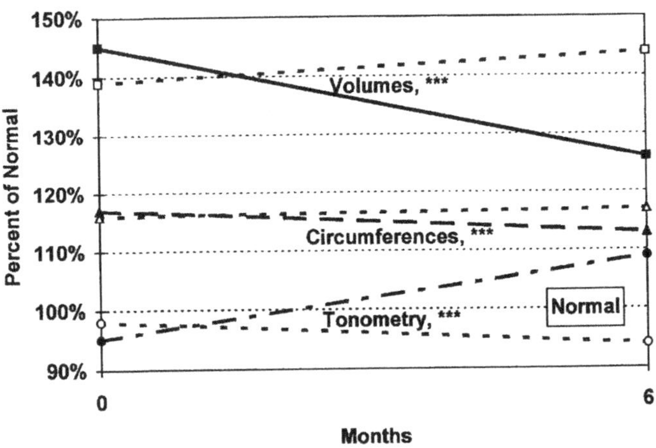

Figure 7.4. Linear regression lines from a double-blind, crossover trial of coumarin (400 mg/day) in 52 arms and legs (shown combined)[51]. Both periods are plotted together. The active drug gave improvements in volumes, (——) circumferences, (— — —) and tonometry (— - —), compared with the placebos (- - -).

Table 7.2. Changes in symptoms for lymphoedema when treated with coumarin for six months.

	Placebo				Coumarin				
	Total	Better	Same	Worse	Total	Better	Same	Worse	Sig. diff.
Bursting pain	26	3	11	12	29	21	6	2	7×10^{-5}
Burning pain	14	2	7	5	11	5	5	1	0.35
Cramps	9	1	5	3	11	4	4	3	0.74
Paraesthesia	7	1	2	4	10	3	5	2	0.68
Feeling of tightness in limb	48	8	22	18	49	30	11	8	2×10^{5}
Feeling of tension in limb	35	4	20	11	37	22	9	6	2×10^{-4}
Feeling of swelling in limb	39	10	15	14	39	24	10	5	0.0063
Feeling of heaviness in limb	41	8	17	16	41	27	7	7	10^{-4}
Feeling of hardness in limb	41	8	18	15	44	31	11	2	8×10^{-6}
Loss of mobility of limb	47	4	31	12	49	32	10	7	10^{-7}
Raised temperature of limb	30	8	15	7	25	10	8	7	0.38
Secondary acute inflammation	14	1	5	8	10	7	3	0	0.010
Dryness of skin	35	7	15	13	33	12	12	9	0.31
Warts	3	0	1	2	3	1	1	1	1.0
General well-being	41	9	23	9	40	27	8	5	2×10^{-4}
Which period was preferred?	3				48				10^{-17}

This was a double-blind, cross-over trial[51]. There were 31 post-mastectomy lymphoedemas and 21 primary and secondary lymphoedemas of legs, all grade 2. A symptom had to be present at some time during a period for it to be included here. Usually they were present at the start, but some developed during a period. If one disappeared during the first period and did not reappear during the second, it was only included for the first.

7.6.3 Filaritic Lymphoedema and Elephantiasis

Primary lymphoedema and that secondary to surgery can lead to elephantiasis, but this condition is seen far more frequently following filariasis. Whether this is because of the infestation itself, or because of the geographical and social circumstances of most of the patients, is a matter of some debate. A few workers consider that frequent infections (SAI) are largely responsible for the elephantiasis. Filaritic lymphoedema and elephantiasis both in size and in the numbers affected[1], is without doubt the major problem of this group of diseases in the world.

A randomised, double-blind, placebo-controlled trial, using proportionally matched groups (56 active, 40 placebo) was conducted in China on filaritic lymphoedema and elephantiasis (grades 1–5) using 400 mg/day of coumarin[57]. The trial was for 12 months, with a 12 month follow-up following the withdrawal of the drug (Figure 7.5). The placebo group did not alter; the active drug reduced the oedema and there were further reductions in the post-treatment year. Mean monthly reductions were: −0.62%, −1.1% and −1.6% of the normal for grades 1, 2 and elephantiasis, respectively. Post-treatment reductions were: −0.18%, −0.54% and −0.87%. Total reductions compared with the oedemas at the start of the first year, were: 100%, 95% and 45% of the oedema, for grades 1, 2 and elephantiasis. Many symptoms and complications were significantly reduced (Table 7.3): feelings of swelling and bursting, increased hardness and thickness of the skin, ulcers and loss of hair, and both the incidence and severity of attacks of SAI.

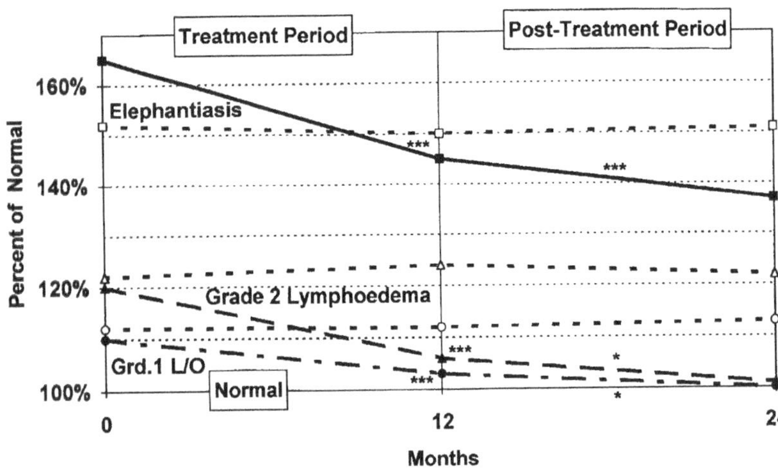

Figure 7.5. Volumes of filaritic lymphoedema (grades 1 and 2) and elephantiasis treated with 400 mg/day of coumarin in a double-blind, matched-groups trial[57]. The active drug is shown with solid symbols, the placebo with open symbols. Reductions during the year when treatment was given are in the left half. Notably, reductions continued after the drug was stopped (right half).

Table 7.3. Alterations in symptoms, signs and complications in filaritic lymphoedema and elephantiasis during treatment with coumarin (first year only).

	Placebo		Coumarin		Sig. diff.
	Start	End	Start	End	
Total numbers of patients			40	56	
Feeling of swelling and bursting of leg					
Severe	10	10	15	5	***
Moderate	6	4	6	10	
Mild	24	23	34	20	
Slight	1	2	1	21	
Feeling of swelling and bursting of foot					
Severe	4	4	6	0	***
Moderate	5	5	6	1	
Mild	10	10	14	11	
Slight	4	4	3	0	
None	17	17	27	44	
Increased hardness of skin					
Severe	6	6	6	2	***
Moderate	8	8	21	2	
Mild	24	26	28	5	
Slight	2	0	1	16	
None	0	0	0	31	
Increased thickness of skin					
Severe	6	6	8	2	***
Moderate	8	8	13	2	
Mild	24	24	22	6	
Slight	2	2	12	17	
None	0	1	1	29	
Secondary acute inflammation (lymphangitis, lymphadenitis)—SAI					
Attacks of SAI in patients prone to it	6	6	7	2	*
Total no. of attacks in these patients	18	18	21	4	***
Attacks of SAI in patients not prone to it	0	5	0	9	NS
Patients with fungal infection	6	6	9	3	**
Patients with ulceration	8	8	16	2	***
Patients with loss of hair	38	38	36	5	***
Patients with gross deformities (including sulci, warts and papillomata)	6	6	8	4	*

A double-blind trial, using proportionally matched groups[57]. The various grades did not differ significantly and are combined.

A similar trial, conducted in India[58] used 400 mg/day of coumarin, with and without 6 mg/kg/day of diethylcarbamasine (DEC). This kills microfilariae and some have thought it reduces lymphoedema and elephantiasis (but it also releases toxins from the dead worms). The format was: randomised, double-blind, matched groups (40–55 per group), single and double placebo-controlled, for 3 to 45 months

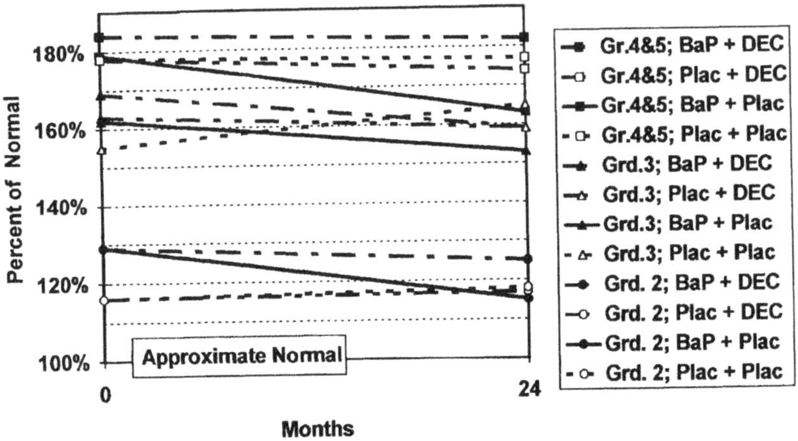

Figure 7.6. A double-blind, matched-groups trial of coumarin (400 mg/day) and diethylcarbamasine (DEC), in filaritic lymphoedema (grade 2) and elephantiasis (grades 3–5), in India[59]. There were slow but significant reductions in all grades with coumarin over two years; these were lessened by the addition of DEC, although there were still some reductions.

(75% for over 24 months). Coumarin slowly reduced all grades of lymphoedema and elephantiasis (Figure 7.6). Over two years grade 2 (lymphoedema) was reduced by 70% of the oedema, grades 3–5 (elephantiasis) by 20%. The greater the initial oedema, the greater was its rate of resolution. Many symptoms and complications were significantly improved (Table 7.4), particularly bursting pains, SAI (bacterial and fungal) and ulcers, and the fever caused by DEC. Giving continuous DEC reduced the effectiveness of coumarin, while coumarin improved that of DEC. Given together, they reduced feelings of swelling and bursting pain, fungal infections, lymphangitis and lymphadenitis more than either alone. Thus, it might be best to use DEC in short bursts every 2–3 months to kill any re-infestation by filariae, rather than continuously.

This trial was then continued as an open trial for up to eight years (mean five years), with coumarin and DEC as before, but giving the active drugs in place of the placebos for those who received the latter (Figure 7.7)[59]. For the 36 taking coumarin this was −0.55% while for the 47 taking coumarin plus DEC it was −0.19%. The difference between these was significant. Both reductions were very significant, compared with zero and also with the double-placebo group in the first trial. The initial amount of oedema differed between the grades, so that coumarin alone returned grade 2 (lymphoedema) to normal after a mean of five years, while at the end of eight years grades 3–5 (elephantiasis) was only 40% reduced. For coumarin plus DEC at the end of eight years, grade 2 was 40% reduced and grades 3–5 was 18% reduced. For this trial, the incidences of complications and symptoms were

Table 7.4. Alterations in symptoms, signs and complications (per three-month period) in filaritic lymphoedema and elephantiasis during treatment with coumarin and/or DEC.

	Results					Significances				
Coumarin	No	Yes	No	Yes		No	No	No	No	Yes
Diethylcarbamazine	No	No	Yes	Yes		No	No	No	Yes	No
								versus		
Number of patients	43	47	39	45	Coum	Yes	No	Yes	Yes	Yes
Total no. of 3-month periods	272	337	305	308	DEC	No	Yes	Yes	Yes	Yes
Feeling of swelling and bursting	47	31	36	15		**	NS	***	**	*
Generalised fever	6	11	24	11		NS	**	NS	*	NS
Secondary acute inflammation (SAI)										
Lymphangitis and lymphadenitis	9	6	2	0		NS	NS	**	NS	NS
Generalised bacterial infection	20	7	4	8		**	***	**	NS	NS
Fungal infection	16	23	14	4		NS	NS	**	*	***
Ulcer	18	4	4	4		***	***	***	NS	NS

A double-blind, matched-group trial for 3–45 months (most patients attended for two years)[58]. Each person was counted once, irrespective of whether their disease was uni- or bilateral. If patients missed an appointment, the preceding three months was not included. The numbers refer to the number of such three-monthly intervals in which the symptom, etc. was present.

not significantly different if DEC was added to coumarin (Table 7.5), but compared with the placebo group, coumarin very significantly reduced their incidence.

A randomised, double-blind, placebo-controlled, matched-group (30 per group) trial was made in China of primary and secondary (mostly filaritic) lymphoedema and elephantiasis, using 400 mg/day of coumarin for six months[60]. In comparison to the placebo (Figure 7.8), coumarin very significantly improved volumes (20% of oedema), circumferences and tonometry. Symptoms such as feelings of swelling, pain and heaviness, and the loss of mobility (Table 7.6) were also markedly improved.

7.6.4 The Question of the Effective Dose

In many diseases, results from the benzopyrones often improve when higher doses are used[1]. It is a great pity that very few dose-efficacy studies have been performed on coumarin[46] or other benzopyrones[61] in high-protein oedemas. This is understandable, given the history of these drugs, but it means that we have to rely on comparing often quite disparate trials in an attempt to guess the most suitable doses. From the trials discussed above, it would seem for lymphoedema that the higher the dose, the

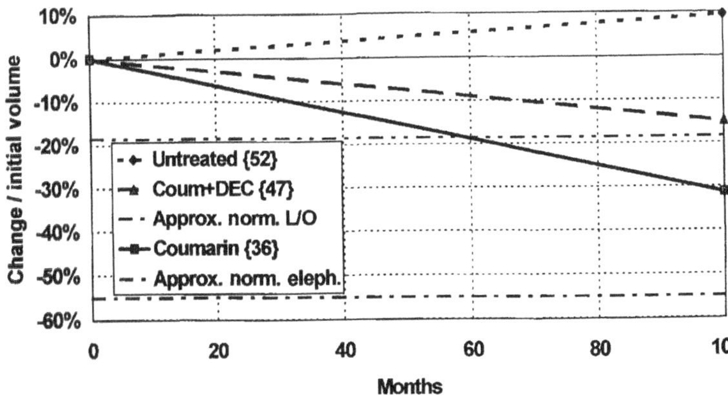

Figure 7.7. The trial shown in Figure 7.6 was continued as an open trial[59]. There were no significant differences between the grades in the change/initial volume values and they are combined. The untreated (double-placebo group) of the first trial are shown (- - -), as are those receiving coumarin and DEC (— — —), and coumarin alone (——). Both reductions were significant, but coumarin alone was greater. Approximate normals are shown for grade 2 (— - - —) and elephantiasis (— - —). The average lymphoedema became normal at 60 months. Elephantiasis was only about half way there after eight years, but at least it was reducing, not increasing.

Table 7.5. Complications of lymphoedema and elephantiasis in eight-year open trial of coumarin and coumarin plus DEC, compared with placebos to both in previous double-blind trial.

	Coumarin ± DEC (%)	Coumarin ± DEC (No.)	Placebos, previous trial (%)	Placebos, previous trial (No.)	Sig. Diff. Coumarin ± DEC vs. placebos
No. of patients		56		43	
No. of quarter-years		1135		272	
Incidents (per 3-month period)					
Feelings of swelling and bursting	0.2	20	17	47	10^{-23}
Lymphangitis and lymphadenitis	0.8	9	3	9	0.003
Generalised bacterial infection	0.6	7	7	20	10^{-12}
Fungal infection	0.5	6	6	16	10^{-9}
Ulcer	0.5	6	7	18	10^{-10}
Generalised fever	0.9	10	2	6	0.13

An open, matched-groups trial for up to eight years (most patients attended for five years)[59] was performed following that shown in Table 7.4. There were no significant differences between coumarin with or without DEC, or between the grades, and they are all combined. They are compared with the double-placebo group of the previous trial[58].

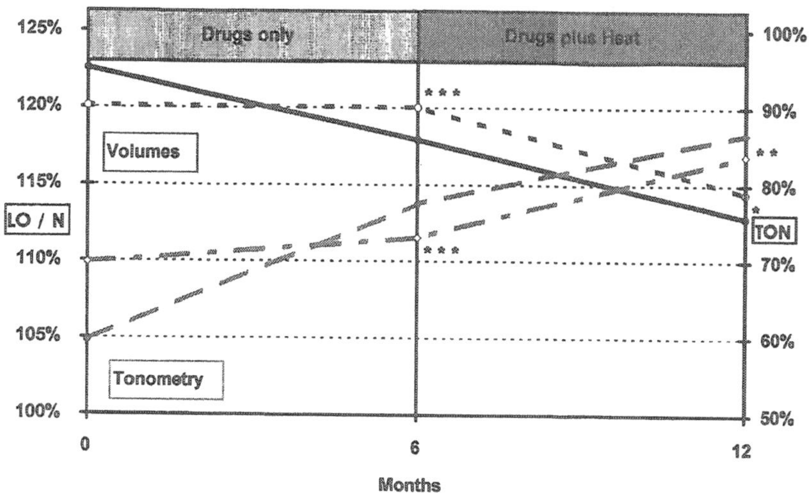

Figure 7.8. Double-blind, matched-group trial of coumarin/placebo alone (left half), followed by these plus heat (right half)[66]. Coumarin (——) improved the volume of lymphoedema/normal (LO/N) compared with the placebo (- - -), and the tonometry (coumarin: (— — —), placebo: (— - —), TON right axis). For the next six months, patients continued with the same drugs and had a series of heat (microwave) treatments and compression bandages/garments. There were improvements from heat, but those from heat plus coumarin were greater.

Table 7.6. Alterations in symptoms in lymphoedema and elephantiasis (mostly filaritic) by coumarin.

	Placebo			Coumarin			
	No.	Improved	Same	No.	Improved	Same	Sig. diff.
Feeling of swelling	29	5	24	30	25	5	10^{-7}
Burning pain	14	3	11	11	8	3	0.031
Feeling of heaviness	26	10	16	28	24	4	0.0009
Restricted mobility	24	4	20	25	23	2	10^{-6}

This was during the drug-only period (0–6 months) of a more complicated double-blind, matched-group trial which in the next period had heat added to the drugs[60]. No patient became worse over this period so these possibilities are omitted from this table. Only those complaining of a symptom are included (30 per group).

better the result (at least for the doses used so far). This hypothesis receives some confirmation from the next two studies cited.

Coumarin (100 mg/day) was studied in open trial (for a mean of 18 months) of 21 patients who had previously undergone treatment with coumarin and troxerutin (two tablets/day, equivalent to 60 mg/day of coumarin) for a year and who were dissatisfied with the speed of their improvements. This was controlled by 71 patients who remained on the previous dose of coumarin and troxerutin[62]. Coumarin significantly improved the circumferences and tonometry compared with conventionally treated patients. These improvements were reported to be 'about twice those with coumarin and troxerutin', but no statistical tests between them were given.

Primary and secondary lymphoedema (6 arms and 14 legs) were studied in an open trial of coumarin, 400 mg/day (for six months) versus 800 mg/day (for 12 months) in the same patients[63]. Circumferences were reduced by 400 mg/day, but more by 800 mg/day—there was a significant difference between them.

7.6.5 Topical Coumarin in Lymphoedema

Two topical coumarin preparations have been used: an ointment of 10% coumarin in polyethylene glycol (PEG) and a 10% mixture in sterile talc (the latter is an unusual manner of applying a drug, but coumarin's good lipid-solubility permits it to penetrate the skin[1,64]). The ointment, in particular, permits a high local dose, but the powder is generally more suitable for use under compression garments and is less sticky. Both sting if applied to mucous membranes, lymphatic fistulae and some ulcers. However, treatment with local anaesthetic spray a few minutes earlier prevents this and permits their use, which is often very effective, because of the high local doses obtained.

An open, but objectively measured trial was conducted for up to 12 months, using complex physical therapy (CPT) with and without oral coumarin (400 mg/day), and with and without topical coumarin, on the first consecutive 628 lymphoedematous limbs treated at 25 centres (from some 50 therapists of very varying expertise)[65,66]. There were 379 legs and 249 arms examined in the trial. The effects of many factors on the reduction of lymphoedema and on the maintenance of this reduction were studied using multivariate analysis (Figures 7.9 and 7.10). At the end of the year, the final reduction of oedema (compared with its initial value) was 43% from CPT alone, 48% from CPT plus oral coumarin, 52% from CPT plus topical coumarin, and 64% from all of these. The effects of oral and topical coumarin were not significantly different from each other, but their combination was more effective than either alone. Each of them combined with CPT was very significantly more effective than CPT alone, particularly in maintaining the reductions.

Pain is not usual in huge elephantitic legs (unless they are infected) but is sometimes quite extreme in the much smaller post-mastectomy and post-pelvic surgery lymphoedemas. This is because these increase far more rapidly[44], the tissues being torn apart. Two randomised, double-blind, matched-group, placebo-controlled, prospective trials were conducted for 21 days using coumarin powder (15 per group) and ointment (20 per group) for painful post-mastectomy lymphoedema (Figure 7.11)[67]. Patients estimated the pain weekly. There was considerable placebo effect, but both topical forms of coumarin reduced pain greatly, even after only seven days. If topical coumarin is combined with physical therapy, pain is reduced much better

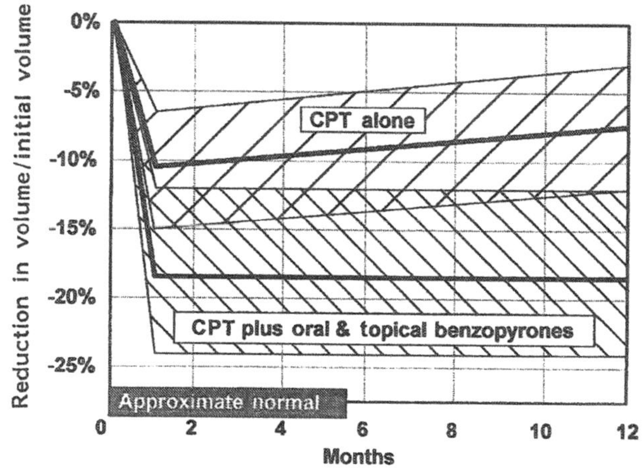

Figure 7.9. The effect of complex physical therapy (CPT) alone and in combination with oral and topical coumarin[65,66]. Numbers of limbs were: CPT alone 356 for the first month (the course of CPT) and 90 for the next year; CPT plus oral and topical coumarin: 92 and 54. The 25%, 50% and 75% quartiles are shown for the reduction in volume of the lymphoedematous limbs/their initial volumes. Coumarin increased the reductions during the course, and helped to maintain these better for the next 11 months.

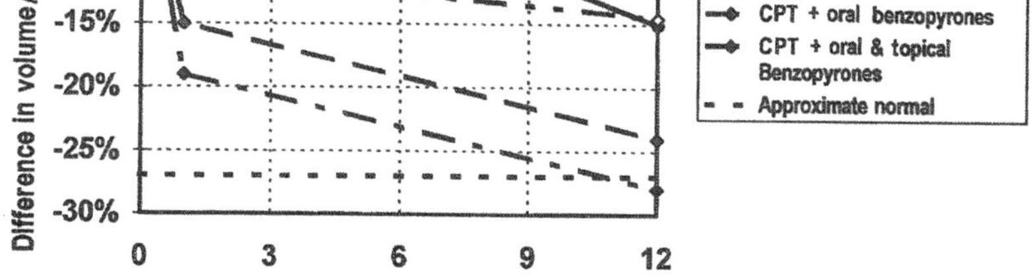

Figure 7.10. Means for various combinations of therapies mostly from the trial[65,66] shown in Figure 7.9, but also with untreated lymphoedema (- - -)[44] and the combined values from a number of trials for legs treated only with oral coumarin (——)[51,53,58,60,78]. Here, only grade 2 legs are shown. Grade 1 legs or arms give similar results, but the effects of oral coumarin alone varies. While CPT gives more rapid results, oral coumarin gives good results over time. It is even better when combined with CPT and topical coumarin.

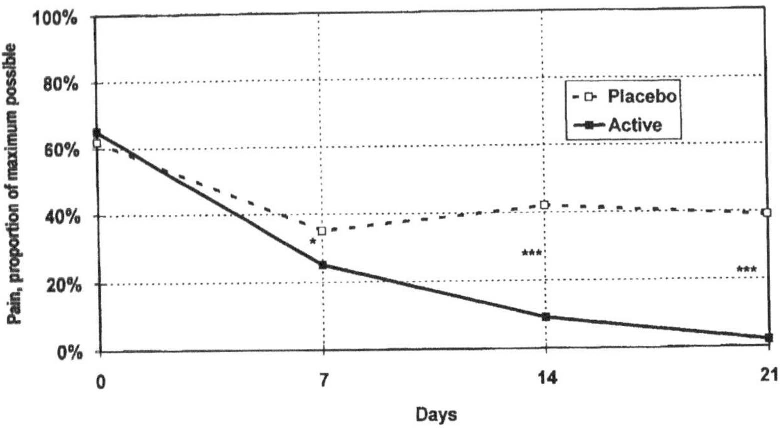

Figure 7.11. A double-blind, matched-group trial of the effect of coumarin ointment on the pain of post-mastectomy lymphoedema[67]. While there was a placebo effect, the active drug reduced the pain much more. A separate, similar trial showed the same for coumarin powder (which is much better under compression garments, if these are used—they were not used in these trials). The relief of pain is much faster if the drugs are combined with CPT. This itself relieves the pain but not nearly as rapidly as the two together.

than by physical therapy alone[67]. Coumarin powder also apparently benefited some cases of ulcers associated with lymphoedema and lymphocutaneous fistulae[68].

7.6.6 Use of Coumarin and Troxerutin Mixture in Lymphoedema

Post-mastectomy lymphoedema was treated with nine tablets of coumarin (135 mg) and troxerutin (810 mg) per day equivalent to 270 mg/day of coumarin, for six months in a trial using a randomised, double-blind, placebo-controlled, matched-group (46 per group) format[69]. Symptoms, SAI and isotope lymphography were improved in the treated group but not in the placebo group. A randomised, double-blind, cross-over (24–28 days/period), placebo-controlled trial of three tablets of coumarin and troxerutin per day (equivalent to 90 mg/day of coumarin) was performed on nine leg lymphoedemas (aetiology not stated)[70]. There was a 6.1% decrease in oedema with active drug and a 1.5% increase with placebo (very significant).

Post-mastectomy lymphoedema was successfully treated for up to 100 months (mean 24) with two tablets of coumarin and troxerutin per day (equivalent to 60 mg/day of coumarin) in 103 patients[71,72]. An incidental finding was a remarkable lack of recurrence of the carcinoma[1]. Primary lymphoedema and that secondary to surgery and trauma responded to two tablets/day of coumarin and troxerutin (equivalent to 60 mg/day of coumarin) in several open trials[73–76].

7.6.7 Coumarin Combined with Other Therapies in Lymphoedema

The benzopyrones provide an alternative pathway for the removal of the excess protein in lymphoedema; most other methods of treating it attempt to restore lymphatic drainage in some way. Complex physical therapy (CPT) attempts to improve the collateral drainage of the lymphostatic tissue. Hence, it was very interesting to note that combining coumarin with other methods improves the efficacy of both (Figures 7.8–7.10)[65–68,77]. These papers point to the fact that coumarin alone is just as useful (and much cheaper and easier) than poorly designed physical methods. This has been further shown from the one-year results produced by oral coumarin alone versus CPT with and without oral and/or topical coumarin (Figure 7.10)[78,79].

7.7 EFFECT OF COUMARIN ON LYMPHATICS

Many benzopyrones, including coumarin, increase pumping by collecting lymphatics[1]. Improved lymphatic functioning has now been found in lymphoedema treated with coumarin[52,80]. This is understandable as there is always some lymphatic function remaining in human lymphoedema, albeit insufficient.

Coumarin and troxerutin also improved the lymph drainage (as measured by radiotracers) from the cavities surrounding artificial eyes[81].

The thoracic duct was cannulated in a controlled, open trial of 10 patients with normal hearts, kidneys and guts[82]. After a control period of 20 min, they were given 4 ml of injectable coumarin and troxerutin (the equivalent of 24 mg of coumarin), followed 10 min later by the start of 10 ml of coumarin and troxerutin (the equivalent of 60 mg of coumarin) given by a drip over the next 60 min. Flow in the thoracic duct increased by 96% 10 min after the first injection of coumarin and troxerutin. It was still significantly elevated by 35%, 20 min later. There was a mean increase of 68% during the infusion.

A similar experiment was performed in 10 normal subjects, using randomised, double-blind, matched groups, in which pancreatic amylase and lipase were estimated during a control period and after the administration of intravenous 10 ml of coumarin and troxerutin (the equivalent to 60 mg of coumarin)[83]. There was no effect on pancreatic secretion (volume, bicarbonates, amylase, lipase), but the thoracic duct lymph content of amylase and lipase significantly increased at 1 h, returning to normal at 2 and 3 h. This indicated that the lymphatic removal of pancreatic enzymes from tissues was increased.

Before lymphangiography, coumarin and troxerutin (1 or 2 ampoules, equivalent to 30 or 60 mg of coumarin) were administered endolymphatically, or intravenously (as a bolus or an infusion) to 71, 160 and 46 patients with dubious lymphatic systems[84]. Oxerutins (HR) were similarly administered to 28, 52 and 33. These were controlled by 100 other patients. Both drugs were reported to improve lymphatic functioning so that the lymphatic trunks were more visible, as were the central lymph nodes.

Lymphangiography was used to observe the effect of 6 ml of coumarin and troxerutin given intravenously (equivalent to 33 mg of coumarin) to 26 subjects (12 normal, 8 hypoplastic and 6 hyperplastic lymphoedema), in an open, controlled trial[85]. With the drug, less injection pressure was necessary to inject the contrast medium into the lymphatics in the foot, and if lymphangiospasm was present, it relaxed. The drug gave better filling of lymph trunks, a more rapid flow of contrast medium and more uniform distribution up to inguinal nodes—attributed to improved pumping by the collecting lymphatics. (This could only be the collecting lymphatics since initial lymphatics do not pump actively.)

7.8 CHRONIC VENOUS INSUFFICIENCY AND VARICOSE VEINS

Contrasting with the oxerutins (hydroxyethyl rutosides, HR) for which many studies have demonstrated their effectiveness in venous disorders[1], there have been few using coumarin alone, although there have been a number with coumarin and troxerutin. These show beneficial effects in various venous disorders, from simple varicose veins to chronic venous insufficiency. It is unfortunate that only one small study has been performed on haemorrhoids[73]. From the benefits given by the oxerutins in this condition[1], it is likely that coumarin would also be useful.

7.8.1 Oral Coumarin

A randomised, double-blind, cross-over, placebo-controlled trial of coumarin (200 mg/day) was performed in 15 normals and 15 patients with chronic venous insufficiency (CVI) for four weeks (with an intermediate four-week wash-out period)[86]. Oedema and most symptoms (Table 7.7) significantly improved in CVI patients (Figure 7.12). There was no difference between patients with CVI and the normals for the clearance of a non-metabolisable tracer ([^{99m}Tc]rhenium colloid). However, the clearance of a metabolisable tracer ([^{51}Cr]γ-globulin) was initially very significantly greater in CVI than in normals. In both groups it increased after the placebo but more after the active. The difference between active and placebo was significant in the patients with CVI, but not in the normals. This suggests that the drug was acting at least partially via increasing proteolysis in the high-protein oedema, but that proteolysis was not important in the normal condition (as is also suggested by model data—Table 7.1[5,78]). However, the ratio of the clearances between the two tracers was not significantly different even in the CVI patients, unlike in similar animal studies[7]—possibly because of greater standard errors in the human study.

7.8.2 Coumarin and Troxerutin Mixture

The efficacy of coumarin/troxerutin has been well-confirmed by the randomised, double-blind, placebo-controlled trials conducted on patients with CVI. They all

Table 7.7. Effect of coumarin on chronic venous insufficiency (four weeks).

	Placebo			Active			
	Total	Improved	Worse	Total	Improved	Worse	Sig. diff.
Clinical signs							
Clinical oedema	13	4	9	13	8.5	4.5	NSD
Clinical tenderness	10	3.25	6.75	10	9	1	*
Symptoms							
Feeling of heaviness in legs	15	4.75	10.25	15	12.5	2.5	*
Feeling of swelling/ tenseness in legs	14	4.25	9.75	12	10.75	3.25	*
Feeling of tiredness in legs	14	4	10	14	12	2	**
Feeling of aching pains in legs	13	4.5	8.5	13	10.5	2.5	*
Restless legs	7	2.25	4.75	7	4.75	2.25	NSD
Paraesthesia	8	3	5	8	5	3	NSD
Nocturnal cramps	8	2.75	5.25	8	5.5	2.5	NSD
General well-being	15	6.75	8.25	15	13.25	1.75	*
Preferred period	1			14			***

A double-blind, cross-over trial[86]. Patients showing no clear alterations in a given symptom were scored as 0.5 in each column; those with mild alterations were scored as 0.75 in the column towards which they tended and 0.25 in the other.

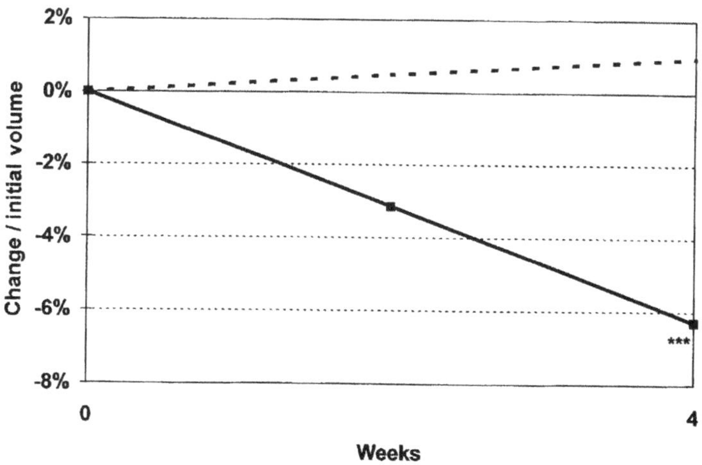

Figure 7.12. A double-blind cross-over trial of coumarin (200 mg/day) in chronic venous insufficiency[86]. Coumarin (—■—) gave a reduction of leg volume, compared with its initial volume, compared with the placebo (----).

Table 7.8. Alterations in feelings of tenseness and sural pain, in chronic venous insufficiency treated with coumarin and troxerutin.

Day	Drug	Number	Worse	Same	Slightly improved	Improved	Much improved
7	Active	30	0	10	4	15	1
	Placebo	28	1	24	5	1	0
14	Active	30	0	9	4	15	2
	Placebo	28	1	20	2	5	0
21	Active	30	0	7	3	18	2
	Placebo	28	0	23	1	4	0

A double-blind, matched-group trial for three weeks[87]. There were 14 per group with grade 3 varicose veins and 16 with post-thrombophlebitis, with two drop-outs in the placebo group. Alterations were significant at each period.

showed excellent results[87–94] (Table 7.8). Similar results are also reported in randomised, open, standard treatment-controlled trials[70,95–97].

Varices, chronic venous insufficiency and lymphangitis were treated with coumarin and troxerutin (3 tablets/day, equivalent to 90 mg/day of coumarin) in an open trial for six months[98]. Only vague generalisations are given concerning the results, which the author says were 'very good'. This trial is mentioned, not because it gives any evidence about the efficacy of the drug, but because 7900 patients were involved—with only occasional, trivial, side-effects.

7.9 ACUTE INFLAMMATION

It is a pity that patients are permitted to suffer the pain, poorer wound-healing, and the many other disabilities associated with oedema, when coumarin treatment would alleviate the problems. The use of coumarin often results in considerable savings in hospitalisation, analgesics, other disabilities and later sequelae—to say nothing of some deaths. The routine use of coumarin in elective surgery (preferably starting a few days before the operation) could considerably reduce post-operative morbidity and time in hospital.

7.9.1 Accidental and Surgical Trauma

Topical Coumarin

Randomised, double-blind experimental studies, using the opposite limbs for the placebo controls, were performed on three groups of 10 men using coumarin ointment (10% in PEG)[17]. Experimental haematomas were produced by injecting 100 ml of the subject's own blood intramuscularly into each thigh, using this format.

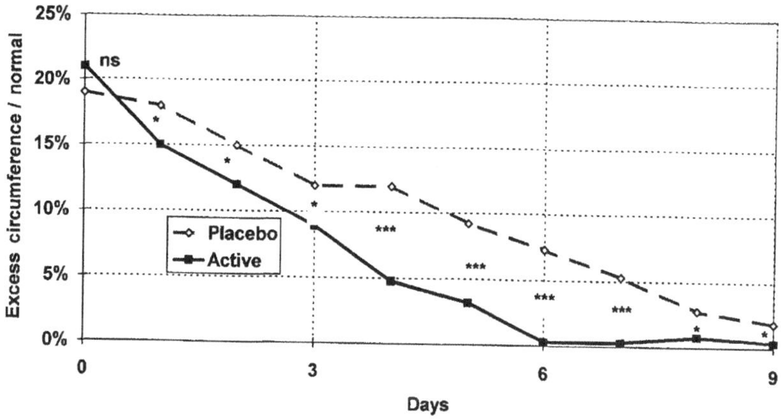

Figure 7.13. Experimental haematoma in humans[17]. After the first day there was reduction in the swelling treated with coumarin ointment (——), compared with the placebo (– – – –).

Haematomas are not often thought of as a high-protein oedema, but they are. Circumferences were reduced over the injection site from day 1 to day 9 compared with the placebos (Figure 7.13). This shows the usefulness of this topical preparation in haematomata and (probably) in crush injuries—which has been demonstrated in rats[17].

An experimental burn was made with a brass rod (100 °C for 2 s) applied to the volar aspects of both forearms using this format. Oedema was measured, using a microscope, by the elevation of the injured site relative to the rest of the skin. It was reduced at 6 h and from then until day 8, but was not different at day 14. The pain was reduced at 6, 12 and 24 h. Oxygen tensions in the site were increased from 12 h.

A milder, longer experimental burn, using this format, was given by placing both little fingers in hot water (56 °C for 60 s)[17]. Oedema was reduced at 5 min and at days 1–5 (Figure 7.14). Pain was reduced at 5 min and at 24 h.

A similarly randomised, double-blind trial was performed on 20 subjects with moderate sunburn, applying the active formulation to one side of the back and the placebo to the other[17]. Pain was markedly reduced from 15 min to 48 h, but there was no significant difference at 96 h.

Oral Coumarin

While experimental haematomas and burns are relatively easy to perform, it is rather more difficult to find a large number of subjects willing to have reasonably large, fairly deep, experimental, incised wounds. In this respect then, it is fortunate that obstetricians perform many episiotomies and that these are fairly constant. They were used in a major, and very important series of studies by Pethö[46] (who performed or observed all of them). Many patients were given coumarin and troxerutin

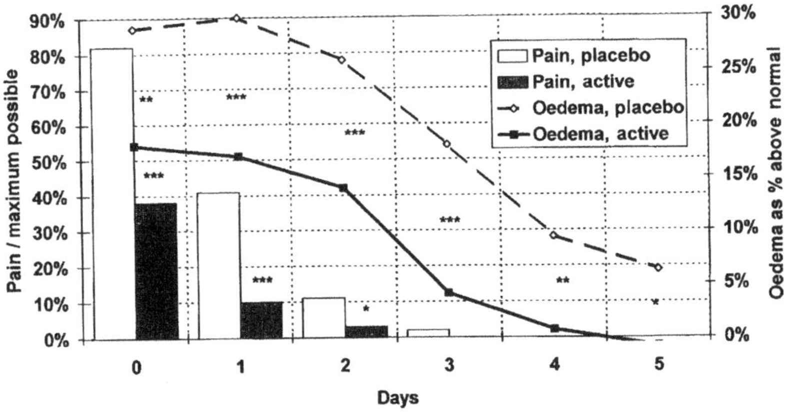

Figure 7.14. Experimental burn (56 °C for 60 s) in humans[17]. From day 0 to day 5 there were reductions in the oedema (right axis) treated with coumarin ointment (——), compared with the placebo (- - - -). Pain was also reduced from day 0 to day 2 (left axis). (This was a double-blind trial.)

and will be discussed in Section 7.9.2. However, one series had coumarin alone and is discussed here. This study involved 60 patients with episiotomies who received coumarin (90 mg/day), 60 who received six tablets per day of coumarin and troxerutin (equivalent to 180 mg/day of coumarin, but containing the same amount of actual coumarin as the above), and 30 who received the placebo (later expanded to 221 since none of the placebo groups for the various subtrials differed significantly). The format was randomised, double-blind, placebo-controlled, with matched groups, studied for seven days. This was a double-blind trial[99]. Hard data were obtained and analysed statistically for: incidence of oedema, incidence of redness and consumption of analgesics. This last was confirmed, but not statistically analysed, by the incidences of spontaneous pain when lying in bed, on walking and on defaecation. Coumarin reduced all of these from day 3 onwards. It was reported that there were less haematomas in the active groups than the placebos, less dyspareunia on coitus, and less scar-pain at six weeks. No statistics were given for these, but they were very impressive clinically.

An open trial of third-molar dental extractions was carried out using 400 mg/day of coumarin (for three days pre-operatively and three days after it) in 12 patients. It was controlled by similar extractions in these subjects without coumarin. Pain and oedema were significantly reduced[100].

7.9.2 Coumarin and Troxerutin Mixture

Pethö performed a dose-efficacy study of coumarin and troxerutin[46] in patients undergoing episiotomy. The incidence of redness was reduced by the equivalent

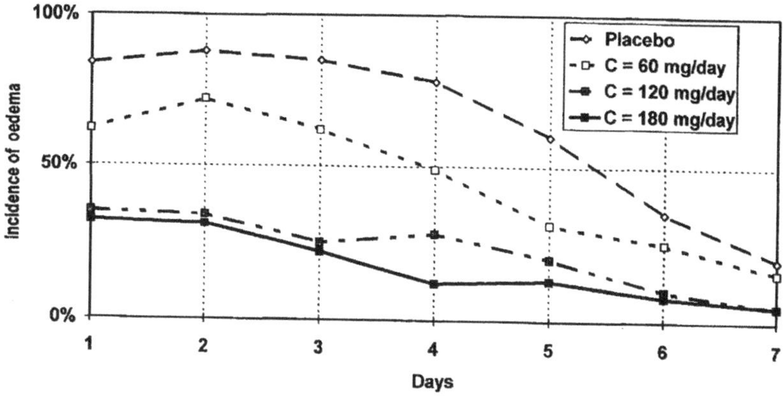

Figure 7.15. Episiotomies were performed on 560 women in a double-blind, matched-group trial[46]. The incidence of oedema is shown here for various doses of coumarin and troxerutin (expressed as mg of coumarin). The reductions increased with each increase in dose.

of 60 mg/day of coumarin, but 120 mg/day reduced it more, and 180 mg did so still further (from the third day onwards). The same all applied to the incidence of oedema (Figure 7.15) and for amount of analgesia used (Figure 7.16). Reductions in pain were confirmed by the incidences of spontaneous pain when lying in bed, on walking and on defaecation. It was also noted that there were less haematomas in the active groups than the placebos, less dyspareunia on coitus, and less scar-pain at six weeks. Thus, the efficacy increased as the dose was increased. It showed no sign of reaching a plateau at these dose levels for analgesia reduction, although it did for oedema.

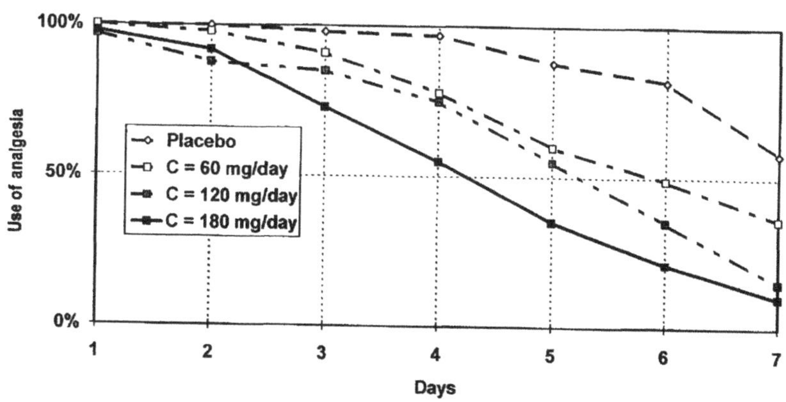

Figure 7.16. As for Figure 7.15, showing the amounts of *ad lib* analgesia (compared with the maximum) taken with varying doses of coumarin and troxerutin (as mg of coumarin)[46]. The analgesia taken reduced with each increase in dose.

Pain, oedema and the formation of keloid were studied after laparotomies in randomised, single-blind, placebo-controlled trial using matched groups (60 per group)[101]. For the first three days the active group received intravenous injections of six ampoules of coumarin and troxerutin per day (equivalent to 180 mg/day of coumarin), then for the next 12 days they received this amount orally as six tablets/ day. On the eighth day the wound was very significantly better in the active group for redness, oedema, leaking of serum, and the start of keloid formation. This last was also notable at the end of the second week, when the rest were not significantly different (it was present only in 10/60, compared with 30/60 in the control group). The active group were also much less sensitive to pressure (6/60, compared with the placebo group 20/60). The placebo group also had other complications (7/60) compared with none in the active group.

Vaginal hysterectomies were studied using a randomised, single-blind, placebo-controlled trial of matched groups of 60 patients[102]. The active group received intravenous injections of six ampoules of coumarin and troxerutin per day (equivalent to 180 mg/day of coumarin) for three days, then for the next 14 days they received this amount orally as six tablets/day. The active group had less redness, oedema, leakage of serum, pain, formation of keloid and problems with micturition.

Another source of standardised 'experimental' incised wounds in humans is circumcision for phimosis. This was used as a model to test 1–5 ml intravenous or intramuscular coumarin and troxerutin pre-operatively (equivalent to 5.5 mg/ml of coumarin) in 52 children of various ages, controlled by 51 given no coumarin[103]. The incidence of oedema was reduced very significantly (Table 7.9). Haematomas were present in 1/52 for the treated group and in 6/51 for the controls, but this was not significant. However, the mean after-care time necessary was 5.5 days for the treated group and 11 days for the controls.

Table 7.9. Prophylaxis against oedema following circumcision, using coumarin and troxerutin.

Amount of oedema	Coumarin	Placebo
None	33	3
Mild	14	25
Moderate	4	20
Severe	1	3

Frequencies of mild oedema versus none

Group	Oedema	No oedema	Sum
Coumarin	14	33	47
Placebo	25	3	28
Sum	39	36	75

Frequencies of severe oedema versus none

Group	Oedema	No oedema	Sum
Coumarin	5	35	38
Placebo	23	3	26
Sum	28	36	64

An open, randomised trial, using consecutive patients given intravenous or intramuscular coumarin and troxerutin before circumcision for phimosis[103]. The differences shown in each of the three subtables are highly significant.

Intramuscular injections of two coumarin and troxerutin ampoules per day (equivalent to 60 mg/day of coumarin) were given one day before and four days after oral and maxillofacial surgery (141 sinus operations, 101 osteotomies and 53 cystectomies)[104]. This was a randomised, double-blind, matched-group, placebo- and oxyphenbutazone-controlled trial. Oedema was reduced by coumarin and troxerutin versus both controls (Figure 7.17). Trismus was reduced by coumarin and troxerutin versus the placebo, but not versus the oxyphenbutazone. The obvious benefit to the patient was reflected by the reduction in the amount of on-demand analgesia (Novalgin®) by coumarin and troxerutin versus both controls.

A variety of dental operations (including extractions) were studied in a randomised, double-blind, oxyphenbutazone-controlled, matched-group (150 active, 50 control), 11-day trial, using six tablets/day of coumarin and troxerutin (equivalent to 180 mg/day of coumarin)[105]. The duration of oedema, analgesic requirement and pain were reduced and healing time was shortened in the coumarin group.

Pain was reduced, and wound-healing improved following tonsillectomy[106].

One trial (randomised, double-blind, matched groups) for five days, was on oedema following arterial reconstruction in the legs because of arterial insufficiency (Figure 7.18)[107]. There were 20 controls, 20 received coumarin and troxerutin as above, 30 received this dose half intravenously and half intramuscularly, and 20 received half the above dose of coumarin and troxerutin intravenously. Leg circumferences were significantly reduced each day in all active groups compared with the controls. There were no significant differences between the various coumarin groups.

The second trial (using the above format) was of brain oedema after reconstructive surgery of the carotid and subclavian arteries, over five days, with 27 in the active group and 29 controls[108]. The active drug reduced neck oedema, and gave

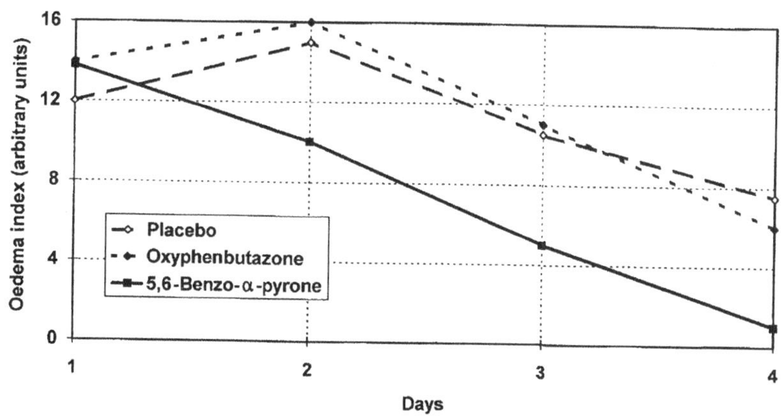

Figure 7.17. A double-blind, matched-group trial of the amount of oedema after oral and maxillofacial surgery[104]. The reductions were considerably greater after coumarin (5,6 benzo-α-pyrone) and troxerutin (expressed here as 180 mg/day of coumarin) than with the placebo or oxyphenbutazone.

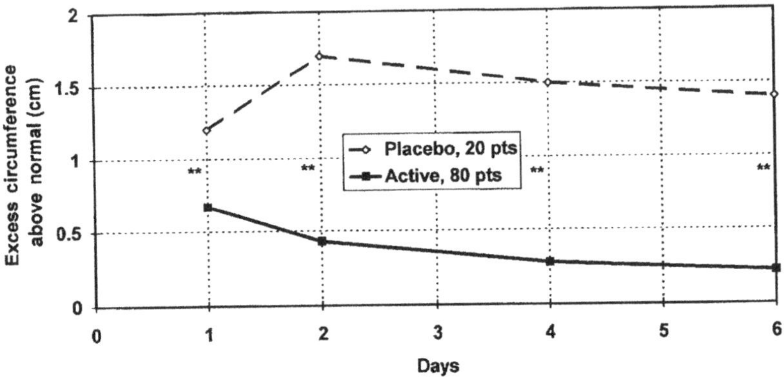

Figure 7.18. Oedema following arterial reconstruction in the legs owing to arterial insufficiency was studied in a double-blind, matched-group trial of patients receiving the equivalent of 120 mg/day of coumarin[107]. From day 1 to day 6 the active drug gave very considerable reductions.

improvements in the state of consciousness, neurological signs, and respiratory acidosis (all monitored in a double-blind manner). The duration of hospitalisation was also reduced (12.2 days to 8.8).

The third of these trials (again with the same format) was of post-operative oedema following gastroenteric anastomosis, for six days, with 25 in the active group and 21 controls[109]. The active drug significantly facilitated patients eating earlier. On day 15, 25/25 of the actives could eat; only 15/21 of the controls could do so. The actives were also able to leave hospital earlier. On day 11, 19/25 of the actives had left, compared with only 11/21 of the controls.

The reduction of post-reconstructive oedema was also observed in a randomised, double-blind, placebo-controlled seven-day trial, using matched groups—27 placebo and 29 receiving six coumarin and troxerutin tablets per day (equivalent to 180 mg/day of coumarin)[110] (a similar suppository was used on the day of operation). Oedema was reduced, but only the severe oedema of a femoro-popliteal bypass. Minor oedema from lesser operations could not be measured accurately enough.

A randomised, double-blind, placebo-controlled, matched-group (23 active, 20 placebo), 14-day trial was conducted, using six tablets/day of coumarin and troxerutin (equivalent to 180 mg/day of coumarin) on closed fractures of the lower leg[111]. The reductions of the mean circumferences (measured at six standardised places) were twice those of the placebo at 7 and at 14 days (Figure 7.19).

The effect of intravenous coumarin and troxerutin (two ampoules per day, equivalent to 60 mg/day of coumarin on the day of operation and for the three subsequent days) was tested (80 active and 300 retrospective controls) on post-operative oedemas in gastroenteric anastomoses (Billroth I or II) in an open trial[112]. For 150 of the 300 controls the stomach tube had to remain for more than 48 h; for 78 of the 80

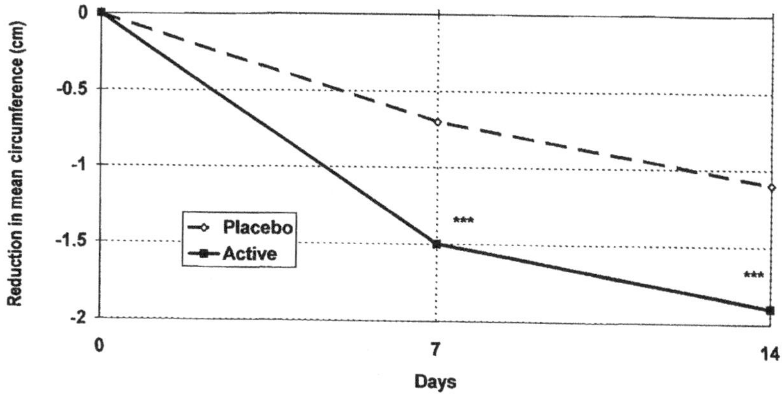

Figure 7.19. A double-blind, matched-group trial was performed on oedema following closed fractures of the lower leg using the equivalent of 180 mg/day of coumarin[111]. This gave great reductions in the oedema.

actives it could be removed by 12–18 h. The risk of oedema from this operation in the coumarin group was calculated to be 2.5% of that of the controls.

Traumatic and post-operative haemophthalmia, retinal haemorrhage, and vitreous opacities treated with coumarin and troxerutin (three tablets/day, equivalent to 90 mg/day of coumarin) in open trials of 149 patients gave considerable improvements[113]. This compares with well-controlled animal experiments[114], illustrating the potential usefulness of this drug in an area which otherwise might be overlooked.

Clinically the need for antibiotics with ischaemic wounds can be avoided by the use of 12 ampoules of injectable coumarin and troxerutin per day (equivalent to 350 mg/day of coumarin), which strongly promotes healing by first intention[115]. This avoidance of gangrene by increased oxygenation produced by coumarin was confirmed in the rabbit-ear model[16]. This action is also valuable in the trophic changes of the skin in diabetes, as has been shown for other benzopyrones in diabetic retinopathy[1].

7.9.3 Acute Pancreatitis—Coumarin and Troxerutin Mixture

Two trials show the benefits of coumarin in acute pancreatitis. In one the death rate was reduced. It is not certain whether these benefits were from increased proteolysis or improved lymphatic functioning, perhaps both. Undoubtedly, there are many other acute high-protein oedemas, minor or life-threatening, which would similarly benefit from coumarin. Unfortunately, no trials have been performed.

Acute pancreatitis with and without disturbed liver function was treated with intravenous coumarin and troxerutin (four ampoules/day, equivalent to 120 mg/day of coumarin), for 4–5 days, in a randomised, double-blind, matched-group, placebo-controlled trial for 10 days[116]. In the treated group urinary and serum amylase was reduced on day 5. Leucocytosis was improved on day 10 and the temperature and the urine and plasma chemistry became normal earlier (by a mean of 9.6 days compared with 16.7).

In the patients with liver damage in addition to pancreatitis, the urine and plasma chemistry became normal earlier (by a mean of 14.4 days compared with 20.6).

Acute haemorrhagic necrotising pancreatitis was the subject of the other randomised, double-blind, matched-group, aprotinin-controlled, 10-day trial, using intravenous coumarin and troxerutin (six ampoules/day, equivalent to 180 mg/day of coumarin)[117]. The active group contained nine grade II, one grade II/III, and two grade III patients; the control group: 7, 3 and 2 respectively. There were no deaths among the treated group and four in the control group (30%). This was significant, as was the difference in their hospitalisation (medians of 24 days for the coumarin-treated group and 34.5 days for the controls).

7.10 CHRONIC INFLAMMATION

There appear to be great benefits in the use of coumarin in chronic inflammation. It is singularly unfortunate, therefore, that many more and better designed trials have not yet been performed.

7.10.1 Oral Coumarin

Thornes used coumarin (25–100 mg/day) in chronic brucellosis over two-year periods[118,119]. He found that such patients usually have low IgA levels, lymphocytopenia and negative skin tests for delayed hypersensitivity. Of a number of drugs, only coumarin increased and maintained IgA levels, in addition to increasing helper T cells and restoring delayed hypersensitivity reactions for long intervals. He treated 1600 cases, previously incurable, with coumarin and antibiotics, with a 75% success rate (R.D. Thornes, 1984, pers. comm.). Contrasting with the general belief that present dose levels are too low, he believes a lower dose (25–50 mg/day) more suitable, at least in such conditions.

Coumarin (400 mg/day) has been used to treat thoracic outlet syndrome (excess fibrosis around the brachial plexus) with success (W. Ellis, 1994, pers. comm.). Of 40 patients with this chronic and most difficult condition, he has achieved excellent results in 11, good in 10, fair in six, equivocal in eight and unaltered in three.

A side-benefit of trials of oral coumarin, 200 or 400 mg/day for other conditions, has been 33 patients who spontaneously reported that their previously chronically bleeding gums ceased to bleed after a few weeks[120].

7.10.2 Topical Coumarin

Chronic or repeated acute tenosynovitis is often quite incapacitating. In an open study of 14 such cases, using coumarin ointment, all were considerably improved over a few days to weeks[120]. Ellis (1994, pers. comm.) has had similar results.

Intractable decubitus ulcers (bed-sores) are extremely difficult for both patients and nursing staff. A number of intractable cases have responded well, using either the coumarin ointment or powder[120].

7.10.3 Coumarin and Troxerutin Mixture

A randomised, double-blind, matched-pairs, placebo-controlled, 21-day trial was performed on chronic aggressive (relapsing) hepatitis and cirrhosis patients[121]. Some had diabetes mellitus (14 patients per group), some did not (13 per group). Coumarin and troxerutin (three tablets/day) were used (equivalent to 90 mg/day of coumarin). Many parameters were not significantly altered, but there were improvements in unsaturated iron-binding capacity, and in thymol turbidity, alkaline phosphatase, γ-globulin and albumin.

Schizophrenia may involve a high-protein oedema of the brain, possibly of viral origin. An accidental finding in one patient, who improved quite remarkably[122], led to a randomised, double-blind, cross-over trial of troxerutin (HR) in this condition[123]. There were considerable improvements in some of the patients, but not in all. Coumarin has not yet been tested in schizophrenia.

In 17 patients with chronic tuberculosis[124], coumarin and troxerutin (three tablets/day, equivalent to 90 mg/day of coumarin) was used, in an open trial, for 100–200 days, in conjunction with normal therapies. The objective was to reduce excess fibrosis around TB lesions and so permit antibiotics to reach the bacteria. After a nine-year follow-up, scar formation had been reduced compared with other non-coumarin-treated patients.

Objective measures were used to assess the efficacy of coumarin and troxerutin in chronic prostatitis, with co-trimoxazole (39 cases) or without (16 cases)[125]. Improvements were found in 75% and 87%, respectively.

Euthyroid endocrine ophthalmopathy was studied in a randomised, standard-treatment controlled, six-month trial, using matched groups[126]. Of these, 47 patients received coumarin and troxerutin (three tablets/day, equivalent to 90 mg/day of coumarin) as well as thyroxine, prednisolone and retrobulbar irradiation. Thirty-seven controls received only thyroxine whilst 43 received this plus prednisolone, while 21 had retrobulbar irradiation as well. The benzopyrone-treated patients had significantly better results than all control groups.

Coumarin and troxerutin (six tablets/day, equivalent to 180 mg/day of coumarin) for six months, reduced the high protein-content of oedema found in two cases of chronic contact dermatitis (complicated by recurrent infections)[127].

High-protein pleural effusions, even those associated with malignancy, have been treated using a coumarin/troxerutin mixture (Jüngst, 1982, pers. comm.). It is unfortunate that more trials have not been made, in view of the actions of these drugs in experimental pleural effusions in animals[1].

A curious phenomenon often associated with chronic high-protein oedemas is the desquamation of endothelial cells. Coumarin (with or without troxerutin) prevents this, as do other benzopyrones[128].

Experimental erysipelas in the rat is often used as a model for rheumatoid arthritis. Coumarin and troxerutin caused reductions of paw volume, corneal oedemas and, even more importantly, mortality (30% compared with 60% in the control group)[129]. Unfortunately, trials in humans have not been performed.

7.11 CONCLUSION

It is evident that coumarin (and indeed the many benzopyrones which had been tested) reduce all high-protein oedemas on which they have been tried. This leads to considerable lessening in the amount of disabilities, and to improvements in wound-healing and in other cellular functions. It is a great pity that such drugs are not used routinely in all these conditions.

There are also many experimental models of acute and chronic inflammations, in which benzopyrones are known to be effective, but which have not yet been tested in clinical trials. These are burns (and keloid), cerebral vascular incidents, various forms of ascites, and adult respiratory distress syndrome, to name just a few. One wonders how many patients have suffered unnecessarily, and have died before their time, because of these neglected opportunities.

REFERENCES

1 Casley-Smith JR, Casley-Smith Judith R. *High-Protein Oedema and the Benzo-pyrones*. Sydney & Baltimore: J.B. Lippincott, 1986: 536. *Copies now only available from the Lymphoedema Association of Australia, University of Adelaide, South Australia 5005, Australia.*

2 Farkas L, Gábor M, Kállay F (eds). *Flavonoids and Bioflavonoids, Current Research Trends*. Amsterdam: Elsevier, 1977.

3 Gábor M. *The Pharmacology of Benzopyrone Derivatives and Related Compounds*. Budapest: Akadémiai Kiadó, 1986: 253.

4 Gábor M. Anti-inflammatory substances of plant origin. In: Vane JR, Ferreira SH eds. *Anti-Inflammatory Drugs*. Berlin: Springer, 1979: 698–739.

5 Casley-Smith JR. A model of the factors affecting interstitial volume in oedema; Part II: Their effects at various abnormal steady-states. *Biorheology* 1993; **30**: 9–30.

6 Casley-Smith JR, Gaffney RM. Excess plasma proteins as a cause of chronic inflammation and lymphoedema. *J Path* 1981; **133**: 243–72.

7 Piller NB, Casley-Smith JR. The effect of coumarin on protein and PVP clearance from rat legs with various high protein oedemas. *Br J Exp Path* 1975; **56**: 439–43.

8 Piller NB. Further evidence of the induction of proteolysis by coumarin in rats with various high protein oedemas. *Arzn-Forsch (Drug Res)* 1977; **27**: 860–4.

9 Piller NB. The influence of various benzopyrones on acid and neutral protease activity levels, the cells from which they may arise and their importance in the resolution of lymphoedema. *Res Exp Med* 1977; **170**: 115–24.

10 Piller NB. Variations in acid and neutral protease activity in rats with thermal oedema together with the influence of various benzopyrones. *Arzn-Forsch (Drug Res)* 1977; **27**: 1069–73.

11 Piller NB. An integration of the modes of action of coumarin. An explanation for its effectiveness as a therapy for thermally injured tissue. *Arzn-Forsch (Drug Res)* 1977; **27**: 1135–38.

12 Szabó G, Magyar Z. Mechanism of action of benzo-pyrones in experimental ascites [trs]. *Arzn-Forsch (Drug Res)* 1977; **27**: 1064–9.

13 Szabó G. Drug influence on acute thrombophlebitic edema [trs]. *Arzn-Forsch (Drug Res)* 1977; **27**: 2332–5.

14 Casley-Smith Judith R, Casley-Smith JR. Effects of varying doses of 7-hydroxy-coumarin and coumarin in acute lymphoedema and other high-protein oedemas. In: Casley-Smith JR, Piller NB eds. *Progress in Lymphology X*. Adelaide: University of Adelaide Press, 1985: 194–6.

15 Casley-Smith JR, Eckert P, Földi-Börcsök E. The fine structure of pulmonary contusion and the effect of various drugs. *Br J Exp Path* 1976; **57**: 487–96.

16 Casley-Smith JR, Clodius L, Földi M. Experimental blood vascular and lymphatic occlusion in the rabbit ear and the effect of benzo-pyrones. *Arzn-Forsch (Drug Res)* 1977; **27**: 379–89.

17 Casley-Smith Judith R, Casley-Smith JR. The reduction of haematomas, burns, crush injuries and chronic high-protein oedemas by coumarin ointment. In: Partsch H ed. *Progress in Lymphology XI*. Amsterdam: Elsevier, Excerpta Med Int Cong Ser 779, 1988: 647–52.

18 Casley-Smith JR, Casley-Smith Judith R. Effect of 7-hydroxy coumarin cream on acute and chronic high-protein oedemas. In: Casley-Smith JR, Piller ND eds. *Progress in Lymphology X*. Adelaide: University of Adelaide Press, 1985: 191–3.

19 Casley-Smith JR, Casley-Smith Judith R. The action of Unguentum lymphaticum on chronic high-protein oedema. In: Casley-Smith JR, Piller ND eds. *Progress in Lymphology X*. Adelaide: University of Adelaide Press, 1985: 197–9.

20 Casley-Smith JR, Clodius L. Földi M. Experimental blood vascular and lymphatic occlusion in the rabbit ear and the effect of benzo-pyrones. *Arzn-Forsch (Drug Res)* 1977; **27**: 379–89.

21 Casley-Smith JR, Clodius L, Piller NB, Földi M. A model of lymphoedema in the rabbit's ear. The effects of the benzo-pyrones. *Chir Plast* 1977; **4**: 5–14.

22 Casley-Smith JR, Földi-Börcsök E, Földi M. Fine structural aspects of the influence of lymphatic blockage on cold injury of the brain and skin, and the effects of the benzo-pyrones. *Arzn-Forsch (Drug Res)* 1977; **27**: 367–78.

23 Casley-Smith JR, Eckert P. Földi-Börcsök E, Földi M. The effect of macrophage poisoning by silica on experimental pulmonary contusion and its benzo-pyrone treatment. *Br J Exp Path* 1977; **58**: 386–90.

24 Casley-Smith JR, Földi-Börcsök E, Földi M. A fine structural study of the removal of the effectiveness of benzo-pyrone treatment of lymphoedema by the destruction of the macrophages by silica. *Br J Exp Path* 1978; **59**: 116–27.

25 Casley-Smith JR, Casley-Smith Judith R. Sodium chromoglycate ('Intal') cream and acute lymphoedema in rat's legs; with and without macrophages. In: Casley-Smith JR, Piller NB eds. *Progress in Lymphology X*. Adelaide: University of Adelaide Press, 1985: 188–90.

26 Casley-Smith JR, Casley-Smith Judith R. The effects of calcium dobesilate on acute lymphoedema (with and without macrophages) and on burn oedema. *Lymphology* 1985; **18**: 37–45.

27 Casley-Smith Judith R, Casley-Smith JR. The effect of 'Unguentum lymphaticum' on acute experimental lymphoedema and other high-protein oedemas. *Lymphology* 1983; **16**: 150–6.

28 Piller NB. A morphological assessment of the stimulatory effect of coumarin on macrophages. *Br J Exp Path* 1978; **59**: 93–96.

29 Piller NB, Clodius L. The effect of coumarin (5,6 benzo-α-pyrone) on elicited members of the mononuclear systems in dogs with chronic secondary lymphoedema. Personal communication, 1984.

30 Piller NB, Clodius L. A morphological assessment of macrophages attaching to subcutaneously implanted coverslips in dogs with chronic lymphoedema. *Br J Exp Path* 1979; **60**: 401–7.

31 Huot J, Hubbes M, Nosal GI, Radouco-Thomas C. Biphasic stimulo-inhibitory effects of flavonoids on cell proliferation *in vitro*. *Archs Int Pharmacodyn Ther* 1974; **209**: 49–65.

32 Bolton T, Casley-Smith JR. The *in vitro* demonstration of proteolysis by macrophages and its increase with coumarin. *Experientia* 1975; **31**: 271–3.

33 Casley-Smith JR. 6-Fluoro-coumarin causes enhanced proteolysis by mouse peritoneal macrophages. In: Casley-Smith JR, Piller NB eds. *Progress in Lymphology X*. Adelaide: University of Adelaide Press, 1985: 106–8.

34 Casley-Smith Judith R, Casley-Smith JR. The effects of *o*-(*β*-hydroxy-ethyl)-rutosides (HR) on acute lymphoedema in rat's thighs, with and without macrophages. *Microcirc Endothelium Lymphatics* 1990; **6**: 457–64.

35 Piller NB. The ineffectiveness of coumarin treatment of thermal oedema of macrophage-free rats. *Br J Exp Path* 1976; **57**: 170–80.

36 Diegelmann RF, Cohen IK, Kaplan AM. The role of macrophages in wound repair. *Plast Reconstr Surg* 1981; **68**: 107–13.

37 Beuscher N, Beuscher H, Otto B, Schäfer B. Øber die medikamentöse Beeinflussung zellulärer und humoraler Resistenzmechanismen in Tierversuch. I. In-vitro-Unter-suchungen an Peritoneal-Leukozyten und Sere der Ratte. *Arzn-Forsch (Drug Res)* 1977; **27**: 1655–60.

38 Beuscher N. Wirkmechanismus und therapeutische Wirksamkeit von Benzopyronen. *Lymphologie* 1994; **11–12**: 464–7.

39 Dunn RF, Koh MS, Willoughby DA, Giroud GP. The value of multifactorial screening for anti-inflammatory activity as shown by coumarin. *J Path* 1977; **122**: 201–7.

40 Hardt TJ, Ritschel WA. The effect of coumarin and 7-hydroxy-coumarin on *in vitro* macrophage phagocytosis of latex particles. *Meth Find Exp Clin Pharmacol* 1983; **5**: 39–43.

41 Weissleder R, Bogdanova A, Weissleder H. Medikamentöse Lymphödemtherapie. *Vasomed* 1994; **6**: 236–9.

42 Mislin H. Die Wirkung von Cumarin aus *Melilotus officinalis* auf die Funktion des Lymphangions. *Arzn-Forsch (Drug Res)* 1971; **21**: 852–3.

43 Casley-Smith Judith R, Casley-Smith JR. The prevention of death in rats with severe acute lymphoedema by the benzo-pyrones. In: Partsch H ed. *Progress in Lymphology XI*. Amsterdam: Elsevier, Excerpta Med Int Cong Ser 779, 1988: 645–7.

44 Casley-Smith JR. Alterations of untreated lymphoedema and its grades with time. *Lymphology* 1995; **28**: 174–85.

45 Findley-Jones J. Lack of effect of coumarin on mouse peritoneal abscesses. Personal communication, 1984.

46 Pethö A. Die Wirksamkeit von Benzopyronen bei der posttraumatischen Entzündung. *Arzn-Forsch (Drug Res)* 1981; **31**: 1303–7.

47 Piller NB, Morgan RG, Casley-Smith JR. A double-blind cross-over trial of *o*-(*β*-hydroxy-ethyl)-rutosides (benzo-pyrones) in the treatment of lymphoedema of the arms and legs. *Br J Plastic Surg* 1988; **41**: 20–7.

48 Casley-Smith JR, Földi M, Ryan TJ *et al.* Lymphedema. Summary of the 10th International Congress of Lymphology working group discussions and recommendations. *Lymphology* 1985; **18**: 175–80.

49 Casley-Smith JR, Casley-Smith Judith R. Frequency of coumarin hepatotoxicity. *Med J Aust* 1995; **162**: 391.

50 Cox D, O'Kennedy R, Thornes RD. The rarity of liver toxicity in patients treated with coumarin (1,2 benzo-pyrone). *Human Toxicol* 1989; **8**: 501–6.

51 Casley-Smith JR, Piller NB, Morgan RG. Treatment of lymphedema of the arms and legs with 5,6 benzo-α-pyrone. *New Engl J Med* 1993; **329**: 1158–63.

52 Cluzan R, Pecking A. Benzopyrone (Lysedem) double blind crossing over study in patients with secondary upper limb edemas. In: Nishi M, Uchino S, Yabuki S eds. *Progress in Lymphology XII*. Amsterdam: Elsevier, Excerpta Med Int Cong Ser 887, 1990: 453–4.

53 Lafontan B de, Dilhuydy JM, Monnier A *et al.* Intérêt de la coumarine associée à la physiothérapie dans le traitement du lymphoedème du membre supérieur après cancer du sein. *But Cancer/Radiother* 1991; **78**: 119–24.

54 Le Pen C. Economic evaluation of medical treatments: the example of lymphology and phlebology. *Phlebology* 1992; Suppl. 3: 15–16.

55 Pecking A, Banal A, Bougon Ph, Lemiere E, Sachwald J, Mahé E. Benzopyrones in the treatment of head and neck edemas. In: Partsch H ed. *Progress in Lymphology XI*. Amsterdam: Elsevier, Excerpta Med Int Cong Ser 779, 1988: 635–7.

56 Nieto S, Grandval C. The use of 5,6-benzo-(α) pyrone for the treatment of lymphoedema. In: Cluzan RV, Peching AP, Lokiec FM eds. *Progress in Lymphology XIII*. Amsterdam: Elsevier, Excerpta Med Int Cong Ser 994, 1992: 357–8.

57 Casley-Smith JR, Wang CT, Casley-Smith Judith R, Cui Zi-hai. Reduction of filaritic lymphoedema and elephantiasis by 5,6 benzo-[alpha]-pyrone. *Br Med J* 1993; **307**: 1037–41.

58 Casley-Smith JR, Jamal S, Casley-Smith Judith R. Reduction of filaritic lymphoedema and elephantiasis by 5,6 benzo-α-pyrone (coumarin) and the effects of diethylcarbamazine (DEC). *Ann Trop Med Parasitol* 1993; **87**: 247–58.

59 Casley-Smith JR, Jamal S, Casley-Smith Judith R. Results and side-effects of coumarin in the treatment of filaritic and other lymphoedemas. In: Witte MH, Witte CL eds. *Progress in Lymphology XV. Lymphology* (Suppl) 1996; **29**: 20–3.

60 Chang T-s, Gan J-l, Fu K-d, Huang W-y. Clinical application of 5,6 benzo-α-pyrone on chronic lymphoedema of the extremities. *Lymphology* 1996; **29**: 106-111, and *J Chinese Med* 1990; **70**: 655–6.

61 Nocker W, Diebschlag W, Lehmacher W. Clinical trials of the dose-related effects of *o*-(*β*-hydroxyethyl)-rutosides in patients with chronic venous insufficiency. *Phlebology* 1990; **5** (Suppl. 1): 23–6.

62 Clodius L, Piller NB. The conservative treatment of post mastectomy in patients with coumarin results in a marked continuous reduction in swelling. In: Bartos V, Davidson JW eds. *Advances in Lymphology.* Prague: Avicenum 1982: 471–4. (*Supplemented by an unpublished study, 1984, with more patients.*)

63 Casley-Smith JR, Casley-Smith Judith R. Increased doses of 5,6 benzo-α-pyrone (coumarin) may improve responses in lymphoedema (in preparation)

64 Ritschel WA, Grummich KW, Kaul S, Hardt TJ. Biopharmaceutical parameters of coumarin and 7-hydroxy-coumarin. *Pharm Ind* 1981; **43**: 271–6.

65 Casley-Smith JR, Casley-Smith Judith R. Lymphoedema therapy in Australia, complex physical therapy and benzo-pyrones in over 600 limbs. In: Witte MH, Witte CL eds. *Progress in Lymphology XIV. Lymphology* 1994; **27** (Suppl): 622–6.

66 Casley-Smith JR, Casley-Smith Judith R. Treatment for lymphoedema by complex physical therapy (C.P.T.), with and without oral and topical benzo-pyrones; what should therapists and patients expect? *Lymphology* 1996; **29**: 76–82.

67 Casley-Smith JR, Casley-Smith Judith R. The effect of topically applied coumarin (5,6 benzo-α-pyrone), as an ointment in P.E.G. or mixed with talc, on the pain of postmastectomy lymphoedema (in preparation)

68 Casley-Smith JR, Casley-Smith Judith R. Four cases treated with complex physical therapy and/or benzo-pyrones. In: Witte MH, Witte CL eds. *Progress in Lymphology XIV. Lymphology* 1994; **27** (Suppl): 618–21.

69 Desprez-Curely JP, Cluzan R, Pecking A. Benzopyrones and postmastectomy lymphoedemas. Double blind trial placebo versus sustained release coumarin with trioxyethylrutin (T.E.R.). In: Casley-Smith JR, Piller NB eds. *Progress in Lymphology X*. Adelaide: University of Adelaide Press, 1985: 203–5.

70 Pflug JJ. Medikamentöse Therapie des Chronisch geschowllenen Beines. *Folia Angiol* 1976; **24**: 78–81.

71 Clodius L, Piller NB. Conservative therapy for postmastectomy lymphoedema. *Chir Plast* 1978; **4**: 193–203.

72 Piller NB, Clodius L. Clinical results of the effectiveness of Venalot in 103 post-lymphoedema patients. In: Bartos V, Davidson JW eds. *Advances in Lymphology.* Prague: Avicenum, 1982: 475–9.

73 Lerut T, Gruwez JA, Vanhoof P. Het gebruik van Cumarin-Rutine bij de Behandeling van peripheer Oedeem, waarbij de Invloed op de Bloedstolling Gecontroleerd Wordt. *Medikon* 1976; **5**: 19–23.

74 Eggert A, Krischner H, Schröder HJ, Wittmann DH. Das lymphostatische Ödem des Armes nach der Brustkrebsbehandlung. *Med Klin* 1977; **72**: 2106–113.

75 El-Shammah A. Behandlung der Armlymphödems nach Mastektomie und Strahlentherapie mit einem Benzopyronpräparat. *Therapiewoche* 1976; **26**: 4915–21.

76 Gregl A, Kutscher K-R, Muschter R. 2. Jahres-kongress der Deutschen Gesellschaft für Lymphologie. *Med Welt* 1979; **30**: 905–8.

77 Casley-Smith Judith R, Casley-Smith JR. Effects of various combinations of complex physical therapy, benzo-pyrones and mercury compression on lymphoedema; which combination works best? In: Cluzan RV, Pecking AP, Lokiec FM eds. *Progress in Lymphology XIII*. Amsterdam: Elsevier, Excerpta Med Int Cong Ser 994, 1992: 537–8.

78 Casley-Smith JR, Casley-Smith Judith R. Volume alterations in lymphoedema, untreated and after Complex Physical Therapy, benzo-pyrones, or both. In: Witte MH, Witte CL eds. *Progress in Lymphology XIV. Lymphology* 1994; **27** (Suppl): 627–31.

79 Casley-Smith JR, Casley-Smith Judith R. Cost-efficiencies of various treatments for lymphoedema. In: Witte MH, Witte CL eds. *Progress in Lymphology XV. Lymphology* (Suppl.) 1996; **29**: 49–55.

80 Lokiec F, Pecking A, Santoni J, Cluzan R. Benzopyrone pharmacokinetic parameters: dose-efficacy of two coumarin dosages in patients with secondary upper limb edemas. In: Nishi M, Uchino S, Yabuki S eds. *Progress in Lymphology XII*. Amsterdam: Elsevier, Excerpta Med Int Cong Ser 887, 1990: 451–2.

81 Grüntzig J, Schicha H, Becker V. Beeinflussung der Lymphdrainage des Glaskörpers durch Kryotherapie, Venalot und Dionin, gemessen mit Lymphpflichtigen radioaktiven Tracern. *Ber Dtsch Ophthalmol Ges* 1979; **76**: 335–42.

82 Bartos V, Brzek V. Das Wirkung von Venalot auf die Lymphaströmungen im Ductus thoracicus des Menschen. *Med Klin* 1970; **39**: 1701–3.

83 Bartos V, Brzek V. Exocrine secretion of the pancreas and enzymes in the lymph after administration of a combination of coumarin and rutin sulphate. *Arzn-Forsch (Drug Res)* 1979; **29**: 548–9.

84 Schmidt F, Keilacker M. Pharmakolymphographie mit Venalot® und Venoruton®. *Radiolidagn* 1985; **26**: 41–51.

85 Collard M. Pharmacoradiologie et pharmacothérapie lymphatiques. *Folia Angiol* 1973; **21**: 293–6.

86 Casley-Smith JR. Behandlung chronischer Veneninsuffizienzen mit 5,6 benzo-α-pyron. *Therapiewoche* 1992; **8**: 671–8.

87 Klüken N. Eine klinische Doppelblindstudie zum Versuch einer Objektivierung der Therapie mit Venopharmaka. *Therapiewoche* 1971; **21**: 72–8.

88 Languillat N. Etude en double aveugle contre placebo de l'activité veinotonique d'Esberiven: évolution de l'hémodynamique musculaire des membres inférieurs appréciée grâce au Xénon 133. *Act Méd Int Angiol* 1989; **5**: 665–8.

89 Blume J. Therapie des venösen Ödems. *Therapiewoche* 1987; **37**: 1362–7.

90 Blume J, de la Haye R. Verstärkte Wirksamkeit der intermittierenden Druckbehandlung durch medikamentöse Begleittherapie. *Phlebol Proktol* 1988; **17**: 53–6.

91 Sandmann W. Cumarin-Kombination bei Durchflußstörungen venöser und lymphatischer Genese. *Ärztl Prax* 1971; **23**: 3449–500.

92 Blume J, de la Haye R, Rühlamm K. Effektivität und Compliance in die Therapie chronisch-venöser Insuffizienz. *Corvas* 1988; **2**: 89–92.

93 Borzeis MG, Angignard J, Dedieu F, Dupont JM, Miloradovich T, Leutenegger E. Effect of a combination of coumarin derivatives and rutosides on venous and lymphatic circulations during severe constriction of the caudal vena cava in rabbits. *Arzn-Forsch (Drug Res)* 1995; **45**: 262–6.

94 Bosse K, Drieschner P, Klose L. Vergleichende Untersuchung über die Wirksamkeit von Venentherapeutika chronisch-venöser Insuffizienz. *Phlebol Proktol* 1985; **14**: 111–14.

95 Simic L, Pirnat L. Behandlung des Ulcus cruris mit einem Benzopyron-Präparat. *Inform Arzt* 1985; **13**: 14–20.

96 Berson I, Geiser J-D. Etude comparative ouverte entre deux phlébotropes chez deux groupes rêpartis au hasard. *Schweiz Rundschau Med (Praxis)* 1980; **69**: 1244–6.

97 Desmons F, Simon M. L'activité d'une préparation associant extrait de Mélilot et rutine hydrosoluble sur les troubles trophique des membres inférieurs d'origin vasculaire. *Phlébologie* 1975; **28**: 155–8.

98 Lacativa AS. Associaçao de cumarina/rutina na rotina do tratamento das flebopatias no Hospital dos Servidores do Esado do Rio de Janeiro. *Med Welt* 1980; **31**: 727–9.

99 Ritschel WA. Therapeutic concentration of coumarin and predicted dosage regimes. *Arnz-Forsch (Drug Res)* 1984; **34**: 907–10.

100 Casley-Smith JR, Casley-Smith Judith R. The effect of oral coumarin on the oedema of dental extractions (in preparation)

101 Pethö A. Nachbehandlung von Laparotomien. *Münch Med Woch* 1984; **126**: 943–5.

102 Pethö A. Nachbehandlung von vaginalen Hysterekotomien mit einen Benzopyron-Präparat. *Fortsch Med* 1984; **17**: 493–6.

103 Mahjoub O. Prophylaxe der Ödementstehund nach Zirkumzision. *Klinikarzt* 1976; **5**: 374–6.

104 Dieckmann J. Möglichkeiten der Schmerzverringerung durch Ödemprophylaxe und Ausschaltung köpereigener schmerzerzeugender Substanzen. *Fortsch Med* 1978; **29**: 1475–80.

105 Khatir M. Clinical-therapeutic experiences with Venalot in dental and surgical procedures. *Quintessence Int* 1976; **7**: 9–11.

106 Banfai P. Die Wirkung eines Benzopyron Präparates auf die Wundheilung und auf die postoperativen Schmerzen bei Tonsillektomie-Patienten. *Arch Arrzneither* 1976; **1**: 77–8.

107 Kriz M. Medikamentöse Verhütung des postischämischen Ödems. *Vasa* 1975; **4**: 63–7.

108 Kriz M. Rekonstruktive Eingriffe an den Halsschlagadern. Prä- und postoperative Behandlung mit einen Benzopyronpräparat. *Klinikarzt* 1977; **6**: 627–9.

109 Kriz M. Gastroentero-Anastomosenödem; Die Präventivbehandlung des Gastroentero-Anastomoseödems mit Benzopyronen. *Zschr Allgemein Med* 1978; **54**: 165–6.

110 Becker HM, Niedermaier G, Orend K-H. Benzopyrone in die Therapie des postrekonstruktiven Ödems. *Fortsch Med* 1985; **21**: 593–6.

111 Imig H, Sinnokrot A. Doppelblindversuch zur medikamentösen Beeinflussung von Schwellungen und Wundödemen nach Extremitätenfrakturen. *Med Welt* 1973; **4**: 547–8.

112 Vida S. Verhütung des Gastroentero-Anastomosen-Ödems durch Benzopyron-behandlung. *Klinikarzt* 1975; **4**: 503–4.

113 Bider E, Anderes W, Dimitrijevic B, Petrovic D. Die Wirkung von Venalot bei traumatischen sowie postoperativen Hämophthalmien, bei Netzhautblutungen und bei Glaskorpertrübungen. *Klin Mbl Augenheilk* 1975; **166**: 403–4.

114 Piller NB. The enhancement of hyphema resolution following benzopyrone therapy in an experimental model. *Exp Eye Res* 1978; **26**: 725–7.

115 Clodius L. Antibiotikaprophylaxe in der plastischen Wiederherstellungs-chirurgie? *Helv Chir Acta* 1978; **45**: 505–10.

116 Preisich H. Zur medikamentösen Therapie der akuten Pankreatitis. *Leber Magen Darm* 1975; **5**: 26–9.

117 Vida S. Über die Wirkung eines Benzopyronpräparates bei der akuten hemorrhagisch-nekrotisierenden Pankreatitis unter chirurgischer Therapie. *Therapiewoche* 1977; **27**: 5476–80.

118 Thornes RD. Coumarins, melanoma and cellular immunity. In: O'Brien DCH, Slater TF eds. *Protective Agents in Cancer*. London: Academic Press, 1983: 43–56.

119 Thornes RD. Acquired immune suppression in chronic brucellosis. *Irish Med J* 1983; **76**: 225.

120 Casley-Smith JR, Casley-Smith Judith R. Coumarin treats chronically bleeding gums (in preparation)

121 Cerlek S. Die Wirkung eines Benzopyron-Präparates auf die Leberfunktion bei chronischer Hepatitis, chronischer, aggressiver Hepatitis und Leberzirrhose. *Arzn-Forsch* 1977; **27**: 611–12.

122 Casley-Smith JR. Benzopyrones in treatment of schizophrenia. *Lancet* 1983: **i**: 421.

123 Casley-Smith JR, Casley-Smith Judith R, Weston F, Johnston PC. A double-blind trial of benzopyrones in chronic schizophrenia. *Psychiatry Res* 1986; **18**: 267–73.

124 Seeliger K. Benzopyrone therapy in severe inflammation; adjuvant effect to reduce tissue fibrosis. In: Partsch H ed. *Progress in Lymphology XI*. Amsterdam: Elsevier, Excerpta Med Int Cong Ser 779, 1988: 623–7.
125 Akiyama T, Kohri K, Iguchi M, Minami K, Yachiku S, Kurita T. Clinical experience with anti-inflammatory agents Esberiven for chronic prostatitis. *Hinyl (kika) kiyo (Act Urol Jap)* 1980; **26**: 71–7.
126 Horster FA, Wildmeister W. Endokrine Orbitopathie. *Dtsch Med Wschr* 1983; **108**: 413–15.
127 Worm A-M, Staberg B, Thomsen K, Hentzer B. Postinflammatory oedema in two patients with contact dermatitis. *Int J Microcirc Clin Exp* 1984; **3**: 41–8.
128 Hladovec J. Vasotropic drugs—survey based on a unifying concept of their mechanism of action. *Arzn-Forsch (Drug Res)* 1977; **27**: 1073–6.
129 Beuscher N. Beeinflussung einer rheumatoiden Modellkrankheit durch ein Benzopron-Präparat. *Z Rheumatol* 1977; **36**: 285–93.

CHAPTER 8

Mode of Action of Coumarin in the Treatment of Thermal Injuries

Neil B. Piller
Department of Public Health, School of Medicine, Flinders University of South Australia

8.1 INTRODUCTION

Coumarin can reduce the severity, duration and damage of mild to moderate thermal wounding. Scalding, which results in the majority of thermal clinical wounds and is the most treatable, best approximates the thermal wounding used in the experimental studies reported here. At the moment there are no preparations which can be given which will have a significant impact on the severity and duration of the oedema. However, there are medications and strategies specifically aimed at healing and reducing the risk of infection of the injured area. Coumarin, or its metabolic products, have the potential to become the mainstay of early treatment of scalds and other forms of thermal wounding.

The mode of action of coumarin is a complex one (Figure 8.1, page 204), involving actions at the cellular and tissue level. Its main effects involve facilitation of the removal of extravasated protein through proteolytic breakdown by stimulated macrophages. Since the entry of protein into the tissues is one of the main reasons why fluid accumulates, its removal means a reduced swelling volume, better tissue oxygenation, a more rapid resolution of the swelling and less tissue damage. In addition to this major action there are many other important aspects of the mode of action of coumarin, such as its effects on red cell deformability, on stabilising labilised membranes and on facilitating glucose uptake.

Coumarins: Biology, Applications and Mode of Action. Edited by R. O'Kennedy and R.D. Thornes.

There is still much to be explored in terms of the mode of action of coumarin, its metabolic products and their potential beneficial effect in resolving thermal wounding. It is recommended that there should be further exploration of various methods for its administration, such as topical application. Given that the events of thermal wounding occur over a number of days, coumarin has great prophylactic potential if administered early in the wounding phase.

The apparent absence of recent research work into coumarins and thermal wounding is of particular concern. The bulk of the known studies were performed in the mid-1970s to mid-1980s. It may now be opportune to revisit this much needed area of research, especially in view of the emergence of antibiotic-resistant organisms.

8.2 GENERAL PICTURE OF THERMAL WOUNDING

The extent of tissue damage subsequent to burns varies greatly and to deal with this is beyond the scope of this chapter. For details the reader is referred to Muir *et al.*[1], Ruberg[2] or Clark[3].

The burns on which coumarin has the greatest potential to have a positive effect are those damaging the dermal and epidermal tissues, that is partial thickness burns. In these, the proteins are coagulated and a range of vasoactive substances are released which permeate through the local (and distant) tissues where they cause a range of changes in the permeability of the lymphatics, blood vasculature and tissues. The end-point is a rapid loss of protein and fluid from the vascular system into the tissues. The subsequent high protein and high flow of lymph fluid exceeds the carrying capacity of the lymphatic system (dynamic insufficiency) and so the wounded tissues become swollen and oedematous. Fluid loss into the tissues will occur for up to about 48 h, after which the capillary and tissue permeability returns to near normal. The lymphatic system now is capable of removing the reduced lymph load and the tissue swelling resolves. The wet and oozing surface of the skin after this time becomes drier and may appear leathery. Scar tissue begins to form if there is no intervention, the granulating tissue and fibrous tissue formation occur in concert with the repair and regeneration of the vascular system. The progress of this stage is strongly dependent on the presence of bacteria and any subsequent infection and associated inflammation. Any bacteria (even non-virulent types) of the local population or any contaminating ones from other sources will significantly impact on the process of reconstruction and repair, and, in the more severely burned patient, may determine the life or death of the patient. From the prophylactic point of view even though the initial thermal wounding may have occurred, the injurious effects proceed and spread for some days after the initial injury. Such 'secondary damage' is related to the inability of the surrounding tissues to bring an adequate supply of nutrients (especially glucose) and oxygen to the tissues around the initial burn area. This means that the thermally injured tissue may be in an unstable condition for up to nine days, during which any factor that can influence various parameters such as immunocompetency, P_{O_2}, lymphatic and tissue clearance of metabolic by-products or other injurious agents, will have a significant effect on the extent of damage beyond the original wounding site.

8.2.1 Thermal Injury on Living Tissues (*in vitro*)

In the early days of our research it became apparent that mammalian cells have a very poor tolerance to thermal wounding[4–6]. The main changes detected by *in vitro* studies are swelling and vasiculation of mitochondria, disorganisation of the cytoplasm with the rough endoplasmic reticulum showing disorganisation and degranulation. Most of these phenomena and eventual cell death can be related to changes in the permeability characteristics of cell and organelle membranes. Aspects of this will be discussed later.

8.2.2 Healing in Thermal Wounding

The amount of oedema affects the distance between adjacent capillaries, and this, combined with the permeability characteristics of the capillaries and the flow through them, determines the oxygen tension in the wounded region. Re-epithelialisation and collagen synthesis are related to oxygen levels, so hypoxia reduces wound-healing and oedema causes hypoxia. Given our knowledge of these events and of the action of coumarin they will be discussed in this chapter.

8.3 THE NATURE OF THERMAL WOUNDING USED IN THE STUDIES

The model of burn injury used in most of these studies has the benefit of reproducibility and precision of control, but there are many differing underlying reactions occurring in each of the vascular, lymphatic and tissue systems. Thus, the picture is complex and can often mask the effects of a drug such as coumarin which has multimode and multisystem effects. In the majority of cases (except where indicated), the model of thermal wounding was the rat limb and the following procedure was used:

Fifteen minutes prior to thermal wounding the animal was anaesthetised with 0.75 ml/100 g body weight of a 10% solution of Sagital (May and Baker, UK) in 10% AR ethyl alcohol. The limbs were shaved. To create the thermal wounding the limbs were immersed in moving water at 54 °C for 60 s. Temperature was maintained within ±0.5 °C. The limb was blotted dry with towelling and the animal allowed to recover. Animals were fed water and food *ad libitum* and maintained at an ambient temperature of 20 ± 1 °C.

8.3.1 Capillary Permeability

This burn is the minimal stimulus necessary to evoke reproducible changes in capillary permeability[7]. In study of the effects of thermal wounding on the capillaries of the diaphragm, Casley-Smith and Window[8] were able to show initially the presence of a high proportion of open junctions. However, they reported that the calculated permeability of the capillaries was much greater than those indicated by the experiment. On this basis, they postulated that it was the connective tissues (and not the endothelium) that was the limiting factor for the movement of small molecules. They also reported large numbers of large vacuoles but no change in small vesicles, indicating that the former play an important role in the transport of large molecules across the cells. While the implications of this will not be directly dealt with in this chapter the possible impact of such events should be kept in mind.

8.4 COUMARIN AND THE MAJOR EVENTS OF THERMAL WOUNDING

Once it was determined, via a series of exploratory studies, that coumarin had potential as a means of treating the high-protein oedema of thermal wounding a series of experiments were designed to elucidate its precise action. The information which follows is meant to take the reader along this pathway.

8.4.1 The Most Effective Dose

Piller[9] compared the effect of various dose levels of benzopyrones, including coumarin, on the rate of formation of the oedema, the maximal swelling volume and the rate of resolution. All dose levels ranging from 3 mg to 100 mg/kg had significant effects on the rate of resolution of the oedema and some effect on the maximal swelling volume. The most effective dose, however, in terms of minimal injurious effects, maximal effect on the volume of oedema and on the rate of resolution, was 25 mg/kg.

A more speculative, but none the less informative, way of calculating the dose required can be based on coumarin/albumin binding figures. Coumarin binds to albumin in a 1 : 1 molar ratio in most species including humans[10]. In rats (200 g) there is about 6.5×10^{-6} mol of albumin, and in an ideal situation for a 1 : 1 molar binding ratio to be attained a minimum of 10×10^{-4} mol of coumarin are required (about 5 mg/kg). However, since not all of the coumarin enters the bloodstream and in consequence of the competition for binding sites etc., to ensure a 1 : 1 ratio about fivefold greater than the above dose, i.e. 25 mg/kg, is required. For humans, there are about 1.7×10^{-3} mol of albumin. Assuming similar conditions as for the rat, then about 1.5×10^{-1} mol (250 mg/70 kg) would be required. A fivefold amount of this would be about 1250 mg/70 kg and if a 10 : 1 human : rat sensitivity is assumed, 125 mg/70 kg would be required. Since the time of these experimental studies the dose level of coumarin used in the treatment of other forms of high-protein oedema such as lymphoedemas has been and continues to be 200 mg/70 kg[11,12]. The sad point is that the full potential of benzopyrones in the treatment of other high-protein oedemas, such as thermal wounding, has not been adequately explored.

8.4.2 The Optimal Frequency of Administration

The standard dose used in these and other studies was 25 mg/kg[13,14]. This dose was normally administered once only and the subsequent outcomes followed. In this study the dose was also varied by administering 1/4 dose every 6 h or 1/12 of the dose every 2 h or by a full dose once per day for the duration of the study. With the exception of one full dose per day, each significantly increased the rate of swelling. All dose levels significantly reduced the maximal swelling volume with the exception of the 1/12 dose, and all but the full dose per day significantly reduced the time at which the maximum swelling was reached. In terms of the resolution of the thermal oedema all dose levels significantly improved the rate of resolution with the best results being achieved with a single full dose of 25 mg/kg. What these findings suggest is that even though coumarin causes some initial vascular injury which results in additional accumulation of proteins into the already damaged tissues, the stimulatory effect on the cells of the mononuclear system (discussed later in this chapter) compensates for this and results in a reduction in the severity of the oedema and more rapid resolution.

8.4.3 Optimal Ambient Temperature for Oedema Resolution

In the older literature there was much debate about the optimal ambient temperature for the facilitation of oedema resolution with some suggesting surface cooling to 10–14 °C[15], while others suggested 5–13 °C[16] as soon as possible after thermal wounding. Ofeigsson[17], however, showed an increased mortality with very cold water (0 °C) and even increased mortality at temperatures of 15 °C. There was also much debate about the rate of entry of fluids into the interstitium, with King and Zimmerman[15] indicating that about 85% entered during the first 30 min while others claimed fluid entry for up to 48 h. There were unanimous indications that some degree of surface cooling was beneficial. This worked by reducing the amount of capillary outflow. A study by Piller[18] indicated that there was a temperature-dependent effect associated with coumarin administration. To explore this the optimal dose level of coumarin was used in combination with four ambient temperature levels of 10, 20, 30 or 35 °C. Coumarin was most effective when moving air of 20 °C was blown over the burnt animals. At this temperature the maximal swelling volume was least, the time at which this volume was reached was shortest, the rate of resolution the greatest and the time taken for the limbs to return to normal was the quickest[18].

8.5 EFFECTS OF COUMARIN ON GROSS PARAMETERS OF THE THERMALLY INJURED LIMB

8.5.1 Rate of Swelling

All dose levels of coumarin increased the rate of swelling of the thermally injured limb over and above that of the thermal injury alone. The dose level of 25 mg/kg had the least effect[9], thus making this dose level the most attractive since, as will be seen later, even though the net result of coumarin administration is to eventually reduce the severity and duration of the wounding it does have an initial minor injurious effect on the vasculature[9,14,19].

8.5.2 Maximum Swelling Volume

Coumarin, at the optimal dose of 25 mg/kg, significantly shortened the time taken to reach the maximum swelling volume of thermally injured limbs by up to 45%[14,18] and significantly reduced the maximal swelling volume by 20% in most studies[9,14].

8.5.3 Rate of Resolution

At the standard dose level with optimal ambient temperature, coumarin resulted in significant improvement in the rate of resolution of the oedema; the rate being four times that of the untreated thermally wounded control group[20]. Other dose levels, however, also facilitated this increased rate of resolution[9,14], as did other members of the benzopyrone group[9,20].

8.6 THE DISTRIBUTION AND MOVEMENT OF ADMINISTERED COUMARIN

8.6.1 Tissue Distribution of Coumarin

The fate of administered coumarin became the next focus of attention. A single dose of radioactive coumarin, at a dose of 25 mg/kg, to uninjured animals[21] showed that 7.4% of the injected dose remained in the tissues after 100 h. During this time 30.9% of the dose was excreted in the faeces and 47% excreted in the urine. At any given time most of the dose remaining was present in the gut, muscular tissues, skin and liver, this averaged 33% for muscular tissues, 28% for skin and 16% for liver. The highest concentrations per gram of tissue were, however, in kidney and liver, which represent the two organs of metabolism and excretion of coumarin. These findings were in a similar range to those of other earlier studies by Furuya[22] and Kaighen and Williams[23] although they differ from the findings of van Sumere and Teuchy[24], in which much lower doses were used.

Piller[21] showed that at the end of 2 h, 92% of the injected dose remained in the tissues and at 24 h it was 38%. There is much biological variation between animal species, of course, and different rates of excretion with various dose levels. However, with the optimal dose administered daily, the tissue concentrations are high enough to maintain an effect, at least in rats. In comparing the metabolic pathways in the rats to those in humans Shilling *et al.*[25] showed that 79% of an orally administered dose of coumarin was excreted in the urine (mainly as 7 hydroxycoumarin) while in the animal study only 45% of the injected dose was excreted, and then only 55% of this in the urine. Ritschel *et al.*[26] found that the major pharmacokinetic parameters for coumarin show a distinct interspecies relationship. This is discussed elsewhere in this book (see Chapter 3).

8.6.2 Binding of Coumarin to Serum Proteins

Why does so much of the coumarin remain in the tissues for so long? The majority of the members of the benzopyrone group of drugs bind to serum proteins[27]. The degree of binding seems to be related to the number of phenolic groups and dependent on the ability to form hydrogen and hydrophobic bonds with the albumins. The extent of this binding obviously has a strong influence on the pharmacological and pharmacokinetic properties and, thus, on the potential duration of action of the drug. In this respect coumarin can remain at pharmacologically active levels for about 24 h (at least in rats)[21] and at the end of this period some 80% of the administered drug remained in the tissues[21].

8.6.3 *In vitro* and *in vivo* Comparisons of the Binding of Coumarin

To confirm serum binding, a study involving [3-^{14}C]coumarin in a range of concentrations and dose levels from 3 to 50 mg/kg or /ml were used[28]. *In vitro* and *in vivo* estimations of serum binding were made 15 min after administration or injection of the radiolabelled coumarin. *In vitro* results showed the percentage of coumarin bound was between 36.7% and 41.5%. There was no correlation between the concentration of coumarin administered and the percentage bound. Intravenous administration showed a range of 30.6–42% bound, again with no correlation between dose and binding percentage. A similar situation occurred for intraperitoneal administration although the range was between 28.3% and 42.6% bound. The fact that this binding occurs rapidly and coumarin rapidly enters damaged tissues means that high levels quickly reach areas where it is needed and is an important factor in its mode of action.

The comparatively high tissue levels of coumarin is probably due to the fact that coumarin can bind to serum proteins[10], although it is likely that coumarin has only a single binding site. Garten and Wosilait[27] reported approximately 40% binding while Piller[28] reported approximately 37% binding to serum proteins. To confirm the specificity of this binding under normal tissue conditions Piller and Schmitt[29] used [3-^{14}C]coumarin in rats at a dose level of 25 mg/kg. The study showed that

36% was bound to the slow and fast globulins, 11% to the post albumins, 10% to globulin and 9% to albumin. This means that large quantities of coumarin can enter the tissues where the target cells (the monocyte–macrophage system) are located. Once concentrated in these areas what did it do? In part, an answer for this was obtained using a rat leg model of high-protein oedema to explore the effect of coumarin on radiolabelled protein (RPP) and radiolabelled povidone (RPVP) clearance from thermally wounded tissues[30–32]. Their clearance was determined at 0, 3, 6, 24 and 48 h after injection. Coumarin administration at the optimal dose of 25 mg/kg not only resulted in a significant enhancement of the removal of RPP from the normal limbs but also from limbs receiving thermal wounding. There was also a significant increase in the removal of RPVP, but, most importantly, in both the normal and thermally wounded tissues coumarin caused the ratio of RPP : RPVP to fall significantly, indicating greater protein clearance. In the same experiment it was shown that the clearance ratios were similar although slightly less, thus providing an insight into another mode of action of coumarin. It did not need a functional lymphatic system to facilitate the protein removal although there was evidence from other studies that it was capable of increasing lymph flow.

8.7 CELLUAR EFFECTS: THE MACROPHAGE

8.7.1 Coumarin as a Stimulator of the Mononuclear Phagocytic System

The liver has one of the most dense concentrations of potentially phagocytic cells of the body (about 30% of its weight is these cells[32]). Some stimulators of the reticuloendothelial system increase the number of phagocytic sites and cells. This results in an increase in liver and spleen weights[33]. Others act by increasing particle engulfment[34,35] and have no effect on liver or spleen weights. Piller[36] was able to show that coumarin at optimal dose levels results in a statistically significant increase (13%) in rat liver weights in both normal and thermally wounded groups within 48 h. Although the actual increase in liver weights was not great, when this is translated to changes in cell numbers it could mean up to 60% increase in phagocytic cell numbers in the liver alone[36]. The number of potentially phagocytic cells in the rat liver was estimated at 3×10^9 but within 48 h coumarin had increased this number by 1.7×10^9 in both the control and thermally injured group. Thermal wounding alone had no effect on liver weights at 48 h.

8.7.2 Coumarin Effects on Macrophages

There are large increases in the number of macrophages in high-protein oedemas[37] and enzymatic studies of serum, lymph and tissue fluids indicated that they play an important role in tissue repair and oedema resolution. However, they are not the only cells involved, as fibroblastic cell lines also play an important role[38]. To assess the importance of the macrophage and the effect coumarin has on it, a macrophage-free animal was required. This was created in a rather primitive but effective manner with tridymite silica which has a selective toxicity towards macrophages[39–41]. It was difficult to actually assess the viability of the macrophages remaining but certainly the population was substantially depleted[41]. Silica is non-metabolisable, so once in the system it will remain an effective destructor of any cell which re-phagocytoses the fragments of a previous victim. The mechanism of destruction is via rupture of lysosomal membranes with the subsequent release of the lysosomal enzymes into the cytoplasm whereupon death of the cell occurs. While the short-term effects of silica on the phagocytes is devastating, in the longer term it becomes sequestered in lymphoid tissue as fibrous granulomatous lesions. This takes about two months[41].

In the silica-treated cases, the normal processes of tissue proteolysis and the enhanced proteolysis stimulated by the coumarin no longer occur. This results in a reduced proteolytic effect on the extravasated proteins, which largely accumulate in the tissues and there is now a tendency to osmotically trap more fluid. Thus, there was a more rapid rate of swelling in the burn group pretreated with silica and coumarin compared with the burn group receiving only coumarin. Likewise, the maximal swelling volumes were significantly greater and lasted significantly longer[42].

8.8 PROTEINS: THE TARGET FOR ACTION

8.8.1 Abnormally Accumulated Proteins

It then appeared that macrophages were one of the major targets for coumarin and that the end-result of this process seemed to be a more rapid resolution of any swelling. The fact that a stimulation of lymphatic outflow was not likely to be the main action was supported by similar rates of resolution in a model of lymphoedema where the lymphatics had been ligated. Since it was certain that proteins, by virtue of their osmotic effect, could hold fluid, it seemed that the more rapid resolution of the oedema was related to the removal of the protein. To gain a better understanding of the effects of coumarin on accumulated proteins and, thus, to be able to determine the nature of its cellular actions, a number of studies[31,45,46] were conducted using radiolabelled protein (RPP) and radiolabelled povidone (RPVP—a non-metabolisable molecule). The effect of coumarin on the clearance of RPP and RPVP from the blood was explored. The results of this study showed a significant reduction in the levels of labelled RPP in the blood at 30 min (42%) and 1.5 h (55%) compared with the thermally wounded but untreated control group. There were no significant differences in the groups at later observation times, viz. 6 and 24 h. For PVP there were similar significant reductions at 30 min (66%) and 1.5 h (71%) and a lesser, although still significant reduction, at 6 h (39%). These results can be explained by an initial uptake of RPP and RPVP by the cells of the mononuclear phagocytic system or its entry into other tissues. Hence, stimulation by coumarin is very important since, in burns, the activity of the mononuclear phagocytic system is depressed.

8.8.2 Tissue Levels of Radiolabelled Protein and Povidone

Coumarin causes a slight additional injury to the vascular endothelium of thermally injured tissues[35]. This injury, which lasts about 30 min, may result from histamine release[43]. Other relatives of coumarin (the rutosides) also have a similar action[44]. The effect is confirmed by the increased rate of swelling of the coumarin-treated thermally wounded tissues over those only thermally wounded. This injurious effect can also be confirmed by the use of ^{51}Cr-labelled plasma protein. Piller[45] indicated a significant 30% increase in tissue level of RPP labelling 30 min after thermal wounding in the coumarin-treated thermally injured group compared to the group which received thermal injury alone. The injurious effect of coumarin was short-lived and by 1.5 h the amount of ^{51}Cr-labelled protein in the tissues was only 20% of that of the untreated thermally wounded group. This reduction continued for the next 24 h, suggesting that the cessation of the injury is due to enhanced phagocytosis, digestion and *egestion* of the fragments, which can then leave the tissues the way they came in (e.g. via the vascular system).

8.9 EVIDENCE FOR PROTEOLYSIS

It is likely that the form of proteolysis induced by the benzopyrone group generally, and coumarin specifically, is very different from that which is normally used by pharmacologists as a measure of inflammation. The usual proteolysis of inflammation represents tissue derangement. However, the proteolysis induced by coumarin and the other benzopyrones lessens the more injurious effects of this derangement. All of the benzopyrones studied to date induce proteolytic activity which seems confined to the abnormally accumulated protein in the extracellular tissue compartment[31,46,49].

Coumarin facilitated a rapid removal of tracer protein from thermally induced wounds[31], but evidence was required to support the hypothesis that it was via proteolysis. Thus, ^{15}Cr-labelled RPP was bound to homologous rat plasma and dialysis was performed *to ensure 100% binding*. Thermally wounded tissues from the limb and body were sampled after 6, 24 and 48 h. Following tissue homogenisation, proteins were precipitated out using trichloroacetic acid (TCA). The supernatant was decanted and its radioactivity counted[47]. TCA precipitates all proteins with molecular weight greater than 1000 and only fragments with a molecular weight of up to 40 000 have been shown to pass through blood vascular junctions[48]. This experiment considerably underestimates the extent of the lysis and the percentage of the fragments which are small enough to return directly to the vascular system. Coumarin alone produced significantly increased leg and body counts of TCA-soluble ^{51}Cr (protein fragments) of 52 and 44%, respectively, over those of the normal tissues. The tissues which received thermal wounding, in addition to coumarin, resulted in a further 22% significant increase in fragments in the wounded tissues but not in the body tissues. Thermal wounding alone had no effect on fragment levels. At 24 h coumarin increased leg counts of soluble ^{51}Cr by 27% and body counts by 46% over the corresponding untreated thermally wounded tissue. Both increases were significant. At 48 h again there was a significant 59% increase in fragments in the thermally wounded tissues but not in those of the body[47].

It was becoming certain then that coumarin had an ability to lyse homologous ^{51}Cr-labelled protein. The exact nature and extent of this lysis remained to be determined. Was it a direct or indirect effect? A simple experiment, however, in which coumarin was incubated directly with ^{51}Cr-labelled protein at levels approximating those of the tissue showed coumarin to have no effect on the proportion of ^{51}Cr fragments, thus indicating that coumarin did not have a direct effect and, in fact, needed cells of the body to do this[47].

8.9.1 Proteolysis Through Stimulation of Macrophages

Additional evidence for this process relates to the changes in levels of induced proteolytic enzymes active at acid and neutral pH. Of greatest interest in these studies are the effects of the various benzopyrones on acid proteinases[49,50]. Since macrophages are a very rich source of cathepsin D, these levels will closely parallel the general acid protease level. For the enzymes which work optimally at the neutral pH range other cells may be responsible, although it is still speculation at the moment as to the exact cell types involved[51].

8.9.2 Enzyme Activity Levels (Acid and Neutral Proteases)

Enzyme activity levels in the various extracellular regions of the body can be used as an approximation of changes that are occurring in nearby or distant regions. Generally as a consequence of digestion of phagocytosed material, the lysosomal enzymes, together with digestion fragments, enter the surrounding extracellular spaces. Hence in this situation, enzyme activity levels are good indicators of the extent of lysis, with higher levels being associated with more rapid and efficient digestion. Tests for proteases are only general measures of enzymatic activity at either acid or neutral pH: the lysosomes of macrophages are a rich source of cathepsin D which is active at acid pH so changes in its levels are mirrored by changes in protease activity at an acid pH[52]. Proteases active at neutral pH probably originate from neutrophils and fibroblasts[38,53].

8.9.3 Serum

In thermally wounded animals treated with coumarins, serum acid proteases showed a significant increase of 12 and 15%, 6 and 24 h, respectively, after the thermal wounding, compared with those not treated. The results for serum neutral proteases showed smaller increases[49].

8.9.4 Oedema Fluid

In thermally wounded animals treated with coumarin, free oedema fluid acid proteases showed very significant increases of 45 and 60%, 6 and 24 h, respectively, after the thermal wounding, compared with controls. Neutral protease levels showed little change in the free oedema fluid[49].

8.9.5 Skin

Skin acid proteases showed significant increases of 52 and 16%, 6 and 24 h, respectively, after thermal wounding in coumarin-treated animals compared with untreated controls. Skin neutral proteases showed lesser increases of 15 and 13%, respectively, for the same time intervals[49].

8.9.6 Muscle

In thermally wounded animals treated with coumarin, only muscle acid proteases showed a significant increase of 15% at 6 h after the thermal wounding in comparison with untreated controls. The results for skin neutral proteases showed a significant decrease of 22% at 24 h[49].

8.10 SIMILARITIES WITH OTHER BENZOPYRONES

Most of the studies referred to in this chapter have tested a range of the benzopyrones and structurally related substances, and in the main, the actions and the results are similar. Piller[48] showed that there were substantial correlations between the effectiveness of various benzopyrones and the activity levels of acid and neutral proteinases. In particular, there were strong correlations between oedema-reducing abilities of the benzopyrones and the acid protease activity in the fluids of the oedema and also between acid protease activity levels and oedema-reducing abilities in the skin of thermally wounded tissues. Similar very strong correlations were found for neutral proteinases in the skin but a smaller correlation was found for the oedema fluids.

8.11 EFFECTS OF COUMARIN ON CELLS OTHER THAN MACROPHAGES

There is plenty of evidence that there are large infiltrates of polymorphs and fibroblasts into injured areas, particularly early on between 2 and 4 h after injury, but as yet there have been no studies specifically on the action of coumarin on these cells in thermal wounding except for some tenuous links between cell types and extracellular enzyme types and levels[49,50]. Higginbottom *et al.*[54] indicated that mast cell granules from degranulated cells subsequent to injury were taken up selectively by fibroblasts.

It is postulated that coumarin might enhance this selective uptake thus, in part, reducing the oedemas because of the removal of a range of vasoactive substances which are contained therein.

8.12 EFFECT OF COUMARIN ON POST BURN PLASMA VOLUME LOSS

There are many pieces of evidence[31,45] that indicate that there is more rapid removal of extravasated proteins and fluids from the tissues and these have been addressed above. Also, while it has been shown that coumarin (and the other benzopyrones) cause some injury to the microvasculature[19], the effect is transient and is far outweighed by the lymphostimulatory and proteolytic enhancement effects[49]. However, the benzopyrones do have protective capabilities[55]. An awareness of this was first indicated by Armentano[56] and followed up by Ambrose and de Eds[51], Bohr *et al.*[57] and Arturson and Jonsson[58]. Hilton[59–61] expressed some concern about the fact that the earlier studies concentrated only on the local effects of the thermal injury and initiated studies to explore the generalised effects of thermal wounding on vascular permeability. While they used β-hydroxyethyl rutosides there is enough evidence about general similarities between these and other benzopyrones to be able to make some suppositions about coumarin. Hilton[59–61] showed a significant effect of the rutosides (with higher doses being more effective) in reducing general plasma volume loss. Then, as now, there was some considerable speculation about how the rutosides produced this permeability effect and how they can have beneficial effects on capillary fragility. The mechanisms proposed range from a reduction in catecholamine metabolism, a direct effect on the 'cement substances' holding the endothelial cells together, alteration of prostaglandins and a possible action on the adrenals. While this mechanism needs investigation it is without doubt that the prophylactic use of the benzopyrones generally (and coumarin in particular) can have a significant protective effect in conditions where subsequent thermal wounding is likely, or in the early stages of thermal wounding.

8.12.1 The Importance of Lymph Drainage

Changes in temperature, such as the immersion of a foot in temperatures even as low as 42 °C, cause an almost immediate increase in lymph flow and the pulse frequency of the lymph collectors[62]. It is postulated that this increase is a consequence of augmented lymph formation.

The measurement of the changes in the composition of the lymph have great potential to help elucidate the pharmacological and biochemical changes occurring in the damaged tissues and, at present, there is certainly a paucity of information relating to these changes. There are great changes in enzymes, which generally reflect the extent of damage to the range of tissue cell types. For instance, at lesser thermal wounding temperatures there are mainly cytoplasmic and mitochondrial enzymes[63]. At the same time, lymph-to-plasma ratios of the more common proteins increase, indicating an increased capillary permeability to proteins of molecular weights up to about 30 000[64].

At higher temperatures of thermal wounding there is also an increased lymph flow, and almost all of the constituents of the plasma are found in this lymph. However, the external drainage of the lymph, containing this cocktail of substances, does have a significant protective effect. It prevents recirculation of a range of kinins and other vasoactive products which would otherwise exacerbate conditions where the permeability at the wounded site is already increased.

8.13 OTHER EVENTS IN THERMALLY WOUNDED TISSUE

8.13.1 Wound Sepsis

Infection is one of the major problems facing the thermally wounded patient. The organisms found in burn wounds are *Staphylococcus aureus*, *Candida albicans*, *Pseudomonas aeruginosa*, *Klebsiella enterobacter* and *Escherichia coli*. There are many and varied strategies to reduce the opportunity that these organisms have to impact on wound-healing and patient mortality and morbidity. Further information may be obtained from the journal *Burns* or from the established literature[65].

The increased risk of infection stems from a marked reduction in oxygen levels in thermal wounds and the cascade effect that this has on other cellular and tissue functions. Polymorphonuclear cells (PMNs) and macrophages accumulate in thermally injured tissues in large numbers. The PMNs are aerobic, while the macrophages can be regarded as facultative anaerobes. In thermal wounding it is most likely that a number of macrophage subpopulations are affected in and around the injured area. The phagocytic activity of macrophages in burns is depressed and it is mainly the fixed members which are affected. Rittenburg and Hanback[66] suggested that in patients (and animals) who died the phagocytic capacity was generally depressed, while in those who survived the capacity was normal or enhanced. The reduced survival and increased morbidity was related to an increased susceptibility to infection resulting from reduced tissue oxygenation[67].

Creagh *et al.*[68] showed that while coumarin did not seem to have a systemic effect on septicaemia it did have a local beneficial effect on the affected tissue areas.

8.13.2 Vasoactive Substances

Lymph fluid contains most of the same constituents as plasma, albeit in slightly different concentrations and levels. Following thermal injury, many studies have demonstrated an increased lymph flow and some have indicated an increase in its kinin-forming activity[69]. Lewis and Wawratschek[70] reported a large increase in kallikin following thermal injury in the lymph. Its origins were postulated to be the blood plasma from which it escaped through the increase in vascular permeability. Once in contact with the thermally damaged cells it was activated. It would thus seem logical that the recirculation of these and other vasoactive substances via the lymphatics from the injury site could lead to further increases in blood capillary permeability at the site but also could explain the generalised changes which have been suggested[59–61]. Although coumarin certainly was shown to increase lymph flow it seems possible that its protective effect in thermal wounding is not through the facilitation of lymphatic drainage. It may be exerted through some effects on (a) the kinin formation system, (b) the levels of prostaglandins (which are increased in thermal wounding), (c) the lability of cell membranes or (d) the vasoactive substances themselves. While there are still no conclusive answers, early work by Foldi-Borcsok[71] showed a significant protective effect of external drainage, suggesting that the removal of vasoactive substances from the body would reduce further permeability effects. She also found that coumarin had, in addition, the strongest protective effect and that this was not mediated by a further increase in lymph flow, thus adding weight to the protective and deactivation aspects of coumarin.

8.13.3 Neovascularisation

It seems clear that macrophages (or at least their products) are able to stimulate new blood vessel formation[72], which would have the advantage of facilitating a more rapid removal of wastes and increased tissue oxygenation. There is no evidence of macrophages being able to stimulate the growth of any lymphatic structures and this is an area in need of urgent exploration.

8.14 EFFECT OF COUMARIN ON SPECIFIC AND NON-SPECIFIC IMMUNITY

Burn wound sepsis is a primary cause of death in patients following thermal wounding. This is related to changes in the functioning and capabilities of the immune system which predispose infection. Bjornson and Alexander[73] reported a reduction in serum opsonic activity and showed that this preceded and followed episodes of sepsis. In 1976 Bjornson *et al.*[74] showed a reduced ability to convert complement (which was distinct from the opsonic abnormality) in thermally wounded patients. Ransjo *et al.*[75] also showed an impaired PMN function in burns patients, with significant reductions in bactericidal activity, decreased chemotactic activity, and the random migration of PMNs. They reported maximal bacterial growth when the PMN function was most disturbed. This, and other evidence of defective phagocytosis, indicate the potential importance of being able to stimulate the PMNs and other parts of the body's specific and non-specific immune functions when they are exhibiting a lesser than normal function. Marshall *et al.*[76] and others have shown that coumarin does possess immunomodulatory activity through an enhancement of DR and DQ antigen expression by the mononuclear cells. The results of Stuhlmeier *et al.*[77] also support this immunomodulatory theory and there is evidence of activity of one of its major metabolic products, viz. 7-hydroxycoumarin[78]. This topic is discussed in greater detail in other chapters (e.g. Chapters 5 and 9).

8.14.1 Vascular Endothelium

Coumarin initially injured vascular endothelium of the deep venules and superficial capillaries over and above that of the first phase of mild burn injury (measured 20 min after burn). The second phase of mild burn injury (2 h after burn) was normally characterised by an increase in capillary permeability[19]. At this time coumarin reduced the labelling of the superficial capillaries but had no effect on the superficial venules. In the second phase, coumarin prevented the burn-induced increase in permeability. However, in the first phase coumarin increased the permeability of normally unaffected vessels. The net effect is a good one[9,14].

The intravenous injection of colloidal carbon has widely been used to detect microvascular leakage[79]. The colloid moves down the open junctions of the leaking vessels where they become trapped by the basement membrane, resulting in intramural labelling of the vessels.

Coumarin and other benzopyrones have a significant effect on the number of open junctions[8,9], possibly as a consequence of antioxidant effects on adrenaline[51]. The increased adrenaline levels result in a greater tonicity of the precapillary sphincters of the dermal vessels, thus reducing blood flow. However, in muscular tissues blood pressure and flow are increased[80]. It is this, along with the direct effect of coumarin on the vascular endothelium[19,81], which allowed leakage via the open junctions adding to the protein and the fluid in the tissues. Generally, while there is a strong action of the benzopyrones on the junctions, there is little or no effect on the vesicle or vacuole numbers or their movement[8].

The administration of coumarin to animals with thermal oedema increased the rate of swelling of the injured part. When silica was administered prior to coumarin the rate of swelling was greater still. This effect can be explained if the rate of swelling is divided into two components: (a) the entry of protein and fluid, and (b) the rate of its removal via the lymphatics and by proteolysis. (Since macrophages are the major proteolytic force when the lymphatic system is overwhelmed and since they are destroyed by the silica, not only is the protein accumulating in the tissues, but so are the various intercellular components of the destroyed phagocytic cells.) These both considerably increase tissue colloidal osmotic forces which attract extra fluid. Thus, pretreatment with silica means that the normal proteolysis of tissue proteins does not occur nor can coumarin have any enhancing effect on the macrophages since they are destroyed. Hence, oedema fluid rapidly accumulates.

8.14.2 Red Cell Deformability

Coumarin and 7-hydroxycoumarin have been shown to increase the deformability of red blood cells. Schachar *et al.*[82] has indicated that red cell deformability is decreased in thermal wounding. As yet there is no strong evidence that coumarin or the benzopyrones can change this in burns. They can do this in conditions such as metabolic acidosis[83] which is one of the changes very likely to occur in thermally injured tissues. However, it is known that benzopyrones do have a desegregating effect[48] on erythrocytes, which is very important from the haemodynamic viewpoint since the critical level at which red cell aggregates are formed is reduced.

8.14.3 Platelet Aggregation

Although some research suggested that burns can inhibit platelet aggregation[84], the general consensus is that thermal wounding enhances aggregation. There are a range of factors that can elicit this process, including ADP, a range of soluble enzymes, collagen and even heat-denatured plasma. The aggregation can be divided in two phases: an initial partial aggregation and a complete aggregation phase[85]. Benzopyrones can inhibit such aggregation[86], although the exact mechanism is still uncertain. The action may be related to their ability to stabilise previously labile biological membranes[87,88], which, as Marcus[85] showed, could prevent the second phase of aggregation.

8.14.4 Labilised Membranes

Benzopyrones have a variety of different effects on cell membranes. Mixtures of coumarin with other rutosides have been shown to be good stabilisers, generally when the membrane has been previously labilised as a consequence of its ligand power[87,89]. The picture is rather complex, however, and for additional details refer to Casley-Smith and Casley-Smith[43].

8.14.5 Capillary Fragility

Gabor[90] found that most of the benzopyrones reduced abnormally increased capillary fragility. The action is believed to be related to changes in the strength of the bonds between the cells at the blood vascular junctions.

8.14.6 Thrombosis

Haas[91] reported that benzopyrones have some prophylactic protective effect against thrombosis and emboli. However, this mechanism is not via anticoagulation[92].

8.14.7 Glucose Uptake and Transport

Dannon and Elazar[93] showed that benzopyrones enhanced glucose uptake and transport by red blood cells. This effect would seem to protect the red blood cells in stagnant areas of the circulation. Thus, not only can the benzopyrones improve tissue circulation and oxygenation by removing the stagnating fluid, they also have the potential to permit the cells to remain at close to optimal functional levels in any remaining areas of relatively hostile extracellular environment.

8.15 OTHER 'BURN-LIKE' INJURIES

Matook *et al.*[94] reported some similarities between burn injuries and the 'burn-like' injuries which occur when tissue is exposed to the evaporative heat loss caused by contact with high-pressure propane. In view of such similarity, coumarin is likely to also have a beneficial effect in this 'burn' area.

8.16 AN INTEGRATION OF THE ACTIONS OF COUMARIN IN THERMAL WOUNDING

While the above experiments provided an indication of the specific actions of coumarin, it remained to integrate them. Although not all of the actions in what follows have been confirmed in thermal wounding studies, all have a sound basis from other studies of other oedemas such as lymphoedema. An integration of the mode of action of coumarin was first presented by Piller in 1977[95] and, to a great extent, the basis of this has not been challenged to date. The following summary and Figure 8.1 draw together these findings.

Upon administration, coumarin binds to the relatively small serum proteins. Since there is an increase in capillary permeability following thermal injury this infers that, if an optimal dose of coumarin is given, then most, if not all, of the protein which enters the injured areas will carry with it the maximal quantity of bound coumarin. In thermally injured tissues there are large infiltrates of monocytes and the resident tissue macrophages, which will rapidly phagocytose this protein and the bound coumarin. How this more rapid phagocytosis is facilitated is still not clear but the net result is an increased uptake of particular matter, cells and other debris. This stimulatory effect of coumarin occurs not only on individual cells but older quiescent cells may be rejuvenated with increasing cell numbers at the injured site and elsewhere.

Following phagocytosis, the ingested material is digested and then egested or it leaves the cell via a range of diffusion-type processes, or perhaps is actively carried across the cell membranes by events such as reverse pinocytosis. The actual mechanism may be of little consequence. The result is that protein fragments of low molecular weight, such as peptide units and even amino acids, can leave the injured tissue

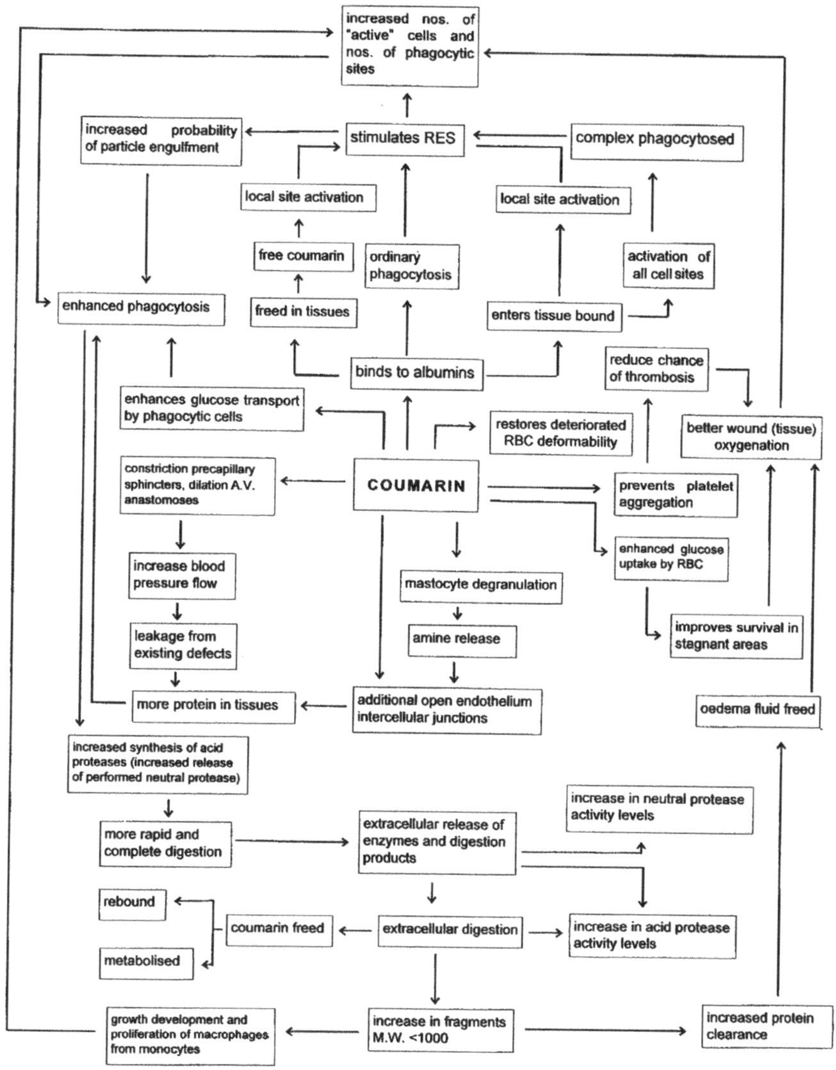

Figure 8.1 A schematic representation of the mode of action of coumarin in thermal wounding. (Reproduced by permission from Piller NB. An integration of the modes of action of coumarin. *Arzneim Forsch* 1977; **27**(1)(6): 1135–8.)

area via back diffusion through the vascular system, which is the way the proteins originally came. Their removal is not dependent on an overloaded lymphatic system. In the early stages of the entry of coumarin into injured tissues, coumarin actually seems to injure the tissues more. This might at first seem potentially deleterious to the tissues, but in fact it seems to be beneficial because it allows higher concentrations of coumarin into the tissues. This facilitates greater phagocytosis and more complete (and rapid) digestion. It seems as if the egested fragments undergo further lysis in the extracellular spaces. These fragments possibly serve as a nutritional base for the large number of monocytes which enter and proliferate in the injured tissues.

The healing of a thermally injured wound is very dependent on oxygen levels and, thus, the vascular supply and also to a lesser extent on the potency of the lymphatic system. Hence, any tendency to improve vascular supply by reducing the distance between the capillaries by reducing the oedema is of crucial importance. This is one of the main strengths of coumarin (and the other benzopyrones). They reduce the oedema by facilitating protein removal from the tissues—once the protein goes, so too does most of the osmotically held water, mainly via the vascular system as it follows the smaller metabolised fragments. Nevertheless, coumarin can also certainly stimulate the functioning of the lymphatic collectors, assisting them to carry more lymph from the area. Less oedema means better oxygenation of the tissues, and more rapid revascularisation. This improves wound strength and tissue remodelling.

The form of proteolysis described here is drug-induced proteolysis and often it is difficult to reconcile proteolysis (which is most often a measure of tissue derangement or damage) with a repair process. However, coumarin-induced proteolysis is a controlled proteolysis, which in net terms is beneficial to the tissues and the cells within.

The induction of proteolysis is the major action of coumarin. There are many other actions such as its effect on platelet aggregation, its effect in stabilising biological membranes, its effect on red cell deformability and its effect on glucose uptake and transport by cells, etc. The end-result is a healthier tissue, less damage, less oedema, less risk of sepsis and more rapid repair process.

REFERENCES

1 Muir IFK, Barklay TL, Settle JAD. *Burns and their Management*, 3rd edn. Oxford: Butterworth Heinemann, 1987.

2 Ruberg RL. Advances in burn care. In: *Clinics in Plastic Surgery*. Philadelphia: W.B. Saunders, 1986: 31.1.

3 Clark JA. *A Color Atlas of Burn Injuries*. London: Chapman & Hall Medical, 1992.

4 Lewis T. *The Blood Vessels of the Human Skin and their Responses*. London: Shaw & Sons, 1927.

5 Moritz AR, Henriques FC. Studies of thermal injury II. The relative importance of time and surface temperature in the causation of cutaneous burns. *Am J Path* 1947; **23**: 695–720.

6 Sevitt S. Inflammatory changes in burned skin: Reversible and irreversible effects and their pathogenesis. In: Thomas I, Uhr JW, Grant L eds. *Injury Inflammation and Immunity.* Baltimore: Williams & Wilkins, 1964: 183.

7 Willem DL, Mason B. Vascular permeability changes in inflammation; the role of endogenous permeability factors in mild thermal injury. *Br J Exp Path* 1960; **41**: 487–506.

8 Casley-Smith JR, Window J. Quantitative morphological correlations of alterations in capillary permeability, following histamine and moderate burning, in the mouse diaphragm and the effects of the benzopyrones. *Microvasc Res* 1976; **11**: 279–305.

9 Piller NB. Benzo-pyrone treatment of mild thermal oedema: Determination of the most effective doses. *Arzneim Forsch* 1977; **27**(1) 6: 1138–41.

10 O'Reilly RA. The interaction of several coumarin compounds with human and canine plasma albumin. *Mol Pharmacol* 1971; **7**: 209–12.

11 Piller NB, Casley-Smith JR, Morgan RG. A double blind cross over trial of *O*-(β-hydroxyethyl) rutosides (benzopyrones) in the treatment of lymphoedema of the arms and legs. *Br J Plast Surg* 1988; **41**: 20–7.

12 Casley-Smith JR, Morgan RG, Piller NB. Treatment of lymphoedema of the arms and legs with 5,6-benzo-alpha-pyrone. *New Engl J Med* 1993; **329**: 1158–63.

13 Piller NB. Lymphoedema, macrophages and benzopyrones. *Lymphology* 1980; **13**(3): 109–19.

14 Piller NB. The effect of benzo-pyrones on the maximal swelling volume and resolution of thermally-induced oedema in the rat: A determination of their optimal frequency of administration. *Folia Angiol* 1976; **XXIV**: 160–5.

15 King TC, Zimmerman JM. First aid cooling of the flesh burn. *Surg Gynecol Obstet* 1965; **120**: 1271–3.

16 Schulman AG. Ice water as primary treatment of burns. *J Am Med Assoc* 1960; **173**: 1916–19.

17 Ofeigsson OJ. Water cooling: First aid treatment for scalds and burns. *Surgery* 1965; **57**: 391–5.

18 Piller NB. The resolution of thermal oedema at various temperatures under coumarin treatment. *Br J Exp Path* 1975; **56**: 83–91.

19 Piller NB. Benzo-pyrones: Their selective injury to rabbit vascular endothelium. *Clin Exp Pharm Physiol* 1976, **3**: 127–39.

20 Piller NB. A comparison of the effectiveness of some anti-inflammatory drugs on thermal oedemas. *Br J Exp Path* 1975; **56**: 554–60.

21 Piller NB. Tissue levels of (3-^{14}C) coumarin in the rat: Distribution and excretion. *Br J Exp Path* 1977; **58**: 28–34.

22 Furuya T. Studies on the metabolism of naturally occurring coumarin. V. Urinary metabolites of coumarin and di-hydro coumarin. *Chem Pharm Bull Tokyo* 1958; **6**: 701–6.

23 Kaighen M, Williams RT. The metabolism of 3-^{14}C coumarin. *J Med Pharm Chem* 1961; **3**: 25–43.

24 Sumere van CF, Teuchy H. The metabolism of 2-^{14}C hydroxycoumarin in the rat. *Arch Int Physiol Bio Chim* 1971; **79**: 968–70.

25 Shilling WN, Crampton RF, Longland RC. The metabolism of coumarin in man. *Nature* (*Lond*) 1969; **221**: 664–5.

26 Ritschel WA, Vachharajani NN, Johnson RD, Hussain AS. Interspecies scaling of the pharmacokinetic parameters of coumarin among six different mammalian species. *Methods Find Exp Clin Pharmacol* 1991; **13**(10): 697–702.

27 Garten S, Wosilait WD. Comparative study of the binding of coumarin anticoagulants and serum albumins. *Biochem Pharmacol* 1976; **20**: 1661–8.

28 Piller NB. The binding of 3-^{14}C coumarin to serum proteins in the rat: A comparison between *in vitro* and *in vivo* estimations. *Exp Path* 1977; **13**(8): 175–9.

29 Piller NB, Schmitt LH. An electrophoretic investigation of the binding of 3-^{14}C coumarin to rat serum proteins. *Experientia* 1977; **33**: 1072–3.

30 Piller NB. 3-^{14}C coumarin distribution in rat tissue after the injection of a single dose. *Res Exp Med* 1977; **171**: 93–100.

31 Piller NB, Casley-Smith JR. The effect of coumarin on protein and PVP clearance from rat legs with various high protein oedemas. *Br J Exp Path* 1975; **56**: 439–43.

32 Daoust R. The cell population of liver tissue and the cytological reference bases. In: Brawer RW ed. *Liver Functions*. Baltimore: Waverley Press, 1958: 3–10.
33 Roos B. Macrophagen: Herkunft, entwicklung und funktion. In: Studer A, Cottier H eds. *Immune Reactions*. Berlin: Springer-Verlag, 1970: 81–93.
34 Hoebeke J, Franchi G. Influence of tetramisole and its optical isomers on the mononuclear phagocytic system. Effect of carbon clearance in mice. *J Reticuloendoth Soc* 1973; **14**: 317–23.
35 Casley-Smith JR, Bolton T. Electron microscopy of the effects of histamine and thermal I injury on the blood and lymphatic endothelium and the mesothelium of the mouse's diaphragm, together with the influence of coumarin and rutin. *Experientia* 1973; **29**: 1386–9.
36 Piller NB. The effect of coumarin on the liver weight of thermally injured rats. *Res Exp Med* 1976; **169**: 29–32.
37 Kalimna TV. The structure and function of intestinal lymphatics and the influence of impaired lymph flow in the ileum of rats. *Scand J Gastroenterol* 1971; **10**(Suppl. 6): 9–87.
38 Houck J, Sharma V. Enzyme induction in skin and fibroblasts by anti-inflammatory drugs. In: Bertelli A, Houck J eds *Inflammation Biochemistry and Drug Interaction*. Amsterdam: Excerpta Medica, 1969: 85–93.
39 Allison AC, Harington JS, Birbeck M. An examination of the cytotoxic effects of silica on macrophages. *J Exp Med* 1966; **124**: 141–54.
40 Nadler S, Goldfischer S. The intracellular release of lysosomal contents in macrophages which have ingested silica. *J Histochem Cytochem* 1970; **18**: 368–71.
41 Pearsall NN, Weiser R. *The Macrophage*. Philadelphia: Lea & Febiger, 1970.
42 Piller NB. The ineffectiveness of coumarin treatment on thermal oedema of macrophage-free rats. *Br J Exp Path* 1976; **57**: 170–8.
43 Casley-Smith JR, Casley-Smith JR. *High Protein Oedemas and the Benzo-pyrones*. Sydney: Lippincott, 1986: 43–93.
44 Lecomte J, Cauwenberge van H, Lallemand G. Sur l'oedema généralise provoque par les ratosides chez la rat. *CR Soc Biol* 1971; **165**: 2462–7.
45 Piller NB. The influence of coumarin on tissue levels of radio-labelled plasma protein and povidone (PVP) in normal and thermally injured rats. *Clin Exp Pharm Physiol* 1976; **3**: 523–30.
46 Casley-Smith JR, Piller NB. The pathogenesis of oedemas and the therapeutic action of coumarin and related compounds. *Folia Angiol* Suppl. 1974; **111**: 33–60.
47 Piller NB. Further evidence for the induction of proteolysis by cumarin in rats with various high protein oedemas. *Arzneim Forsch* 1977; **27**(1):4, 860–4.
48 Renkin EM, Carter RD, Joyner M. Mechanism of the sustained action of histamine and bradykinin on transport of large molecules across capillary walls in the dog paw. *Microvasc Res* 1974; **7**: 49–60.
49 Piller NB. A comparison of the effect of benzopyrones and other drugs with anti-inflammatory properties on acid and neutral protease activity levels in various tissues after thermal injury. *Br J Exp Path* 1976; **57**: 411–18.
50 Piller NB. Variations in acid and neutral protease activity of rats with thermal oedema together with the influence of the various benzopyrones. *Arzneim Forsch* 1977; **27**(1): 5, 1069–73.
51 Ambrose AM, de Eds F. Further observations on the effect of rutin and related compounds on cutaneous capillaries. *J Pharmacol* 1949; **97**: 115–19.
52 Woessner JF. Proteolytic control of collagen breakdown. In: Ribbons DW, Brew K eds. *Proteolysis and Physiological Regulation*. New York: Academic Press, 1976: 35.
53 Ignarro LJ. Non phagocytic release of neutral protease and beta glucuronidase from human neutrophils. *Arthr Rheum* 1974; **17**: 25–32.
54 Higginbottom RD, Dougherty T, Jee WS. Fate of shed mast cell granules. *Proc Soc Exp Biol Med* 1956; **92**: 256–61.
55 Egan D, O'Kennedy R, Moran E, Cox D, Prosser E, Thornes RD. The pharmacology, metabolism, analysis and applications of coumarin and coumarin-related compounds. *Drug Metab Rev* 1990; **22**(5): 503–29.
56 Armentano L. Über den einfluss von substanzer der flavongruppe auf die permeabilitat der kapillaren. *Dtsche Med Wochenschr* 1936; **62**: 1325–31.

57 Bohr DF, McEvor BC, Rinehart JF. Effect of various glucosides on the passage of Evans blue through the damaged capillary wall. *J Pharmacol Exp Ther* 1949; **97**: 243.

58 Arturson G, Jonsson CE. Effects of *O*-beta-hydroxyethyl rutosides (HR) and indomethacin on transcapillary macromolecular transport and prostaglandins following scalding injury. *Bibl Anat* 1973; **12**: 465–70.

59 Hilton JG. Effects of H-R upon post burn plasma volume loss in the non-resuscitated dog. *Burns* 1981; **7**: 211–14.

60 Hilton JG. Dose related inhibition by IV hydroxyethyl rutosides on plasma volume loss after thermal injury. *JR Soc Med* 1981; **42**: 15.

61 Hilton JG. Effects of β-hydroxyethyl rutosides (HR) administered post burn after thermal-injury induced plasma volume loss in the non-resuscitated dogs. *Burns* 1982; **8**: 391–4.

62 Olszewski WL. Injury of lymphatics. In: Olszewski V ed. *Lymph Stasis: Pathophysiology Diagnosis and Treatment.* Boca Raton: CRC Press, 1991: 287–93.

63 Lewis GP. Changes in the composition of rabbit hind limb lymph after thermal injury. *J Physiol (Lond)* 1969; **205**: 619–26.

64 Ganrot K, Jacobsson S, Rothman U. Transcapillary passage of plasma proteins in experimental burns. *Acta Physiol Scand* 1974; **91**: 497–506.

65 Luterman A, Dasco CC, Curreri PW. Infection in burns patients. *Am J Med* 1986; **81**(1a): 45–56.

66 Rittenburg MS, Hanback LD. Phagocytic depression in thermal injuries. *J Trauma* 1967; **7**: 523–40.

67 Hunt TK, Sheldon G, Fuchs R. Physiological mechanisms in repair of burns. *Burns* 1975; **1**: 212.

68 Creagh TA, Leahy AL, McNamara E, Bouchier-Hayes DJ. Local beneficial effect of coumarin in experimental peritonitis. *Ir J Med Sci* 1991; **160**(12): 385–6.

69 Edery H, Lewis GP. Kinin forming activity and histamine in lymph after tissue injury. *J Physiol* 1963; **169**: 568–83.

70 Lewis GP, Wawratschek WA. Effect of thermal injury on the kinin system in rabbit hind limb. *Br J Pharmac* 1971; **43**: 127–39.

71 Foldi-Borcsok E. Effect of external lymph drainage and of coumarin treatment on thermal injury in the rat leg. *Br J Pharmacol* 1972; **46**: 245–59.

72 Thakral KK, Godson NW, Hunt TK. Stimulation of wound blood vessel growth by wound macrophages. *J Surg Res* 1979; **26**: 430–40.

73 Bjornson AB, Alexander JW. Opsonic activity of sera from patients after thermal injury. *Surgical Forum* 1973; **24**: 44–5.

74 Bjornson AB, Altemeier WA, Bjornson HS. Reduction in C3 conversion in patients with severe thermal injury. *J Trauma* 1976; **16**(11): 905–12.

75 Ransjo UA, Forsgren A, Arturson G. Neutrophil leucocyte functions and wound bacteria in burn patients. *Burns* 1977; **3**: 171–8.

76 Marshall ME, Rhoades JL, Mattingly C, Jennings CD. Coumarin enhances DR and DQ antigen expressions by peripheral blood mononuclear cells *in vitro*. *Molec Biother* 1991; **3**(4): 204–6.

77 Stuhlmeier K, Theyer G, Baumgartner G, Zlabinger GJ. Synergistic effect of coumarin (1,2-benzopyrone) and endotoxin in the induction of human interleukin-1. *Clin Exp Immunol* 1991; **84**(2): 317–23.

78 Marshall ME, Mohler JL, Edmonds K, Williams B, Butler K, Ryles M, Weiss L, Urban D, Beuschen A, Markiewicz M *et al.* An updated review of the clinical development of coumarin (1,2-benzopyrone) and 7 hydroxycoumarin. *J Cancer Res Clin Oncol* 1994; **120**: S39–42.

79 Hurley J, Ham K, Ryan G. The mechanism of the delayed prolonged phase of increased vascular permeability in mild thermal injury in the rat. *J Path Bact* 1967; **94**: 1–12.

80 Rose S. A report on the use of an hydroxyethyl rutoside in symptoms due to venous back pressure and allied conditions in the lower limbs. *Br J Clin Pract* 1970; **24**: 161–4.

81 Casley-Smith JR, Foldi-Borcsok E, Foeldi M. The electron microscopy of the effects of treatment with coumarin and by thoracic duct cannulation on thermal injuries. *Br J Exp Path* 1973; **54**(1): 1–5.

82 Schachar NS, Jay AW, Rowlands S. Decreased red cell deformability in burn patients. *Can J Surg* 1974; **17**: 239–43.

83 Schmid-Schonbein H, Volger E, Weiss J, Brandhuber M. Effect of *o*-β-hydroxyethyl rutosides on the microrheology of human blood under defined flow. *Vasa* 1975; **4**: 263–70.
84 Eurenius E, Rothenburg J. Platelet aggregation after thermal injury. *J Lab Clin Med* 1974; **83**: 355–63.
85 Marcus AJ. Platelet function. *New Engl J Med* 1969; **280**: 1213–20.
86 van Haerington NJ, Glasias E, Tencate JW, Gerritsen J, van Geet J. Effect of *o*-β-hydroxyethyl rutosides on red cell and platelet function in man. *Bibl Anat* 1973; **12**: 459–64.
87 Canegham van P. Influence of some hydrosoluble substances with vitamin P activity on the fragility of lysosomes *in vitro*. *Biochem Pharmacol* 1972; **21**: 1543–8.
88 de Ruiter N. Inhibition of heat denaturation of albumin by a combination of coumarin and rutin derivatives. *Arzneim Forsch* 1974; **24**: 126–31.
89 Douwes Fr. Influence of various anti-inflammatory drugs on human leukocytes, their function and release of lysosomal enzymes. In: Foeldi M ed. *Ergebnisse der Angiologie*, vol. 13. Stuttgart: Schattauer, 1976: 135–46.
90 Gabor M. Pharmacologic effects of flavonoids on blood vessels. *Angiologica* 1972; **9**: 355–74.
91 Haas E. Esberivan als Thrombose-Embolie-Prophylacticum. *Med Welt* 1964; **15**: 1783–84.
92 Shimamoto T, Takaori S. Pharmakologische Untersuchungen mit einem Melilotus Extrakt. *Arzneim Forsch* 1965; **15**: 897–9.
93 Dannon D, Elazar E. Glucose utilisation in stagnated blood *in vitro* and in plastic bags *in vivo* and the effect of *o*-β-hydroxy-ethyl rutosides on its rate. *Bibl Anat* 1969; **10**: 431–7.
94 Matook GM, Sasken H, Akehnan E. Propane thermal injuries: Case report and review of the literature. *J Trauma* 1994; **37**(2): 318–21.
95 Piller NB. An integration of the mode of action of coumarin. *Arzneim Forsch (Drug Res)* 1977; **27**(6): 1135–8.

CHAPTER 9

Renal Cell Carcinoma: The Background, Rationale and Current Development of Coumarin (1,2-benzopyrone) as a Potential Therapeutic Agent

Scot W. Ebbinghaus[1], James L. Mohler[2] and M. Ernest Marshall[1]

[1]*Division of Hematology/Oncology, University of Alabama at Birmingham, Birmingham, AL, USA*

[2]*Division of Urology and the University of North Carolina Lineberger Comprehensive Cancer Center, Chapel Hill, NC, USA*

9.1 INTRODUCTION

The United States and Canada share the highest incidence of renal cell carcinoma in the world. There are approximately 25 000 new cases of renal cell carcinoma per year in the United States accounting for nearly 10 000 deaths. While surgery (radical nephrectomy) remains the standard care for patients whose tumor is confined to the kidney clinically, many of these patients develop recurrent or metastatic disease within months to years following surgery. Many patients who have disease clinically confined to the kidney are found to have microscopically residual disease following surgery. This may take the form of microscopic extension of tumor through the capsule of the kidney into the perinephric fat, metastatic involvement of regional lymph nodes or tumor thrombus into the renal vein and/or vena cava. While such

Coumarins: Biology, Applications and Mode of Action. Edited by R. O'Kennedy and R.D. Thornes.

patients are surgically resectable from the technical standpoint, it is clear that they are at very high risk of developing recurrent or metastatic disease and, thus, they were not cured by nephrectomy. The most common sites of metastases from renal cell carcinoma are the lungs, liver and bone. Abdominal, hilar and/or mediastinal adenopathy, and adrenal metastases are also quite common.

Currently, there is no curative therapy for renal cell carcinoma once it is metastatic. While the clinical course of renal cell carcinoma is highly variable, survival among patients with metastatic (stage IV) disease tends to be short, usually from six to eight months. No chemotherapeutic agent has been identified which has an acceptably high response rate against metastatic renal cell carcinoma. The drugs which have shown the greatest promise for the treatment of this disease are those known categorically as biologic response modifiers, or, formerly, as immunotherapies. While it is beyond the scope of this chapter to review the details of all the biologic therapies which have been applied to renal cell carcinoma, two deserve some mention because they comprise an important background to the present work with coumarin. The two most important biologic therapies which have been developed for the treatment of renal cell carcinoma are interferon α (IFN-α) and interleukin-2 (IL-2).

In the United States, the Food and Drug Administration has approved IFN-α for the treatment of hairy cell leukemia and for the AIDS-related Kaposi's sarcoma. In spite of this, there is a substantial body of literature that demonstrates the clinical activity of this drug against metastatic renal cell carcinoma. The response rates reported within the literature vary widely but, in the larger experience, it appears that the objective antitumor response rate for IFN-α, given as a single agent, against renal cell carcinoma is approximately 17%. Interleukin-2 has been approved in the United States for the treatment of metastatic renal cell carcinoma. Using the recommended high-dose intravenous bolus dose and schedule, the response rate for IL-2 as a single agent is reported to be 15%. It can be stated with some certainty that the optimal way to use these drugs has not been identified. Past and present (and future) trials explore various doses, schedules, routes of administration for each of the drugs and there is a growing tendency to attempt to improve upon the single-agent response rates by combining IFN with IL-2 and to combine either or both drugs with other biologic agents. It can be argued, also, that an optimal dose and schedule has never been identified for either IFN or IL-2 given as single agents. A wide range of doses has been reported for these drugs[1–11] and, while responses have been observed at almost any dose used, there has never been a single dose which has emerged as superior to another. The low response rate observed from these drugs can be balanced, however, against the fact that many of the responses are quite durable: in some patients the antitumor responses have lasted for years.

In the following discussions, 'coumarin' will refer specifically to 1,2-benzopyrone. Strictly speaking, 'coumarin' refers to a class of related substances[12]. Because 1,2-benzopyrone is the structurally least complex member of this group of substances, it has acquired the universally accepted trivial name of coumarin. Other coumarins will be referred to by their specific scientific nomenclatures.

Interest in coumarins for the treatment of renal cell carcinoma arose from a need to identify other drugs which, alone or in combination with other agents, might render higher response rates and longer durations of response than those observed with other drugs. It was hoped also that this could be achieved with acceptable toxicities. The results of clinical trials using coumarin for the treatment of renal cell carcinoma mirror some of the experiences noted for IFN and IL-2, above. Specifically, a clearly defined clinical dose–response relationship has not been identified—responses have been observed over a wide range of doses—and, while the response rates tend to be low, the duration of response can be quite long in some patients.

Initial interest in coumarin for the treatment of renal cell carcinoma derives from early reports that coumarin possessed immunomodulatory activity and because Thornes *et al.*[13,14] had reported some clinical activity of coumarin when given to patients with malignant melanoma. Because renal cell carcinoma is one of the few malignancies with a clear track record of response to immunotherapies, it appeared natural to explore the possibility that coumarin would have some level of activity. In addition to its immunomodulatory activities, coumarin has been shown to possess direct (non-immunologically mediated) antiproliferative activity. Which property of coumarin is dominant (or even important) in producing clinical responses is not known. The remainder of this dissertation will review what is known about the immunologic and antiproliferative properties of coumarin, review the results of completed clinical trials, mention on-going trials and highlight what work remains to be done before a decision can be made as to the ultimate clinical utility of this drug.

9.2 IMMUNOLOGIC EFFECTS OF COUMARIN

Early interest in coumarin as an anticancer agent derived from observations that certain anticoagulants, such as coumadin (sodium warfarin), were capable of inhibiting the development of metastases in animal tumor models. While coumadin could inhibit metastasis formation, it was not clear whether this activity of the drug was due to its anticoagulant properties or to some other (unidentified) property of the drug. While coumarin comprises part of the substructure of the coumadin molecule, coumarin itself does not possess anticoagulant activity. Thornes *et al.*[15] demonstrated that both coumadin and coumarin possessed immunomodulatory activity *in vivo* in an animal model. In the rabbit, using an ear chamber model to monitor an implantable tumor, they demonstrated that administration of these drugs would result in the recruitment of macrophages into the tumor site. They showed further that these drugs would inhibit tumor cell locomotion. This was the beginning of an observed duality of drug effects from coumarin, i.e. a potentially positive effect of the drug on the cellular immune response to a tumor and a direct effect of the drug on tumor cells.

Since the observations of Thornes *et al.*, other investigators have examined the effects of coumarin on various components of the cellular immune system. Berkarda *et al.*[16] investigated the effects of coumarin on peripheral blood lymphocytes from cancer patients who were given coumarin. Their patients were treated with coumarin as a single daily oral dose of 1 g/m^2. They attempted to detect any quantitative alterations in peripheral blood lymphocytes through the use of E-rosetting techniques and to detect qualitative changes in lymphocyte function with mitogenesis assays. Coumarin therapy did not produce any alteration in the number of E-rosette-forming cells in the peripheral blood. However, coumarin did produce a significant augmentation of lymphocyte responsiveness to phytohemagglutinin (PHA).

Marshall *et al.*[17] examined the effects of coumarin on peripheral blood immune cells using patients with various advanced malignancies. These patients were given coumarin as a single daily oral dose of 100 mg. Peripheral blood was obtained prior to the initiation of therapy, at two weeks, four weeks and eight weeks on therapy. From the whole blood sample, the mononuclear cell fraction was isolated. Using monoclonal antibody labeling techniques and florescence-activated cell sorting, quantitative alterations in various immune cells were monitored during the course of therapy. Therapy with coumarin did not produce any quantitative alterations in the CD3 + (pan-T-cell) lymphocytes, CD4 + (helper/inducer T-cell) lymphocytes, CD8 + (cytotoxic/suppressor T-cell) lymphocytes, CD24 + or CD9 + B lymphocytes, Ia243 + lymphocytes, CD16 + (natural killer) cells or Leu7 + (natural killer) cells[17]. There was, however, a significant increase in the percentage of M5E2 + peripheral blood monocytes and DR + monocytes. This effect was seen after two weeks of coumarin therapy and it persisted over the eight weeks of monitoring. Subsequently, Marshall *et al.*[18] extended these observations by examining the effects of coumarin on monocytes *in vitro*. Using normal human donors, it was shown that coumarin enhanced the expression of both DR and DQ antigens by peripheral blood mononuclear cells. As above, this work was performed with monoclonal antibody labelling and florescence-activated cell sorting.

Subsequent to the above studies, Marshall *et al.*[19] examined the effects of coumarin on various immune cell functions. This work was done with peripheral blood cells obtained from normal human donors and relied upon standard *in vitro* assays of recognized cellular immune functions. Because various diseases, numerous drugs and cigarette smoking can produce effects on cellular and humoral immunity, care was taken during these studies to use cells from only normal, healthy, non-smoking individuals who were taking no medications. When the effects of coumarin on T-cell mitogenesis were examined *in vitro*, it was shown that coumarin caused an augmentation of T-cell mitogenesis to PHA while no effect was seen on the response to concanavalin-A or to pokeweed mitogen. Using a standard cytotoxicity assay (against radiolabeled K562 target cells), it was shown that coumarin has no significant effect on natural killer cell function *in vitro*. As discussed above, there was some indication that coumarin might 'activate' monocytes, at least by the parameter of increased expression of DR and DQ antigens. When other monocyte functions were investigated, it was shown that coumarin produced a modest suppression of superoxide anion and hydrogen peroxide generation. However, no such effect was observed on macrophages. Coumarin was shown also to enhance the antigen processing of soluble tetanus toxoid by cells from some, but not all, normal human donors. The reason for the individual variations is not known. The data to this point in the chronology of investigations into the immunologic effects of coumarin would indicate that coumarin's immunomodulatory effects were mediated through activation of monocytes/macrophages and the monocyte-dependent functions, such as T-cell mitogenesis.

Stuhlmeier *et al.*[20] have shown that coumarin and 7-hydroxycoumarin can have an influence on cytokine production. Specifically, they demonstrated that the production of interleukin-1 by human mononuclear cells in response to lipopolysaccharide was augmented synergistically by coumarin and by 7-hydroxycoumarin. In the conduct of these studies, the investigators were able to demonstrate that the effect which was seen was due to the drugs and not to endotoxin contamination. Further, they were able to demonstrate that the augmented IL-1 response was due to newly synthesized IL-1 (as opposed to the release of pre-formed cytokine) because the effect was blocked by the addition of protein synthesis inhibitors. The synergistic relationship between coumarin and endotoxin with regard to IL-1 production was seen in all normal donors except one. While the synergism between endotoxin and coumarin with regard to IL-1 production was the predominant effect observed, they found that there was also a synergistic relationship between coumarin and endotoxin with regard to IL-6 and tumor necrosis factor α production in some human subjects. A synergistic effect with regard to IL-6 production was observed in 53% of the normal donors while it was observed in only 32% of the donors for tumor necrosis factor. It is noteworthy that both Marshall *et al.* and Stuhlmeier *et al.* noted variability in measured immunologic parameters among normal donors in response to coumarin. These variations may account, at least in part, for the differences in clinical responses that are observed from coumarin when given to patients with advanced malignancies. At present, however, this is mere speculation without substantiating data.

Extending the work of Stuhlmeier *et al.*[20], above, Zlabinger *et al.*[21] attempted to separate the effects of coumarin from those of lipopolysaccharide. Instead of treating the mononuclear cells simultaneously with coumarin and lipopolysaccharide, they treated them sequentialy, first with lipopolysaccharide then with coumarin. In this sequential model, the synergistic relationship for IL-1 production was seen as in the above model. The authors felt that coumarin probably exerts its effect on IL-1 production through signaling pathways similar to those resulting from the primary stimulus (endotoxin). The following evidence for this was given. They conducted experiments in which mononuclear cells were pre-incubated in lipopolysaccharide inhibitors of particular signaling pathways. Specifically, they used staurosporine, H-7, nordihydroguarietic acid, indomethacin, and forskolin. Almost no effect on lipopolysaccharide-induced IL-1 production was seen under these conditions. Inhibitors of protein kinase C, staurosporine and H-7, and the lipoxygenase inhibitor nordihydroguarietic acid blocked the synergistic response. Neither indomethacin nor forskolin produced any inhibitory effect on either lipopolysaccharide or its synergism with coumarin.

Historically, it has been interesting to explore the effects of coumarin on various components of the cellular immune system. This has been especially so with regard to the use of coumarin for the treatment of renal cell carcinoma because this tumor has been known to be responsive to immunologic therapies. There has been a vast experience of treating this malignancy with IFN-α and with IL-2. While the antitumor effect of interferon can be mediated through augmentation of the immune response, it can also result from the direct cytotoxic effects of interferon. Current knowledge, however, indicates that the antitumor effects of IL-2 are mediated exclusively through immune modulation with no direct antitumor activity from the drug. A number of immunologic effects of coumarin have been described above. It must be remembered, however, that there are major problems with the interpretation of these data. First, many of the experiments have been conducted with immune cells *in vitro*. It is not certain that the phenomena observed from immunologic assays *in vitro* have a direct correlation to events which occur *in vivo*. Further, the studies described above have been conducted with immune cells obtained from the peripheral blood. Observations made from immunologic assays of peripheral blood immune cells may not reflect what is happening immunologically at the tissue sites of tumor. Further experimentation and knowledge are needed in order to determine if, and to what degree, the immunologic properties of coumarin contribute to the antitumor responses which are observed in patients with renal cell carcinoma.

9.3 ANTIPROLIFERATIVE EFFECTS OF COUMARIN

In separate bodies of work, investigators have explored the direct antiproliferative (non-immunologically mediated) effects of coumarin on malignant cells. Both coumarin and 7-hydroxycoumarin, the major human metabolite of coumarin, have been shown to inhibit the growth of a number of malignant cell lines *in vitro*. Table 9.1 is a partial list of the human malignant cell lines which have been used in *in vitro* growth experiments with comarin and 7-hydroxycoumarin by various investigators.

Several investigators have shown that coumarin and 7-hydroxycoumarin inhibit the growth of a number of malignant cell lines *in vitro* and that this growth inhibition is of a cytostatic nature[22–27]. Coumarin and 7-hydroxycoumarin are growth-inhibitory in a dose- and time-dependent fashion. It has been shown further that this growth inhibition is reversible. This phenomenon is illustrated in Figure 9.1, which represents a growth curve experiment in which LNCaP (prostate carcinoma) cells are grown in the presence of various concentrations of 7-hydroxycoumarin. The cells were inoculated on day 1 and allowed to form a monolayer and establish growth for four days. Subsequently, appropriate cells were harvested and counted on days 5–9. As would be expected from the above comments, the growth curve shows a dose- and time-dependent inhibition of growth of the LNCaP cells. On day 9, matching sets of cells grown in the various concentrations of 7-hydroxycoumarin were harvested, washed free of drug and reinoculated into culture in medium lacking

Table 9.1. Some of the human malignant cell lines used in experiments with coumarin and 7-hydroxycoumarin.

Cell line	Type
ACHN	Renal cell carcinoma
Caki-2	Renal cell carcinoma
A549	Lung carcinoma
HCT-15	Colon carcinoma
COLO-232	Colon carcinoma
Caco-2	Colon carcinoma
LNCaP	Prostate carcinoma
PC-3	Prostate carcinoma
Du 145	Prostate carcinoma
H727	Malignant carcinoid tumor
MCF-7	Breast carcinoma
EJ	Bladder carcinoma
HepG2	Hepatoma
G-CCM	Astrocytoma
G-UVW	Astrocytoma
K562	Erythroleukemia
Dakiki	Myeloma (IgA)
HS-Sultan	Myeloma (IgG)
RP-1788	Myeloma (IgM)
CCRF CEM	Acute lymphatic leukemia
HL60	Lymphoma

drug. As before, four days were allowed for adherence and establishment of monolayer growth. Subsequently, cells were harvested on days 12–15 and counted. As shown, in cells which had been grown in all concentrations of 7-hydroxycoumarin recovered from the effects of the drug and resumed growth that was identical to the control cells grown in drug-free medium.

The LNCaP cell line is known to produce prostate specific antigen (PSA) *in vitro*. Figure 9.2 shows the results of an experiment to determine the effects of 7-hydroxycoumarin on the intracellular production of PSA by LNCaP cells. This experiment was done in parallel with the growth curve shown in Figure 9.1. The results of this experiment demonstrate that, in the presence of 7-hydroxycoumarin, there is a dose- and time-dependent suppression of intracellular PSA production which parallels the growth inhibition. Further, the suppression of PSA production is shown to be reversible when the cells are removed from drug-containing medium.

Having demonstrated that coumarin and 7-hydroxycoumarin are growth inhibitory for a number of malignant cell lines, some fairly crude probing experiments were performed in an attempt to begin to investigate the subcellular mechanisms of action of these drugs. In these experiments, the effects of coumarin and

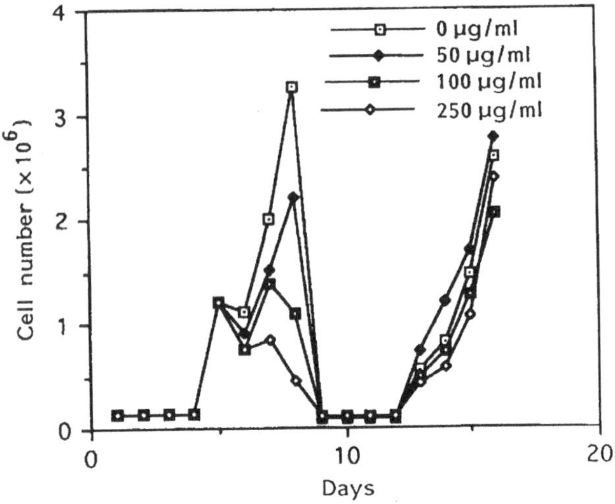

Figure 9.1. Results of a growth curve experiment in which LNCaP cells were grown in various concentrations of 7-hydroxycoumarin. After allowing four days of growth for the establishment of a monolayer, viable cell counts were obtained on days 5–9. The results of this first part of the experiment show a concentration- and time-dependent inhibition of the growth of the LNCaP cells. On day 9, matching sets of cells which had been grown for nine days in various concentrations of 7-hydroxycoumarin were harvested, washed free of drug and reinoculated into medium which was free of drug. After four days to establish growth, cells were harvested for viable cell counts of days 12–15. The results of the second portion of this experiment demonstrate that the growth inhibition from 7-hydroxycoumarin is reversible. (Reproduced by permission from *Journal of Cancer Research and Clinical Oncology*[22].)

7-hydroxycoumarin on DNA, RNA and protein synthesis were examined using radiolabeled precursors[22]. In separate experiments, using malignant cells grown *in vitro*, the effects of drug were examined on the incorporation of tritiated thymidine, tritiated uridine and tritiated leucine. The details of these experimental methods are described elsewhere[23]. Figure 9.3 shows the effects of coumarin and 7-hydroxycoumarin on the incorporation of tritiated thymidine by ACHN (renal carcinoma) cells. As shown, both substances, in a dose-dependent fashion, caused a reduction in the incorporation of tritiated thymidine.

The effect of coumarin and 7-hydroxycoumarin on RNA synthesis by ACHN cells is shown in Figure 9.4. In this experiment, both drugs were shown to inhibit the incorporation of tritiated uridine in a concentration-dependent fashion.

In a fashion similar to that above, coumarin and 7-hydroxycoumarin were shown to inhibit protein synthesis by ACHN cells. Figure 9.5 shows the effects of these drugs on tritiated leucine incorporation by ACHN cells.

As the effects of coumarin and 7-hydroxycoumarin were examined for the list of malignant cell lines shown above it was found that some types of cells were relatively more sensitive to these drugs than other lines (i.e. greater suppression of

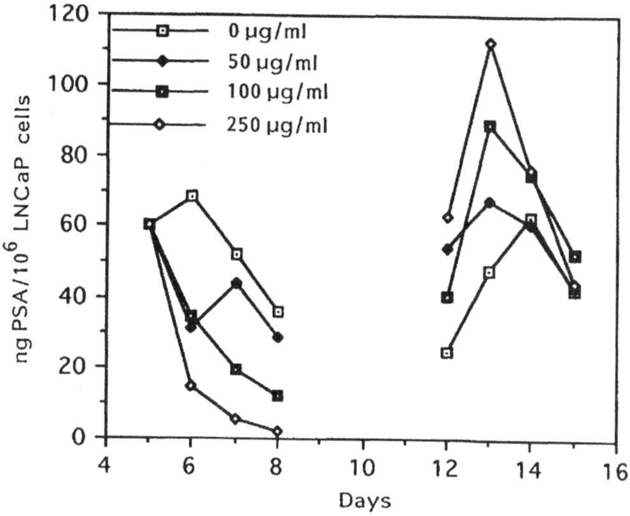

Figure 9.2. The effect of 7-hydroxycoumarin on the intracellular production of prostate specific antigen (PSA) by LNCaP cells *in vitro*. In the presence of 7-hydroxycoumarin, PSA production is suppressed in a dose- and time-dependent fashion. This reduction parallels the inhibition of growth shown in Figure 9.1. Further, just as the growth inhibition from 7-hydroxycoumarin was reversible, the suppression of PSA production is shown to be reversible when the cells are removed from medium containing the drug. (Reproduced by permission from *Journal of Cancer Research and Clinical Oncology*[22].)

growth at lower concentrations). Similarly, the suppression of DNA, RNA and protein synthesis was more profound in certain lines compared to others (although this was not examined for all cell lines listed). For example, the effects of these agents were more profound on HL60 cells than on ACHN cells.

It has been speculated that one of the mechanisms of antitumor action of coumarin and 7-hydroxycoumarin may be mediated through a direct inhibition of oncogene expression. A few investigators have examined the effects of 7-hydroxycoumarin on the expression of several classes of oncogenes including the growth factor receptor oncogene, *HER-2/neu*, the GTPase signal transduction oncogene Ha-*ras* and the nuclear oncogene, c-*myc*. Of these, the c-*myc* oncogene plays a critical role in normal cell function and programmed cell death (apoptosis). Loss of c-*myc* regulation plays a role in the initiation or progression of a number of human malignancies. The rapid induction of c-*myc* expression appears to be necessary for quiescent cells to enter the cell cycle while terminal differentiation of cells is associated with loss of c-*myc* expression. An increased level of the normal c-*myc* protein is observed in a wide variety of malignancies and is caused by amplification, translocation or proviral insertion upstream of the c-*myc* gene. The c-*myc* protein localizes to the nucleus and probably acts through heterodimerization with the *max*

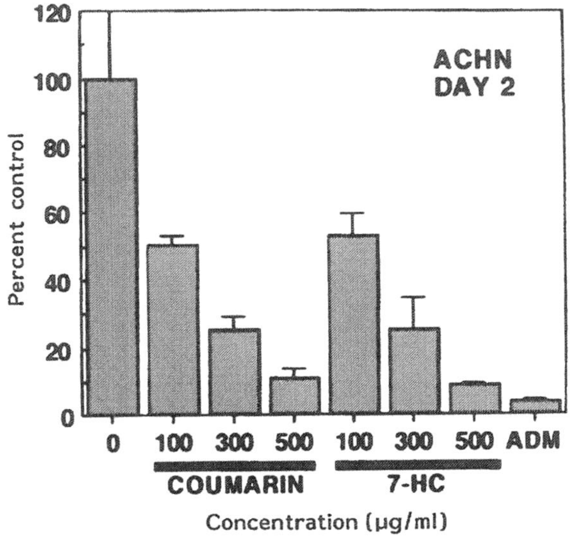

Figure 9.3. The inhibition of DNA synthesis by ACHN cells as measured by incorporation of tritiated thymidine. Results are presented as percentage of control incorporation (by cells in the absence of drug) and as a function of drug concentration. Adriamycin (ADM) served as the positive control for these experiments. (Reproduced by permission from *Journal of Cancer Research and Clinical Oncology*[22].)

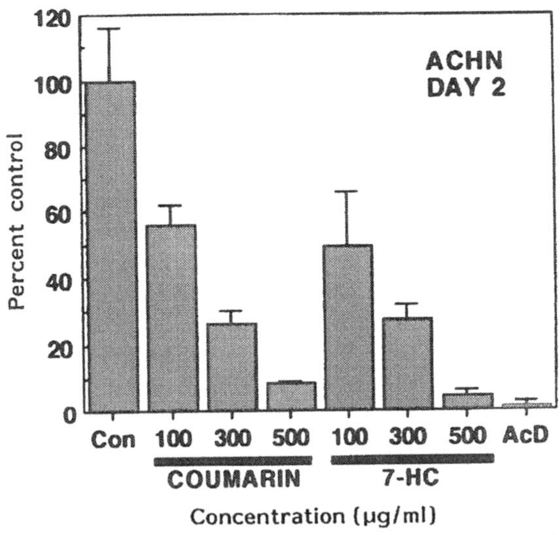

Figure 9.4. Coumarin and 7-hydroxycoumarin inhibit the incorporation of tritiated-uridine by ACHN cells *in vitro* in a concentration-dependent fashion. Actinomycin D (AcD) served as the positive control for these experiments. (Reproduced by permission from *Journal of Cancer Research and Clinical Oncology*[22].)

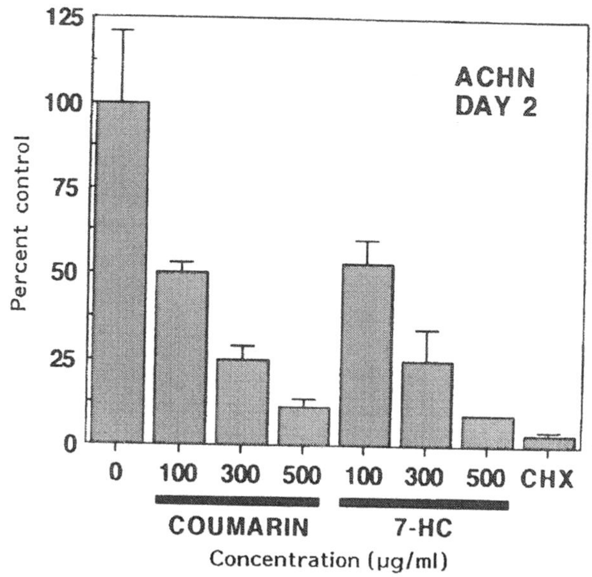

Figure 9.5. In a concentration-dependent fashion, coumarin and 7-hydroxycoumarin were shown to inhibit tritiated leucine incorporation by ACHN cells *in vitro.* In these experiments, cycloheximide (CHX) served as the positive control. (Reproduced by permission from *Journal of Cancer Research and Clinical Oncology*[22].)

protein to form complexes that serve as a transcription factor to induce the expression of other genes involved in cell proliferation. The c-*myc* gene is regulated at both the transcriptional level principally through *cis*-acting promoter elements that are the sites of transcription factor binding and at the post-transcriptional level, principally through the stability of the mRNA. The c-*myc* oncogene may be an intracellular target for the coumarins.

Zanker[25] investigated the mechanisms of antiproliferative action using the DUKX and RBA cell lines. The DUKX cell line is known to overexpress c-*myc* and the RBA line is known to overexpress Ha-*ras*. They demonstrated that 7-hydroxycoumarin was growth inhibitory for both of these cell lines, and they demonstrated that the growth inhibition was associated with decreased levels of mRNAs for both c-*myc* and Ha-*ras* in the DUKX and RBA cells, respectively.

Working with LNCaP (prostate carcinoma) cells, Marshall *et al.* (unpublished data) have shown that 7-hydroxycoumarin, in a concentration-dependent fashion, produces a rapid and potent reduction in the level of c-*myc* RNA. In similar experiments, however, 7-hydroxycoumarin had no effect on the level of mRNA for *HER-2/neu* in LNCaP cells. Ebbinghaus and Kim (unpublished data) confirmed this observation using HeLa (cervical cancer) cells which had been stably transfected to express luciferase under the control of the c-*myc* promoter elements. These cells showed a concentration-dependent inhibition of luciferase activity after treatment with 7-hydroxycoumarin. However, SK-BR-3 (breast cancer) cells stably transfected to express luciferase under the control of the Ha-*ras* oncogene promotor showed slight induction or no change in luciferase activity in response to 7-hydroxycoumarin. The studies suggest a relatively selective effect of 7-hydroxycoumarin on c-*myc* expression in cell culture that must take place in part at the transcriptional level (since the luciferase reporter constructs contain only the c-*myc* promoter elements). It is tempting at this point to speculate that the coumarins are a class of DNA binding compounds that interact directly with the c-*myc* promoter to inhibit transcription factor binding to critical regulatory elements resulting in the loss of c-*myc* transcription. However, at the present time, this is mere speculation awaiting further data to confirm or reject this hypothesis.

Adding further refinement to the understanding of the biologic effects of 7-hydroxycoumarin, Seliger and Petterson[27] explored the effects of this drug on two malignant glioblastoma cell lines. It was known that the growth of these cells was affected by at least two putative autocrine loops affecting the platelet-derived growth factor (PDGF) and epidermal growth factor (EGF) receptor pathway. As was the case with the cell lines mentioned above, 7-hydroxycoumarin produced a time- and concentration-dependent growth inhibition of these cells which was reversible. Further, this growth inhibition by 7-hydroxycoumarin was associated with a rapid and selective downregulation of PDGF-A and PDGF-B mRNA levels.

Marshall *et al.*[22] investigated the possibility that 7-hydroxycoumarin could induce programmed cell death (apoptosis). For these investigations, the following cell lines were used: HL60, LNCaP, PC-3, Du 145, Caki-2, ACHN and K562. The assay for apoptosis depended on detection of the characteristic DNA 'ladder' formation on gel electrophoresis. Apoptotic death could be demonstrated only for the HL60 cells (a cell line known to undergo apoptosis in response to a number of other stimuli). While apoptotic death was not demonstrated for any of the other cell lines listed above, the investigators did not feel that the findings were definitive. It was considered possible that the appropriate experimental conditions were not met in all cases to exclude definitively that apoptotic death was possible (e.g. selection of proper drug concentrations, times of exposure to drug, etc.).

Because coumarin and 7-hydroxycoumarin have direct antiproliferative effects against malignant cell lines, it might be expected that the antiproliferative properties of these drugs would be associated with certain toxicities when they are given to human patients. For example, it might have been predicted that these drugs would be myelosuppressive. However, in none of the phase I or phase II clinical trials using these drugs has there been any bone marrow suppression. These observations were pursued by examining the effects of coumarin on murine and human bone marrow progenitor cells in *in vitro* stem cell culture[28]. It was found that, over a wide range of drug concentrations, there was no suppression of stem cell activity in either species of bone marrow. It is not known why tumor cell growth is inhibited by these drugs while normal proliferating bone marrow stem cells are not affected.

While all of the above studies have contributed to our understanding of the biologic effects of these drugs, they have not resolved definitively the precise mechanism(s) of action of these drugs. It is not known whether the antitumor activity is due to the immunologic effects of the drugs or to the direct antiproliferative effects, or whether both effects may contribute *in vivo*. With regard to the direct antiproliferative effects, it is not known if there is a predominant effect which leads to cell death or whether there are multiple non-specific events at the subcellular level which all contribute ultimately to cell death.

9.4 CLINICAL DEVELOPMENT OF COUMARIN AND 7-HYDROXYCOUMARIN FOR THE TREATMENT OF ADVANCED HUMAN MALIGNANCIES TO INCLUDE RENAL CELL CARCINOMA

Marshall *et al.* have pointed out repeatedly that coumarin has not followed the normal gestalt of new drug development in the clinical setting. While it would be nice to believe that the clinical development of coumarin has been along a logical path, born from basic laboratory research, fed into logical, stepwise, sequentially developed clinical trials, such is not the case. Most of what has been done to date has been born of empiricism. The first clinical experience with coumarin for the treatment of metastatic renal cell carcinoma resulted from publications by Thornes *et al.*[13,14] reporting that coumarin, when given in combination with cimetidine, produced clinical responses among patients with advanced malignant melanoma. As mentioned earlier, Thornes *et al.* had demonstrated that coumarin possessed immunomodulatory effects in an animal model and they were working clinically on the premise that it was these immunologic effects which resulted in antitumor responses *in vivo*. They added cimetidine to the regimen because it was known that cimetidine had immunomodulatory effects of its own which, theoretically, would augment the action of coumarin. Cimetidine had been shown to inhibit suppressor T-cell function[29,30] and it was believed that this activity of cimetidine would be a beneficial adjunct to coumarin. Thornes *et al.*[13,14] reported treating patients with malignant melanoma with coumarin at a dose of 100 mg orally daily continuously. After 14 days of coumarin alone, cimetidine 300 mg four times daily was added and both drugs were continued together thereafter. Because this regimen was reported to render antitumor responses in patients with malignant melanoma, Marshall *et al.*[31] empirically applied this same treatment regimen to patients with metastatic renal cell carcinoma. In this pilot study, there were some complete and partial responses among the patients with metastatic renal cell carcinoma with almost no toxicity from the regimen. This was the first report that coumarin had some degree of activity against renal cell carcinoma. In retrospect, it is unfortunate that this trial was done at all because several investigators rushed to duplicate this study with the same treatment regimen but with lesser results[32–34]. In retrospect, the pilot study should not have been done because (a) it examined a two-drug regimen without first knowing the effectiveness of each drug as a single agent and (b) the dose and schedule of the coumarin were derived empirically based upon limited experience in malignant melanoma. There were no phase I human toxicity data from single-dose and/or multiple-dose schedules to give any idea of an optimal or even effective dose and schedule. To this date, all clinical studies with coumarin have been conducted with the drug given on a single daily oral schedule. From what is known of the toxicity and safety of this drug, it is quite possible to administer it safely on a schedule of multiple daily doses.

Following their pilot study, Marshall *et al.* were eager to move this regimen into an adjuvant setting for the treatment of renal cell carcinoma. (It should be noted that the term adjuvant refers to the treatment of microscopic residual disease following the removal of the kidney containing the primary renal cell carcinoma. Thus, adjuvant therapy would be given to produce cure or delay the time to recurrence of clinically detectable cancer and to prolong survival from the disease.) Plans for an adjuvant trial were held in abeyance in order to conduct a formal phase I trial of the drug. An adjuvant trial would have required a large number of patients, followed for a number of years, and before embarking upon such a large-scale project it was necessary to obtain further data concerning dosing and toxicities.

A phase I trial of coumarin, in combination with cimetidine, was conducted for the treatment of any malignancy which was refractory to standard therapies or for which no standard therapy existed[35]. In this study, the dose of cimetidine was held constant for all patients at 300 mg orally four times daily. Coumarin was given as a single daily oral dose. Three patients were treated on each dose level of coumarin and the dose was held constant for each patient (i.e. there was no dose escalation schema for a given patient). Patients were required to have adequate baseline organ functions and to have a good performance status (Eastern Cooperative Oncology Group performance status of 2 or better). Additional eligibility criteria for this phase I trial included adequate baseline organ functions defined as: total leukocyte count $\geq 3500/mm^3$, granulocyte count $\geq 2000/mm^3$, platelet count $> 100\,000/mm^3$, hemoglobin ≥ 10 gm/dl, hematocrit $\geq 30\%$, serum bilirubin ≤ 2.0 mg%, SGOT and SGPT ≤ 1.5 times the upper limits of normal, alkaline phosphatase ≤ 1.5 times the upper limits of normal, normal prothrombin time and normal thromboplastin time, normal urinalysis, blood urea nitrogen ≤ 20 mg%, serum cretinine ≤ 1.5 mg%, creatinine clearance ≥ 70 ml/min, normal serum electrolytes, normal electrocardiogram and the absence of any cardiac disease. All of these tests of organ function were obtained at baseline to determine eligibility for the study and they were repeated weekly for eight consecutive weeks while on therapy, then monthly for six months and every eight weeks thereafter as long as the patient remained on treatment. Patients were required also to have disease which was either bidimensionally measurable (e.g. by physical examination, radiograph or computerized tomographic [CT] scan) or evaluable (e.g. by radionuclide bone scan, radiographic disease which was detectable but not strictly bidimensionally measurable).

Tumor evaluations were performed with appropriate radiographic studies every month while on therapy for the first six months, then every other month while on study. Patients remained on treatment until there was evidence of progression of disease. Standard response criteria were used. Basically, a complete response was defined as the disappearance of all detectable cancer from physical examination and radiographic studies. A partial response was defined as a reduction by 50% or more from baseline of the sum of the products of all bidimensionally measurable disease. Progression of disease was defined as an increase of 25% or more of the sum of the products of bidimensionally measurable disease when compared to baseline or the occurrence of any new metastatic lesion. Stable disease was defined as anything between a partial response and progressive disease.

The phase I trial started with coumarin given as a single daily dose of 400 mg. All patients were treated with cimetidine 300 mg orally four times daily beginning on day 15 of treatment. The intent was to escalate the dose level of the coumarin, with three patients treated at each dose level, until the maximally tolerated dose and the dose-limiting toxicity of the drug were defined. At the appearance of any significant toxicity, the number of patients treated at any dose level was to increase from three to six. In addition to determining the toxicity profile of coumarin, the investigators were interested in any antitumor activity which might result from this treatment. Fifty-four patients with a variety of malignancies were treated in this trial at coumarin doses ranging from the starting dose of 400 mg daily to 7 g daily, the ultimate dose level for the study. Table 9.2 is a summary of the pretreatment characteristics of the patients treated on this phase I trial.

Table 9.2. Pretreatment characteristics of the patients treated on a phase I trial of coumarin and cimetidine (Marshall *et al. Cancer Biotherapy* 3: 170–178, 1991).

	Number of patients
Total number of patients	54
Male:female	38:16
Age (years): mean (range)	51.8 (32–79)
ECOG performance status[a]	
0	26
1	27
2	1
Diagnosis	
Renal cell carcinoma	37
Malignant melanoma	6
Colon carcinoma	3
Breast carcinoma	2
Prostate carcinoma	2
Malignant carcinoid tumor	1
Sarcoma	1
Small cell carcinoma of lung	1
Adenoid cystic carcinoma	1

[a] ECOG = Eastern Cooperative Oncology Group.

In this phase I trial, the coumarin was tolerated very well over a wide range of doses. Indeed, it was so well tolerated that the true maximally tolerated dose and dose-limiting toxicity from a single daily oral dose could not be defined. Part of the problem was that the coumarin was available only as a 100 mg tablet. Thus, all dosing needed to be in multiples of 100 mg tablets. Consequently, patients receiving the highest doses of coumarin were ingesting 70 tablets as a single daily dose. The study was terminated at this level because it was felt that the limits of practicality had been reached, or exceeded. However, some side-effects were encountered. For the most part, the side-effects were mild and self-limited. Side-effects were almost non-existent below 2 g daily. Two patients (3.6%) receiving doses below this reported side-effects. Each of these patients (one receiving 400 mg and the other 600 mg daily) reported insomnia during the first few weeks of treatment. In both cases this side-effect resolved with continued administration of drug. Nausea was encountered at doses in excess of 2 g daily. Twenty patients (36.4%) reported some degree of nausea at doses higher than 2 g daily. Four of the five patients treated at 7 g daily experienced nausea. Four patients (7.3%) who were taking coumarin at doses in excess of 3 g daily reported episodes of vomiting. With regard to the nausea and/or vomiting, there was a common report that these symptoms had their onset from 2.5 to 3 h following a dose of coumarin. Coumarin (1,2-benzopyrone) is an aromatic substance. In most cases, the side-effects of nausea and vomiting were related to this property of the drug. Especially at higher doses, patients reported that it was the combination of taste and aroma from the large number of tablets taken as a single daily oral dose that led to the nausea and/or vomiting. (Note that this problem would be circumvented with the development of 7-hydroxycoumarin as a drug since it lacks the aromatic properties of coumarin). Typically, the nausea occurred during the first few weeks of therapy and abated with continued administration of drug. Two patients, however, receiving 3.5 g and 7 g daily, respectively, discontinued therapy because of persistent nausea and vomiting which occurred on nearly a daily basis while taking the drug. Two patients (3.6%) who were receiving doses ≥ 3.5 g daily developed diarrhea. The diarrhea did not require therapeutic intervention and was self-limited, abating after a couple of weeks on coumarin. Four patients (7.3%) who were taking doses ≥ 3.5 g daily reported dizziness associated with coumarin which had its onset typically about 2.5 to 3 h following an oral dose. As with the nausea, this side-effect tended to be self-limiting, abating after 1–2 weeks of continued treatment.

Because of what is known of the metabolism and excretion of coumarin, there was keen interest in the effects of coumarin on renal and liver functions, especially at the higher doses. Thirty-seven of the 54 patients lacked any evidence of hepatic metastases from their primary tumors and had pretreatment serum transaminases which were within normal limits. Among these patients, only one showed any evidence of chemical hepatotoxicity. This was manifested as the asymptomatic elevation of hepatic transaminases without hyperbilirubinemia. The patient had renal cell carcinoma for which he had undergone radical nephrectomy. At the time he was entered onto the phase I trial of coumarin and cimetidine, he had pulmonary metastases as his only site of disease and he had an ECOG performance status of 0 (normal activity). He entered the study at the 5 g/day dose level of coumarin. While on therapy at this dose, he achieved a complete response radiographically with regard to his lung metastases. Over the first 12 months of coumarin therapy, he displayed mild elevations of his SGOT but these abnormal elevations were < 1.5 times the upper limits of institutional normal value. At 14 months, however, the SGOT rose to over twice the normal value. The SGGT rose to over twice the upper limits of normal by four months on therapy but the bilirubin remained normal. The peak values for the serum transaminases were as follows: SGOT 120 U/l (normal values, 15–40 U/l); SGPT 333 U/l (normal values, 15–35 U/l); and SGGT 69 U/l (normal values, 10–50 U/l). Once the transaminases rose to these levels, the coumarin was discontinued. Three weeks after treatment was interrupted, the transaminases were all normal. Repeated CT scans of the liver failed to demonstrate any evidence of metastatic malignancy, confirming the impression that the transasminase elevations were due to the drug. Because the patient had achieved a complete radiographic response to his lung metastases and he was otherwise healthy, there was a keen interest in continuing the coumarin at a dose that would be subtoxic for this individual. Once his transaminases returned to normal, coumarin therapy was reinstituted at half dose (2.5 g daily). At this dose, there was no further elevation of transaminases and the patient continued to respond to therapy.

It should be mentioned here that, in a separate study from the phase I trial mentioned above, there was another episode of significant hepatotoxicity. A patient with no liver metastases and normal pretreatment transaminases was treated with coumarin at a dose of 1 g orally daily. This patient developed marked elevations of his transaminases and bilirubin, and became jaundiced clinically. Therapy was discontinued and a liver biopsy was performed as part of the evaluation of liver dysfunction. The biopsy revealed hemorrhagic necrosis. With interruption of therapy, the transaminase and bilirubin abnormalities resolved completely with no adverse sequelae. The patient did not receive further coumarin.

There was no convincing evidence of nephrotoxicity in any patient treated on the phase I trial of coumarin and cimetidine. In the article which reported the results of this phase I trial[35], it was pointed out that there were cyclic variations in the values for the serum creatinine among the patients with renal cell carcinoma who were treated on this trial. As shown in Table 9.2, there were 37 patients with renal cell carcinoma on the study, all of whom had undergone a prior nephrectomy for treatment of the primary tumor. Cyclic variations in serum creatinine were noted in 17 (45.9%) of these patients with creatinine values reaching as high as 2.1. Such fluctuations in serum creatinine are known to occur naturally in patients who had a unilateral nephrectomy and, thus, these aberrations were not attributed to the investigational drug. Only one patient with a diagnosis other then renal cell carcinoma had an elevation of creatinine while on therapy. This occurred in a patient with small cell lung cancer who had a prior history of treatment with *cis*-platinum. In this patient, the creatinine varied between 1.5 and 1.7 mg/dl.

While the goal of a phase I trial is to define the toxicity of a new drug, any evidence of antitumor activity is welcomed because this may give information concerning the potential range of clinical use of the drug. In the phase I trial of coumarin and cimetidine, antitumor responses were seen only among the patients with renal cell carcinoma. It is unusual to have a phase I trial in which as many as 37 patients have a diagnosis of renal cell carcinoma (a relatively uncommon disease), but in this case the disproportionate number of patients with this diagnosis reflected a physician referral bias and not a selection bias. In any event, this skewed situation proved fortuitous because some interesting observations resulted. Table 9.3 summarizes the characteristics of the 37 patients with renal cell carcinoma who were treated on this trial.

By all of the parameters listed in Table 9.3, this was a very typical group of patients with renal cell carcinoma, reflective of the disease as a whole. Among these

Table 9.3. Pretreatment characteristics of the patients with renal cell carcinoma who were treated on the phase I trial of coumarin and cimetidine.

Characteristic	Number of patients
Total number of patients	37
Age (years): mean (range)	52.8 (32–70)
Male:female	26:11
Performance status (ECOG)	
0	23
1	13
2	1
Nephrectomized	37
Prior therapy	
Radiation therapy	6
Chemotherapy	2
Hormonal therapy	3
IL-2	1
IL-2 + LAK cells	7
IL-2 + IFN-α	1
IFN-α	5
IFN-β	1
No prior therapy	16
Sites of metastatic disease	
Lung	28
Liver	6
Bone	13
Abdominal nodes	11
Hilar/mediastinal nodes	8
Renal bed	2
Adrenal gland	1
Spleen	1
Contralateral kidney	1
Soft tissue	1

37 patients, there were seven objective antitumor responses. These included two complete responses, four partial responses and one mixed response. Five of the responders had received no prior therapy except for nephrectomy and one responder had received only radiation therapy in addition to nephrectomy. It was true generally, that the responses occurred only among patients with small or modest tumor burdens and no one with a large tumor burden or bulky masses showed a response. No responses were observed among patients who had received and failed prior systemic therapies such as IL-2 with or without lymphokine activated killer (LAK) cells or the interferons. While these observations are reported in the interest of completeness, it is recognized that there were insufficient numbers of patients to allow for a statistically significant subset analysis of predictive parameters. Such descriptions await larger trials devoted solely to the treatment of renal cell carcinoma and, perhaps, with designed stratifications of patients at the time of entry onto study. In addition to the objective responses, there were 11 patients who experienced stabilization of disease which lasted for at least four months. The mean duration of stable disease among these patients was 9.9 months with a range of 4.5 to over 24 months.

Sites of response among these patients with renal cell carcinoma included lung and bone metastases. It was not surprising that responses were observed in the lung, for this is the single most frequent site of response observed for any therapeutic agent used for this disease. It was something of a surprise that two patients with painful bone metastases had symptomatic and objective responses. With so many patients on study with renal cell carcinoma, it was hoped that some information would be gathered regarding a dose–response relationship for coumarin. Patients with renal cell carcinoma were treated at dose levels ranging from 400 mg daily to 7 g daily. Responses were observed in patients taking coumarin at doses ranging from 600 mg to 5 g daily. The responses were spread out evenly over this dose range. Specifically, responses were seen at the following doses: 600 mg, 700 mg, 900 mg, 1200 mg, 4500 mg and 5000 mg. This would lead to some empiricism in the selection of a dose for a larger trial for renal cell carcinoma to follow this one.

Photographs of the radiographic responses observed in the phase I trial of coumarin and cimetidine were published with the paper describing this trial[35]. In the context of the present discussion, it is illustrative to show here some of the responses which were seen with some description. Figure 9.6A shows the pretreatment chest radiograph of a patient with metastatic renal cell carcinoma. This patient had undergone a radical nephrectomy but had pulmonary metastases at the time of entry onto this study. As shown in Figure 9.6A there were multiple pleural-based pulmonary metastases. Although visible on chest radiograph, they are illustrated better by the computerized tomographic (CT) scan of the chest (Figure 9.6C). On chest CT scan, the tumor in the right chest appears more voluminous and is marked in the photograph with the letter 'T'. This patient was treated with coumarin at a dose of 600 mg daily along with cimetidine 300 mg four times daily. Figure 9.6B shows the post-treatment chest radiograph which illustrates complete radiographic resolution of the pulmonary nodules. The complete radiographic response was documented with a post-treatment CT scan of the chest (Figure 9.6D). This patient remains in complete response with no side-effects from therapy and at a normal performance status at

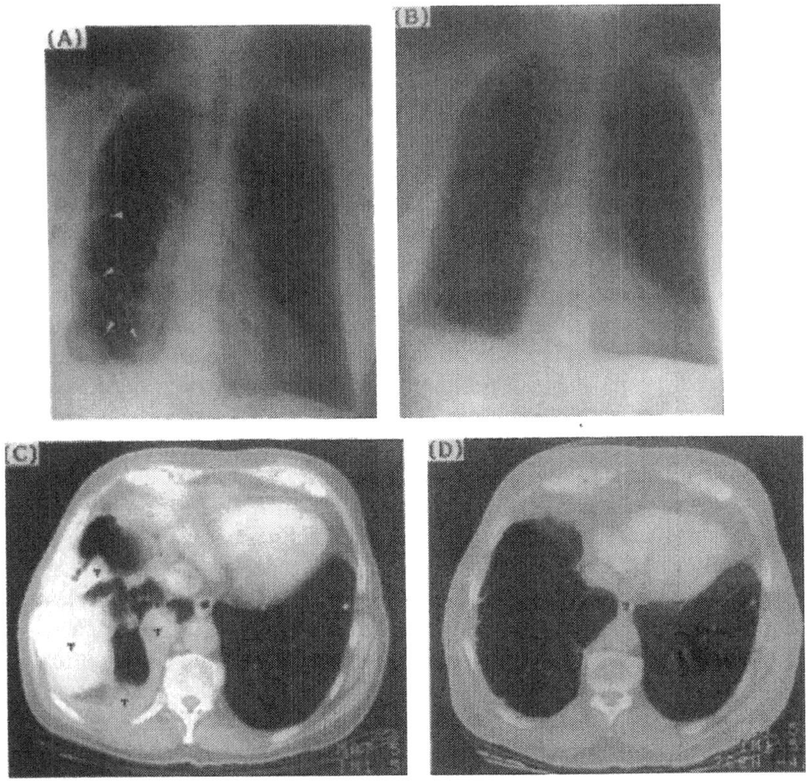

Figure 9.6. (A) Pretreatment chest radiograph of a patient with metastatic renal cell carcinoma. There are multiple pleural-based pulmonary metastases illustrated by the arrows. (B) Chest radiograph of the patient shown in (A) after treatment with coumarin 600 mg daily and cimetidine. This view shows complete resolution of the pulmonary metastases. (C) Pretreatment computerized tomographic scan of the chest of the patient shown in (A). Fairly bulky tumor is shown in the left chest illustrated by the 'T's. (D) Computerized tomographic scan of the chest corresponding to the post-treatment chest radiograph shown in (B). The CT scan confirms the complete radiographic response, showing disappearance of metastatic lesions. (Reproduced by permission from *Molecular Biotherapy*[35].)

greater than seven years on therapy. He has been the longest responder documented thus far from coumarin and cimetidine. Further, the tumor burden which is illustrated by his radiographic studies is the largest tumor bulk known to respond to this therapy.

Figure 9.7 shows the more typical metastatic picture to respond to this therapy. This is the pretreatment chest radiograph of a woman with a metastatic renal cell carcinoma whose disease relapsed following a radical nephrectomy in the form of pulmonary metastases. The arrows in the figures point to some of her multiple small pulmonary metastases as seen on posterior–anterior chest radiograph (Figure 9.7A)

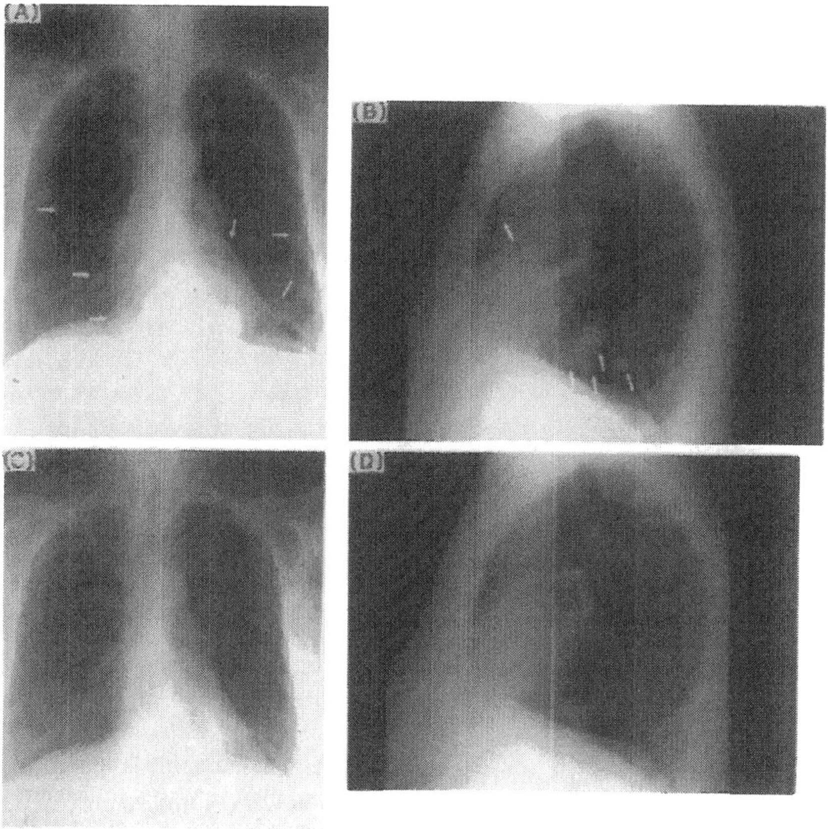

Figure 9.7. Pretreatment posterior–anterior (A) and lateral (B) chest radiographs of a patient with multiple pulmonary metastases from renal cell carcinoma (arrows). (C) Posterior–anterior chest radiograph of the patient shown in (A, B) following treatment with coumarin and cimetidine. This view shows apparent complete resolution of the pulmonary metastases. (D) Lateral view of the chest following treatment showing resolution of all but part of one pulmonary lesion in the patient shown in (A, B). (Reproduced by permission from *Molecular Biotherapy*[35].)

and lateral view (Figure 9.7B), respectively. After treatment with coumarin at a dose of 4500 mg daily with cimetidine, she experienced a nearly complete response with disappearance of all nodules from the posterior–anterior view (Figure 9.7C) and with only a remnant of one nodule seen on the lateral view (Figure 7D). This response was scored for study purposes as a partial response because of the remnant seen on the lateral view.

Two patients on the phase I trial of coumarin and cimetidine exhibited responses at osseous sites of metastases. Figure 9.8A shows a detail from a pretreatment radionuclide bone scan. There is intense abnormal accumulation of radionuclide

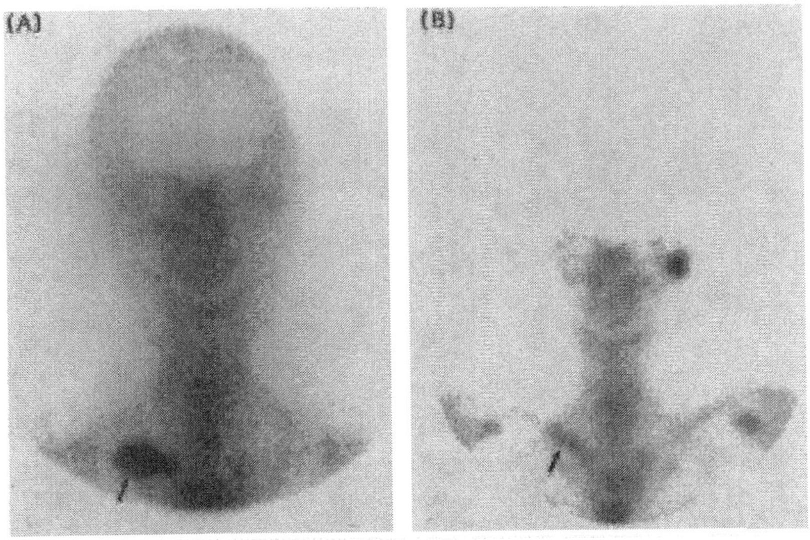

Figure 9.8. (A) Pretreatment radionuclide bone scan of a patient with metastatic renal cell carcinoma. There is an area of abnormal accumulation of the radionuclide in the right clavicle. This bone metastasis was associated with a palpable exophytic mass arising from the clavicle and extending into the soft tissues. (B) Bone scan following treatment with coumarin and cimetidine. This post-treatment scan demonstrates significant improvement in the clavicular metastasis. This radiographic improvement was associated with resolution of the exophytic mass and with improvement in pain[35].

over the right clavicle at the site of a metastatic lesion. This lesion in the bone was associated with a palpable exophytic mass and the lesion was sufficiently painful to require frequent narcotic analgesics. On coumarin and cimetidine, the palpable exophytic mass disappeared from the physical examination and the bone scan improved remarkably (Figure 9.8B). These changes were associated with improvement in pain and diminished need for analgesics. Another patient had a destructive lytic metastatic lesion in the bony pelvis (Figure 9.9A). This lesion produced significant pain requiring narcotic analgesics. Figure 9.9B shows the same area following treatment with coumarin and cimetidine. The arrows point to areas of reossification of bone. This reossification process never became complete but it was consistent with healing bone and improving metastasis. The improvement in the radiographic appearance was accompanied by improvement in bone pain, decreased analgesic use and improvement in performance status.

Some patients on the phase I trial of coumarin and cimetidine with an objective response or stable disease continued to do well clinically for some time following publication of the results. Subsequently, for publication in the proceedings of a meeting symposium, Marshall *et al.* reviewed the survival data for the patients with renal cell carcinoma who had been treated on this phase I trial. An attempt was made to obtain survival information for all 37 patients with renal cell carcinoma

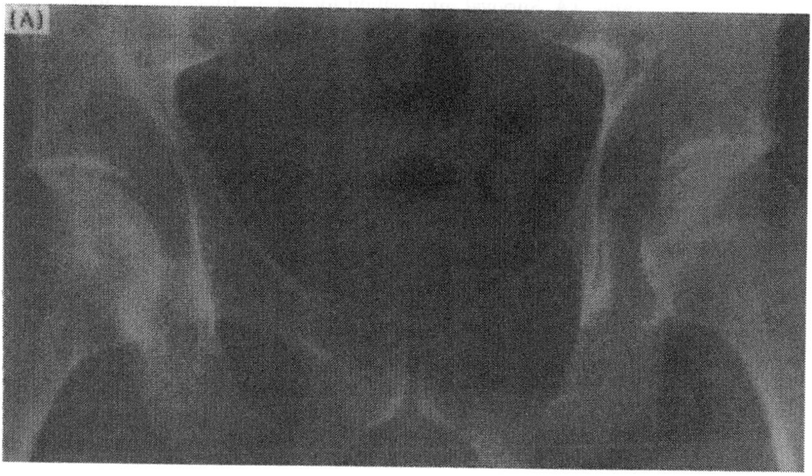

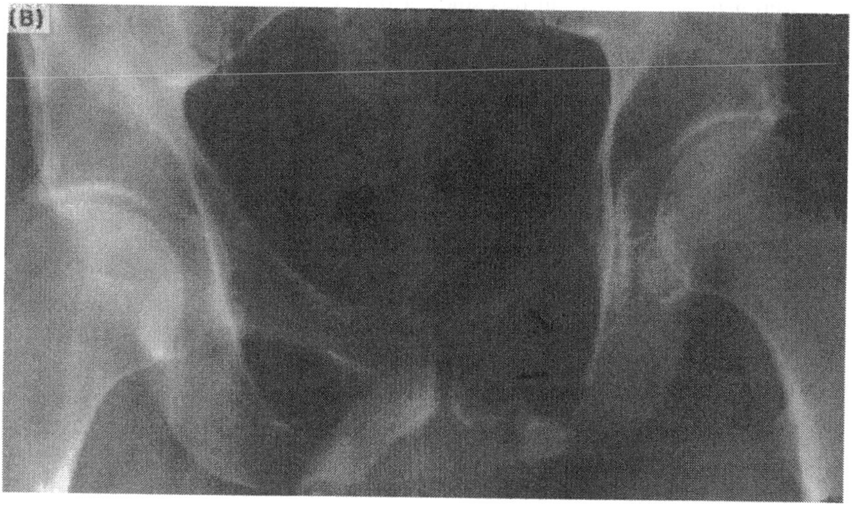

Figure 9.9. (A) Pretreatment radiograph of the pelvis in a patient with metastatic renal cell carcinoma. The film shows a large lytic metastasis which was associated with significant pain and impairment of performance status. (B) Marked improvement in the lytic lesion shown in (A) following treatment with coumarin and cimetidine. The arrows point to areas of re-ossification of bone. These changes were associated with improvement in bone pain and improvement in performance status[35].

Table 9.4. Survival information for the 37 patients with renal cell carcinoma who were treated on a phase I trial of coumarin and cimetidine.

Parameter	Number
Number of patients treated	37
Number evaluable for survival	31
Median survival (months)	28.8
Minimum survival (months)	0.9
Maximum survival (months)	77.8[a]
Number of patients alive	6
Number of responders	7
Number of responders evaluable for survival	7
Median survival for responders (months)	41.3
Minimumn survival for responders (months)	28.8
Maximum survival for responders (months)	77.8[a]

[a] Indicates that the patient is still alive and/or response continues.

who had been treated on this study. However, because a few patients were lost to long-term follow-up, survival information could only be determined for 31 of the patients. Table 9.4 is a summary of the survival information which was gathered and Figure 9.10 is a Kaplan–Meier plot of the survival data.

The 18.9% response rate observed among the 37 patients treated on this phase I trial is interesting and suggests a response rate similar to that of other biologic agents used to treat this disease. A randomized trial of coumarin to another biological agent would be the optimum way to determine relative response rates between the two agents and to compare toxicities. Such a trial has not been done.

Since the report of the pilot study of Marshall *et al.*[31] using coumarin (100 mg daily) and cimetidine for the treatment of renal cell carcinoma, and before the initiation or completion of the phase I trial of coumarin and cimetidine[35], other investigators explored the utility of coumarin for the treatment of this disease. All of these trials were based upon the low-dose regimen of the pilot study and were intended to quickly confirm or reject the findings of the pilot study. Kokron *et al.*[34] used the coumarin and cimetidine regimen, with coumarin at 100 mg daily, to treat 26 patients with renal cell carcinoma. They observed two complete responses, lasting greater than 12 months and 16 months, respectively, two partial responses lasting 12 and 15 months, respectively, three minor responses lasting from one to eight months and stable disease which persisted for up to 10 months. As was the case in the pilot study of Marshall *et al.*[31], Kokron *et al.* found that the features which tended to be associated with response were prior nephrectomy status, good performance status, smaller, non-bulky tumor burdens and lung as the site of metastatic disease. Herrman *et al.*[32], using a similar regimen, reported only two responses among 24 patients treated but he did confirm the safety profile of the drug. Once again using the low-dose coumarin regimen with cimetidine, Dexeus *et al.*[33] reported a 6% response rate for the regimen among 50 patients with advanced

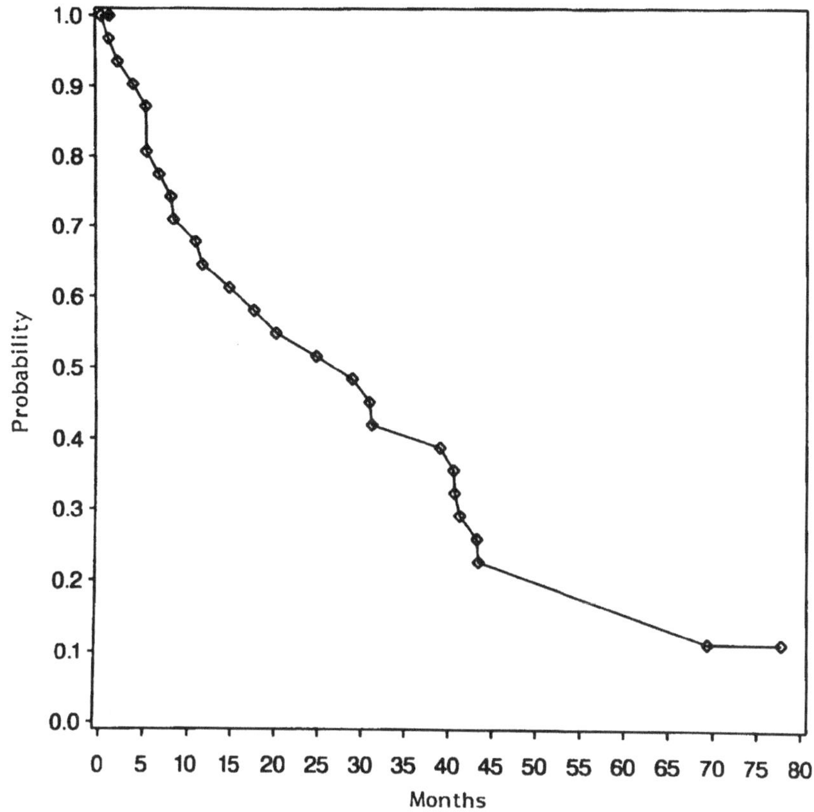

Figure 9.10. Kaplan–Meier plot of the survival data for the 37 patients with renal cell carcinoma who were treated on the phase I trial of coumarin and cimetidine. (Reproduced by permission from *Journal of Cancer Research and Clinical Oncology*[22].)

renal cell carcinoma. In this trial, the investigators reported that 15 of the 50 patients experienced grade 1 nausea/vomiting. It should be noted that Dexeus *et al.* used a preparation of coumarin which was different from that used by the other investigators mentioned above. While this difference in preparation probably had nothing to do with the differences in response rate, it could have accounted for the difference in toxicity. Venook *et al.*[36] used the same regimen to treat patients with renal cell carcinoma who had been treated previously with, and failed, interleukin-2 and lymphokine activated killer cells. They observed no objective responses among these patients. This study was ill-designed for it pre-selected a poor prognostic subset of patients to treat. The data of Marshall *et al.*, as discussed above, indicated that there were no responders among patients treated previously with interleukin-2.

While useful toxicity information resulted from the studies mentioned above, the series of phase II trials which was stimulated by the pilot report of Marshall *et al.*[31] have contributed nothing of significance to the understanding of the potential therapeutic benefit of coumarin for renal cell carcinoma. In some ways, it is unfortunate that these studies were done at all. Marshall has continued to be self-critical for having published the results of the pilot study, or for even having done the study. The usual path for the development of a new drug dictates that phase I dose and toxicity data guide phase II type trials and no such data existed at the time the pilot study was initiated and conducted. It was fortuitous that any responses at all were seen. It is for these reasons that Marshall *et al.*, subsequent to the pilot study, proceeded to conduct the phase I trial discussed above. No other such studies have been done as of this writing. It must be pointed out that there is a great deal of preliminary information lacking in the development of this drug. As mentioned repeatedly above, all trials of coumarin have administered the drug as a single daily oral dose. There are no pharmacologic data or toxicity data to indicate if coumarin can be given safely in multiple daily doses (in order to increase the total daily dose) or to suggest that multiple daily doses might be associated with higher response rates. Currently, this is a major deficiency in the development of the drug.

Subsequent to the completion of the phase I trial of coumarin and cimetidine mentioned above, Marshall *et al.* proceeded to conduct a trial of coumarin alone for the treatment of advanced renal cell carcinoma. As mentioned above, in the phase I trial responses were observed among patients taking coumarin over a wide range of doses. Because there was no clustering of responses around a single dose or dose range, empiricism guided the choice of a single daily dose to be used. Thus, a dose of 1 g daily was chosen (approximately the middle of the dose range among the responders). In the conduct of this next generation trial, Marshall *et al.* wished to compare coumarin to some other recognized agent for the treatment of renal cell carcinoma. The options were limited. It was felt that it would be difficult to attract and accrue patients to a study which randomized them to receive a relatively non-toxic oral agent (coumarin) or to a definitely more toxic agent such as IFN-α (given subcutaneously) or IL-2 (given intravenously or subcutaneously). It would be impossible also to conduct a double-blind study in which one drug is given orally and one is given subcutaneously or intravenously. This presents the risk of introducing investigator bias into the study. Further, because the response rates for these two drugs are relatively low, it is difficult to consider these to be standard of care although they are used frequently to treat this disease. Many patients, by virtue of age or debilitation, are not able to tolerate the side-effects of interferon or IL-2 and are followed simply with palliative, symptomatic supportive care. Thus, it was decided to accept palliative supportive care as the standard of care for patients with metastatic renal cell carcinoma. This arm of the randomized trial would be represented by a matching placebo, which also allowed the study to be double-blinded.

Thus, a trial was designed to treat patients with coumarin or placebo in a randomized, double-blinded fashion. In fairness to the patients, there was a cross-over design which allowed patients failing placebo to receive coumarin on an open label basis. Patients were randomized to receive coumarin 1 g (10 tablets) orally daily continually or placebo (10 tablets daily). Patients were evaluated for antitumor response monthly while on the study and they remained on the assigned (blinded) study drug until there was evidence of progression of disease. At such time, the drug status was 'unblinded'. Patients receiving coumarin were removed from the study except to follow for survival information and patients receiving placebo were crossed-over to receive coumarin until there was progression of disease. The results of this study have no been published as of the time of this writing. The study has reached a major interval statistical analysis point but the results are not available. It can be stated here that there were no responses among patients radomized to receive placebo but there were responses among the patients randomized to receive coumarin, including complete and partial responses. The results of this study will be published elsewhere.

As mentioned above, during the conduct of the phase I trial of coumarin and cimetidine, the toxicities which were encountered seemed to be related to the aromatic properties of coumarin. This was true especially when large numbers of these aromatic tablets were ingested as a single daily dose. Most investigators have felt that coumarin (1,2-benzopyrone) is a pro-drug and that the active form of the drug is 7-hydroxycoumarin. As mentioned earlier, 7-hydroxycoumarin is the major human metabolite of coumarin. This form of the drug lacks the strong aromatic properties of the parent substance. Theoretically, larger does of 7-hydroxycoumarin could be given to patients with fewer side-effects. A phase I trial of 7-hydroxycoumarin has been conducted in humans with advanced malignancies which are refractory to standard therapies or for which no effective therapy exists. The 7-hydroxycoumarin was produced as 100 mg tablets and provided for this study by Schaper & Brümmer (Salzgitter, Germany) pharmaceutical firm. (A word of caution: The metabolism of coumarin is highly species-dependent. 7-Hydroxycoumarin is the major metabolite of coumarin only for humans and baboons. Special caution must be taken in interpreting studies of coumarin conducted in rodents, which produce no 7-hydroxycoumarin, and extending the observations to the human setting.)

The phase I trial of 7-hydroxycoumarin was designed in a fashion similar to the one with coumarin (without the cimetidine). Three patients were treated at each dose level. Patients received 7-hydroxycoumarin at doses ranging from 100 mg to 7 g as a single daily dose. While the results of this study have not been published as of this writing, it can be stated with confidence that 7-hydroxycoumarin was tolerated much better than coumarin. The study ended at 7 g daily because of practical considerations and not because of any toxicity which was encountered. It became impractical for patients to ingest 70 or more tablets as a single daily dose, toxicity notwithstanding. Pharmacokinetic studies were conducted on day 1 and day 15 of treatment and these will be reported in due course. It remains for a separate phase I study to be conducted of 7-hydroxycoumarin given in multiple daily doses. Subsequently, phase II efficacy studies can be performed for various malignancies.

9.5 SUMMARY

As mentioned earlier, coumarin (1,2-benzopyrone) is the structurally least complex member of a large class of naturally occurring substances known generically as the coumarins. For a full appreciation of the biochemistry of this class of substances, the reader is referred to the text of Murray *et al.*[12]. With regard to the range of biologic effects, antitumor effects, immunologic effects, clinical toxicities, pharmacokinetics and clinical responses, it is logical to begin with a study of the simplest of these compounds, 1,2-benzopyrone. These studies will lead naturally to the development of an understanding of structure–function relationships among these substances as further work is done. At the time of this writing, it is clear that a number of things need to be done in the short term to enhance our understanding of coumarin and 7-hydroxycoumarin. For simplicity, the following itemized list is offered.

(1) A formal phase I pharmacokinetic study of coumarin given as a single daily dose needs to be done. This study can build upon the phase I trial which has been reported already[35], guiding the selection of doses from what is known of the toxicity of the drug from that study. Such a study would measure the basic metabolism and pharmacokinetics of the drug and simultaneously look at the usual issues of protein binding of the pro-drug and metabolite as well as the degree of glucuronidation of the metabolite.

(2) The knowledge gained from the pharmacologic study of the single daily dose/schedule should be applied to a separate study to define the pharmacokinetics and toxicity of coumarin when given in multiple daily doses (i.e. define the maximally tolerated dose and dose-limiting toxicity of coumarin given in multiple doses per day). This may allow for more drug to be given within a 24 h period and, perhaps, lead to enhanced antitumor activity.

(3) Following the second item, above, a phase II (efficacy) trial of coumarin given in multiple daily doses should be conducted to determine if this may be superior to the single daily dose regimen. This could be done as a single-arm study or, for more information, it could be done with a placebo control arm in a randomized, double-blind fashion.

(4) Once an optimal or, more accurately, preferred, dose/schedule of coumarin is identified, it should be applied to the treatment of renal cell carcinoma in the adjuvant setting, that is, for the treatment of microscopically residual disease following nephrectomy. Of necessity, this would require a randomized, double-blind, placebo-controlled trial where time to recurrence of disease is the end-point. Because this would require large numbers of patients, this study would need to be conducted through one or more of the large cooperative oncology groups. If coumarin is effective in this setting, it would have an advantage over IL-2 or IFN-α as an adjuvant therapy. Because of the toxicity profile of these latter drugs, they can be administered only for relatively short periods of time following nephrectomy, especially since the toxicities can be cumulative with these agents. Since coumarin has very little toxicity, it can be administered indefinitely until such time as relapse might occur without any cumulative side-effects (i.e. for years). The ability to offer adjuvant therapy for protracted periods following nephrectomy may have some advantage. Of course, at this time, all of this is speculative and the data are lacking.

(5) If 7-hydroxycoumarin is to be developed further as a therapeutic agent, it needs to be subjected to the same steps outlined above.
(6) Because of the low toxicity of coumarin, and 7-hydroxycoumarin, there is the potential of combining either of these drugs with interferon, IL-2 or other biologics. It is very likely that coumarin can be added to a multi-drug regimen in an attempt to increase clinical activity without appreciably increasing toxicity.
(7) The data collected from the preclinical studies (activity and mechanism of action) and the clinical trials (pharmacology and metabolism, phase I toxicity studies and phase II efficacy studies) should be extrapolated to a study of the other substances included within the group known collectively as the coumarins. Structure–function relationships and toxicity profiles may lead to the identification of other drugs with activities against renal cell carcinoma and/or other targeted malignancies.

DEDICATION

This chapter is dedicated to the memory of Mr Hans-Henning Schaper, past owner of Schaper & Brümmer GmbH, whose scientific interest in the coumarins and whose support for research made possible much of what we presently know concerning these substances.

REFERENCES

1 Muss HB. Interferon therapy for renal cell carcinoma. *Semin Oncol* 1987; **14** (Suppl. 2): 36–42.
2 Kirkwood JM, Harris JE, Vera R *et al.* A randomized study of low and high doses of leukocyte alpha-interferon in metastatic renal cell carcinoma. The American Cancer Society Collaborative Trial. *Cancer Res* 1985; **45**: 863–71.
3 Rosenberg SA, Lotz MT, Muul LM *et al.* Observations on the systemic administration of autologous lymphokine-activated killer cells and recombinant interleukin-2 to patients with metastatic cancer. *N Eng J Med* 1985; **313**: 145–92.
4 Lotze MT, Chang AE, Seipp CA *et al.* High-dose recombinant interleukin-2 in the treatment of patients with disseminated cancer. *JAMA* 1986; **256**: 3117–24.
5 Rosenberg SA, Lotze MT, Muul LM *et al.* A progress report on the treatment of 157 patients with advanced cancer using lymphokine-activated killer cells and interleukin-2 or high-dose interleukin-2 alone. *N Eng J Med* 1987; **316**: 889–97.
6 Thompson JA, Lee DJ, Lindgren CG *et al.* Influence of schedule of interleukin-2 adminstration on therapy with interleukin-2 and lymphokine-activated killer cells. *Cancer Res* 1989; **49**: 235–40.
7 Clerk JW, Smith JW, Steis RG *et al.* Interleukin-2 and lymphokine-activated killer cell therapy: Analysis of a bolus interleukin-2 and a continuous infusion interleukin-2 regimen. *Cancer Res* 1990; **50**: 7343–50.
8 Weiss GR, Margolin KA, Aronson FR *et al.* A randomized phase II trial of continuous infusion interleukin-2 or bolus injection interleukin-2 plus lymphokine-activated killer cells for advanced renal cell carcinoma. *J Clin Oncol* 1992; **10**: 275–81.
9 Rosenberg SA, Lotze MT, Yang JC *et al.* Combination therapy with interleukin-2 and alpha-interferon for the treatment of patients with advanced cancer. *J Clin Oncol* 1989; **7**: 1863–74.

10 Whitehead RP, Ward D, Hemingway L *et al.* Subcutaneous recombinant interleukin-2 in a dose escalating regimen in patients with metastatic renal cell adenocarcinoma. *Cancer Res* 1990; **50**: 6708–15.

11 Sleijfer DT, Janssen RA, Butler J *et al.* Phase II study of subcutaneous interleukin-2 in unselected patients with advanced renal cell cancer on an outpatient basis. *J Clin Oncol* 1992; **10**: 1119–23.

12 Murray RDH, Mendez J, Brown SA *The Natural Coumarins: Occurrence, Chemistry and Biochemistry.* New York: Wiley, 1982.

13 Thornes RD, Lynch G, Sheehan MW. Cimetidine and coumarin therapy of melanoma. *Lancet* 1982; **II**: 328.

14 Thornes RD, Lynch G. Combination of cimetidine with other drugs for treatment of cancer. *N Eng J Med* 1983; **308**: 591–2.

15 Thornes RD, Edlow DW, Wood S Jr. Inhibition of locomotion of cancer cells *in vivo* by anticoagulant therapy. Effects of sodium warfarin on V2 cancer cells, granulocytes and macrophages in rabbits. *Johns Hopkins Med J* 1968; **123**: 305–16.

16 Berkarda B, Bouffard-Eyoboglu H, Derman U. The effect of coumarin derivatives on the immunological system of man. *Agents Actions* 1983; **13**: 50–2.

17 Marshall ME, Riley LK, Rhoades J *et al.* Effects of coumarin (1,2-benzopyrone) on peripheral blood lymphocytes, natural killer cells and monocytes in patients with advanced malignancies. *J Biol Resp Mod* 1989; **8**: 62–9.

18 Marshall ME, Rhoades JL, Mattingly C *et al.* Coumarin (1,2-benzopyrone) enhances DR and DQ antigen expressions by peripheral blood mononuclear cells *in vitro. Molec Biother* 1991; **3**: 204–6.

19 Marshall ME, Conley D, Hollingsworth P *et al.* Effects of coumarin (1,2-benzopyrone) on lymphocyte, natural killer cell and monocyte functions *in vitro. J. Biol Resp Mod* 1989; **8**: 70–85.

20 Stuhlmeier K, Theyer G, Baumgartner G *et al.* Synergistic effect of coumarin (1,2-benzopyrone) and endotoxin in the induction of human interleukin-1. *Clin Exp Immunol* 1991; **84**: 317.

21 Zlabinger GJ, Bohmig G, Rosenkranz AR *et al.* Intracelluar signalling in coumarin-enhanced IL-1 production induced by LPS. Proceedings of the International Society for Coumarin Investigators, Dublin, Ireland. *J Ir Coll Physicians Surg* 1993; **22** (Suppl. 1): 38–40.

22 Marshall ME, Kervin K, Benefield C *et al.* Growth inhibitory effects of coumarin (1,2-benzopyrone) and 7-hydroxycoumarin on human malignant cell lines *in vitro. J Cancer Res Clin Oncol* 1994; **120** (Suppl.): 3–10.

23 Moran E, Prosser E, O'Kennedy R *et al.* The effect of coumarin and 7-hydroxycoumarin on the growth of human tumor cell lines. *J Ir Coll Physicians Surg* 1993; **22** (Suppl. 1): 41–3.

24 Siegers CP, Bostelmann HC. Effect of coumarin on cell proliferation in human tumor cell lines. *J. Ir Coll Physicians Surg* 1993; **22** (Suppl. 1); 47–8.

25 Zanker KS. Coumarin(s) interface between cell growth inhibition, expression of oncogenes, tumor cell locomotion and cell-to-cell communication. *J Ir Coll Physicians Surg* 1993; **22** (Suppl. 1): 51–5.

26 Myers RB, Parker M, Grizzle WE. The effects of coumarin and suramin on the growth of malignant renal and prostatic cell lines. *J Cancer Res Clin Oncol* 1994; **120** (Suppl.): 11–13.

27 Seliger B, Petterson H. 7-Hydroxycoumarin inhibits oncogene-induced transformation of murine fibroblasts. *J Cancer Res Clin Oncol* 1994; **120** (Suppl.): 23–7.

28 Gallicchio VS, Hulette BC, Marshall ME. Toxicity of coumarin (1,2-benzopyrone) on human peripheral blood mononuclear cells and human and murine bone marrow (GM) progenitor stem cells. *J Biol Resp Mod* 1989; **8**: 116–21.

29 Friedman H. Immunomodulating effects of cimetidine. In: Hersh EM *et al.* eds. *Augmenting Agents in Cancer Therapy.* New York: Raven, 1981: 417–25.

30 Sakasrabudke DM, McCune CS, O'Donnell RW *et al.* Inhibition of suppressor T-lymphocytes (Ts) by cimetidine. *Proc Am Assoc Cancer Res* 1986; **27**: 328.

31 Marshall ME, Mendelsohn L, Butler K *et al.* Treatment of metastatic renal cell carcinoma with coumarin (1,2-benzopyrone) and cimetidine: a pilot study. *J Clin Oncol* 1987; **5**: 862–6.

32 Hermann R, Maurer B, Manegold C *et al.* Phase II trial of coumarin and cimetidine in renal cell cancer. (In press)

33 Dexeus FH, Logothetis CJ, Sella A *et al.* Phase II study of coumarin and cimetidine in patients with metastatic renal-cell carcinoma. *J Clin Oncol* 1990; **8**: 325–9.

34 Kokron O, Maca S, Gasser G *et al.* Cimetidine and coumarin therapy of renal-cell carcinoma. *Oncology* 1991; **48**: 102–6.

35 Marshall ME, Butler K, Fried A. Phase I evaluation of coumarin (1,2-benzopyrone) and cimetidine in patients with advanced malignancies. *Molec Biother* 1991; **3**: 170–8.

36 Venook AP, Davenport Y, Tseng A Jr. Activity of coumarin and cimetidine in metastatic renal cell carcinoma. *J Clin Oncol* 1989; **7**: 402–3.

CHAPTER 10

The Potential Role of Coumarins in the Therapy of Prostate Cancer

James L. Mohler, Bernadette T. Williams, John A. Freeman
Division of Urology and the University of North Carolina Lineberger Comprehensive Cancer Center, University of North Carolina at Chapel Hill, Chapel Hill, NC, USA

and

M. Ernest Marshall
Division of Hematology/Oncology, University of Alabama at Birmingham, Birmingham, AL, USA

10.1 INTRODUCTION

The lack of an effective treatment for metastatic androgen-independent and clinically localized but pathologically unfavorable prostate cancer warrants trial of new and promising treatments. Preliminary studies in patients with metastatic prostate cancer have shown (a) subjective but no objective responses to 100 mg coumarin and cimetidine daily; (b) objective responses in 3 of 40 patients treated with 3 g coumarin daily, all of whom had normal performance status and one of whom remains with three resolved bone metastases and stable prostate-specific antigen levels after seven years; and (c) toxicity only in bedridden patients. We recently initiated two multicenter trials of 1 g coumarin daily. Metastatic prostate cancer patients of normal performance status were treated in a phase II trial. Patients who had been treated by radical prostatectomy, but had surgical margin, seminal vesicle or lymph node involvement or detectable prostate-specific antigen post-operatively, were randomized to coumarin or placebo.

Coumarins: Biology, Applications and Mode of Action. Edited by R. O'Kennedy and R.D. Thornes.

10.2 PROSTATE CANCER: GENERAL DESCRIPTION

The prostate gland is located beneath the base of the bladder and surrounds the proximal portion of the urethra. In the young adult, it weighs an average of 20 g and is composed of alveoli lined by tall columnar secretory epithelial cells which are surrounded by a dense stroma. The prostate has a rich vascular supply, depends for growth and development upon testicular androgens and undergoes involution with androgen deprivation. The prostate gland has no known function although it produces secretions rich in fructose, calcium, citric acid and phosphoryl choline which enhance, but are not necessary for, sperm motility and fertility. In addition, it produces marker substances that include prostatic acid phosphatase, prostate specific antigen (PSA) and prostatin. PSA is a serine protease which, in the absence of pathology, is present at low levels in the serum. Inflammation, infarction, urinary retention and cancer cause elevation of serum PSA levels.

The most common disease of the prostate, benign prostatic hyperplasia, can be found pathologically in 81% of men over 50 years of age. Its etiology is unknown. Obstruction of the bladder neck produces voiding dysfunction which can be reversed by medical or surgical means in most patients. Benign prostatic hyperplasia occurs in the transitional zone of the prostate which is centrally located, in contrast to prostate cancer which occurs most commonly in the prostate periphery.

Like benign prostatic hyperplasia, the etiology of prostate cancer is unknown. At autopsy, prostate cancer is present in 30% of men in the sixth decade of life and increases to nearly 100% of men beyond 90 years of age. Prostate cancer rarely produces symptoms due to its peripheral location within the prostate until such time that the cancer becomes locally advanced (obstruction) or metastatic (bone pain and anemia). Prostate cancer grows slowly with tumor doubling times of approximately 2–6 years until metastatic at which time growth may result in tumor doubling times of 6–12 months. Incidentally discovered cancer at autopsy, or after prostatectomy for benign prostatic hyperplasia, may have a tumor doubling time as long as eight years. This cancer is most commonly well-differentiated, focal and microscopic (tumor volume less than $0.5\ cm^3$). Clinically significant prostate cancer is usually moderately to poorly differentiated, multifocal, and larger volume. Prostate cancer may extend to the peri-prostatic tissues including the seminal vesicles and bladder base and spread via the lymphatics to the vesical, sacral, external iliac, and lumber lymph nodes and via the vasculature to the bones of the pelvis, femurs and lumbar spine. Diagnosis requires biopsy and a staging evaluation to estimate the extent of disease. Patients with incidentally found prostate cancer which is focal and well-differentiated may be expectantly followed using serial digital rectal examination and measurement of PSA. Patients with clinically significant prostate cancer may be observed or treated by radical prostatectomy or radiation therapy. Cure is evidenced by organ-confinement of the cancer upon pathological examination of the prostatectomy specimen combined with an undetectable post-operative PSA. Since the prostate gland is the only source of PSA, persistence of PSA after radical prostatectomy is indicative of treatment failure in all cases. Patients with metastatic disease may be palliated by androgen deprivation and irradiation of painful metastases. Chemotherapy has proven unsuccessful due to the prolonged tumor doubling times which allow prostate cancer cells to repair damage induced by treatment prior to cell division.

10.3 TWO SPECIFIC CLINICAL PROBLEMS PRESENTED BY PROSTATE CANCER

The aging of our population, greater public awareness, and more frequent use of PSA for early detection has led to a dramatic increase in the incidence of prostate cancer. In 1992, prostate cancer surpassed lung cancer as the most frequently diagnosed non-skin malignancy in American men[1]. In 1995, American men suffered 40 400 deaths and 244 000 new diagnoses of prostate cancer[2]. Prior to widespread use of PSA, approximately 50% of men had bone metastases, and 30% of men had metastases to regional lymph nodes or extraprostatic extension at the time of diagnosis[3]. In the absence of intercurrent disease, most of these men progress to bone metastases within five years. Dr Charles Huggins received the Nobel Prize for work performed in the 1940s that demonstrated the androgen responsiveness of benign and malignant prostate tissue. Since this discovery, little progress has been made in the treatment of metastatic prostate cancer although many alternatives for androgen deprivation have been developed. Approximately 80% of men with metastatic symptomatic prostate cancer respond to androgen deprivation with an average life expectancy of 43 months. Altogether 20% of men fail to respond and survive an average of 7–9 months[4].

Since advanced prostate cancer may only be palliated, efforts to improve survival have focused upon earlier detection. Performance of digital rectal examination, measurement of PSA, and transrectal ultrasound-guided systematic biopsies have led to an increased rate of detection of prostate cancer. Increased detection has resulted in a fourfold increase in the number of radical prostatectomies performed for clinically localized disease from 1984 to 1990[5]. Prostate cancer detected by PSA-based screening and treated by radical prostatectomy has been found in an organ-confined state in 71% of cases[6], of whom approximately 90% are cured. Many believe that aggressive case finding has resulted in overdetection of prostate cancer and unnecessary treatment of many patients. However, criteria for the pathologic identification of significant cancers have been established[7,8]. Review of prostatectomy specimens from patients with PSA-detected tumors revealed that 84–93% of these tumors are clinically significant and not latent[8].

Patients with clinically organ-confined prostate cancer who are treated by radical prostatectomy may have a high risk of treatment failure for three reasons. First, lymph nodes may contain metastatic cancer upon staging pelvic lymphadenectomy in spite of a negative pre-operative evaluation. Secondly, pathological examination of the prostatectomy specimen may reveal seminal vesicle invasion or positive surgical margins. Lastly, in spite of favorable clinical and pathological findings, patients may have persistent disease indicated by detectable levels of PSA.

Patients with positive lymph nodes upon staging pelvic lymphadenopathy have been treated with radiation therapy, androgen deprivation, or observation with or without radical prostatectomy. When the primary tumor is left in place, the value of radiation therapy for local control as well as survival appears minimal[9–11]. Three retrospective analyses have shown no apparent difference in progression of disease between patients treated with pelvic irradiation and those treated with delayed hormonal therapy only[12–14]. In the only randomized trial of extended-field irradiation versus delayed hormonal therapy, Paulson and associates[15] first reported a longer disease-free interval in patients treated with irradiation, but in a later report of a similar series of patients[16] they reported that regional megavoltage radiotherapy did not delay progression. More recently, patients with prostate cancer metastatic to regional lymph nodes have been treated with radical prostatectomy, with and without androgen deprivation, although survival benefit appears unlikely[17,18].

Seminal vesicle invasion and margin positivity following radical prostatectomy are regarded as poor prognostic indicators[19]. By contrast, penetration through the capsule alone, especially when associated with other than high-grade cancer, appears to be less ominous[20]. Survival depends upon development of metastatic disease, which appears unaffected by delivery of radiation therapy to the pelvis although radiation may decrease local recurrence[21–23]. Metastases probably result from the continued growth of micrometastases not detected at the time of operation and only rarely may originate from secondary metastasis from pelvic local recurrence.

Radiation therapy has been employed for both palliation and cure in patients with prostate cancer. The enthusiasm for radiation therapy has diminished due to the inability of radiation therapy to sterilize intraprostatic tumor which has been documented by rates of post-treatment biopsy positivity ranging from 61%[24] to as high as 91%, even in the presence of a normal post-treatment digital rectal examination[25]. Clinical follow-up of irradiated patients relies upon radionuclide bone scans and digital rectal examinations and has been compromised by short durations of follow-up. Monitoring of disease status by PSA has produced the alarming finding that 80% of patients treated for clinically localized prostate cancer with radiation therapy had increasing PSA five years later[26]. In locally advanced prostate cancer, where radiation therapy has been employed for palliation, benefit is equally difficult to demonstrate. Smith[27] and Steinberg[28] and colleagues reported no differences in local progression when irradiated patients were compared with non-irradiated controls. Paulson and colleagues[16] reported no difference in the time to first evidence of distant metastases in patients randomly assigned to external beam irradiation and those expectantly treated with hormonal therapy. The poor results obtained after radiation therapy may be due to failure to sterilize local disease, but also may result from micrometastatic disease that is unrecognized at the time of initial treatment. This is particularly true in patients with clinically advanced prostate cancer in whom pelvic lymph node dissection has revealed lymph node metastases in approximately 50% of patients[29,30].

Palliation of prostate cancer can also be achieved by androgen deprivation. However, prostate cancers contain a subset of androgen-independent cells that grow even in the absence of androgen. Isaacs and Coffey have demonstrated that subpopulations of androgen-independent tumor cells exist before androgen deprivation treatment has begun[31]. Hence, it is not surprising that the Veterans' Administration Cooperative Urological Research Group studies 1 and 2 demonstrated no difference in average survival rates between patients with advanced prostate cancer treated immediately and patients who were observed and treated upon progression of their prostate cancer. Earlier application of androgen deprivation made possible by PSA detection of treatment failure, as well as in combination with new agents to trigger apoptosis, inhibit growth factors, stimulate immunity, or induce cytotoxicity in slowly proliferating tumors may increase the efficacy of treatments that have shown minimal or no benefits in advanced prostate cancer. For example, in the Dunning R-3327 hormone-sensitive H rat prostatic adenocarcinoma, androgen ablation as a single treatment or androgen ablation combined with cytoxan chemotherapy, were both more effective when given as early as possible[32]. When administered simultaneously and early, it was possible to increase survival beyond that found for either treatment administered alone. Patients with minimal disease recognizable only by detectable PSA may provide a situation analogous to the studies performed in animal models. They may benefit from early androgen ablation coupled with multiple modality therapy, an approach not yet tried in prostate cancer. Stimulation of the immune system to augment the normal defense mechanisms of the body to deal with micrometastatic disease after eradication of bulky primary disease may prove effective.

10.4 RATIONALE AND EVIDENCE FOR THE USE OF COUMARIN IN PROSTATE CANCER

Prior to 1970 the prostate was thought to be an immunologically privileged organ. However, the finding of lymphatics within the prostate and the occurrence of granulomatous prostatitis led many to search for an immune response to prostate cancer. Prostate cancer may evoke a humoral response with the production of autoantibodies[33,34] and elicit non-specific cellular responses as demonstrated by phytohemagglutinin blastogenesis impairment[35] and cell-specific responses as demonstrated by a skin reaction to autologous tumor extract[36]. Attempts to actively immunize patients with prostate cancer have shown little or no therapeutic benefit[37,38]. In addition, the only two randomized clinical trials of adoptive immunotherapy used bacillus Calmette–Guérin and produced conflicting results[39,40]. Thus, it appears that prostate cancer, like most other human neoplasms, is weakly immunogenic and effective therapies will require augmentation of the patient's immune system or genetic alteration of his tumor cells to secondarily effect enhanced immune surveillance.

Design of effective strategies for immunotherapy in prostate cancer depends upon an improved understanding of antigen processing and the potential pathways for immune evasion. Reliance upon Class I major histocompatibility complex (MHC)-mediated tumor cell recognition by cytotoxic T lymphocytes has been shown to be undesirable by two groups of investigators. As prostate cancer becomes progressively more poorly differentiated, the expression of HLA Class I antigens declines[41]. Absence of HLA Class I antigens may occur due to defects in Class I MHC assembly due to abnormalities of heavy-chain β_2-microglobulin and/or the androgen transporter gene product TAP-2, such as occurs in human prostate cancer cell lines[42]. Immunotherapy via gene transfer of interferon γ or tumor necrosis factor α which rely upon interaction of cytotoxic T cells and Class I MHC assembly and function may be unsuccessful.

In contrast, immunotherapy based upon granulocyte and macrophage colony stimulating factor (GM-CSF) or gene transfer which depends upon Class II MHC expression or natural killer-mediated cytolysis may prove more successful[42]. In the Dunning R-3327 MAT-LyLu anaplastic androgen-independent rat prostate cancer, cytokine-secreting tumor cell preparations were evaluated as therapeutic vaccines for the treatment of advanced prostate cancer. IL-2-secreting irradiated tumor cell preparations cured animals with subcutaneously established tumors and induced immunological memory that protected animals from subsequent tumor challenge. GM-CSF-secreting tumor cell preparations were less effective and interferon γ-secreting cells had only marginal effect[43]. IL-2 need not be secreted by the tumor cells themselves as mixtures of parental and IL-2-transfected cells were similarly rejected. The absence of T-cell mediation was confirmed by rejection of tumor cells in athymic mice. Subcutaneous injection of IL-2-transfected cells conferred protection only against small tumor cell inoculants[44]. The clinical feasibility of gene therapy for prostate cancer has been demonstrated by efficiently transferring reporter genes and GM-CSF in prostate cancer cells harvested from radical prostatectomy specimens in 10 consecutive patients[45]. The first human prostate cancer gene therapy trials were approved by the National Cancer Institute and began in late 1995.

The activity of coumarin has also been investigated in the non-immunogenic sublines of the Dunning R-3327 model. Omarbasha and associates demonstrated activity for coumarin in the highly metastatic MAT-LyLu subline and the slow-growing androgen-dependent non-metastatic H subline. In the MAT-LyLu subline, the incidence of lung and lymph node metastases was diminished and primary tumor volume was less in the treated than control groups[46]. However, the differences found in tumor volume translated into a five-day survival advantage in the tumor model where average survival without treatment is 39 days, which is of questionable significance. In addition, an apparent anti-androgen effect of coumarin or the co-administered cimetidine was noted which complicated the assessment of H tumor responsiveness to coumarin therapy. However, Maucher and associates found no evidence of anti-androgenic activity of coumarin in rats[47]. In addition, coumarin is metabolized differently in rats and humans and has a much greater bioavailability of different metabolites at equivalent doses[48].

Maucher and associates provided additional evidence for activity of coumarin in the Dunning model. When studying the androgen-sensitive but non-metastatic G-tumor and the androgen-independent and highly metastatic MAT-Lu tumor, no effect upon tumor growth rates or, in the case of MAT-Lu, lung metastases, was observed when coumarin therapy was administered after tumor growth was established. However, in the inoculation and resection model, where MAT-Lu tumor was excised five days after implantation, coumarin produced a 40% reduction in the number of lung metastases but was ineffective as a pre-treatment[47].

10.5 CLINICAL TRIALS OF COUMARIN IN PROSTATE CANCER

The unavailability of effective treatment for metastatic androgen-independent prostate cancer warranted a clinical trial of coumarin[49]. Fourteen patients with metastatic androgen-independent prostate cancer received 100 mg coumarin orally as a single daily dose for 14 days. On day 15, 300 mg cimetidine four times daily was added and both drugs were continued until progression of the disease. No objective evidence of tumor regression was observed; however, three patients experienced significant improvement in bone pain with decreased analgesic use that persisted until disease progression occurred at three, nine and 12 months. The patients all had extensive bone metastases and were heavily pretreated. The improvement in bone pain suggested some minimal activity for coumarin and cimetidine, which prompted a second study to evaluate a higher dosage of coumarin.

A group of 48 patients with metastatic hormone-naive (5 stage N+ and 10 stage M1) or androgen-independent (33 stage M1) prostate cancer, of average age 68 years (range 46–86 years) and Eastern Cooperative Oncology Group (ECOG) performance status of two or better, was given 3 g coumarin daily by mouth and evaluated monthly for toxicity and response by rigid criteria in a multicenter trial[50]. Toxicity was limited to asymptomatic liver transaminase elevations in three patients and nausea and vomiting in four patients, requiring cessation of therapy in two. Eligibility and protocol violations removed six additional patients from response evaluation. There were no complete responses. Partial responses occurred in two patients with bidimensionally measurable disease and one patient with disease evaluable by bone scan and elevated PSA and prostatic acid phosphatase (three of 40 patients, 8%). The remaining patients progressed after 1–12 (average 4.4) months. A 73-year-old man with asymptomatic bone metastases had resolution of three metastases detected by bone scan and the prostate primary visualized by magnetic resonance imaging (MRI) disappeared at three months. He remains with stable PSA levels, a normal bone scan and reappearance of the primary at six months without subsequent change, as shown by MRI or digital examination, after seven years (Figure 1).

These three responses contrast with the natural history of metastatic prostate cancer. The three responses occurred among patients with favorable performance status and relatively small tumor burdens. Prior studies of immunotherapy in many different neoplasms indicate that responses are most frequent when small tumor burdens are treated. Thus, favorable results have been achieved when coumarin

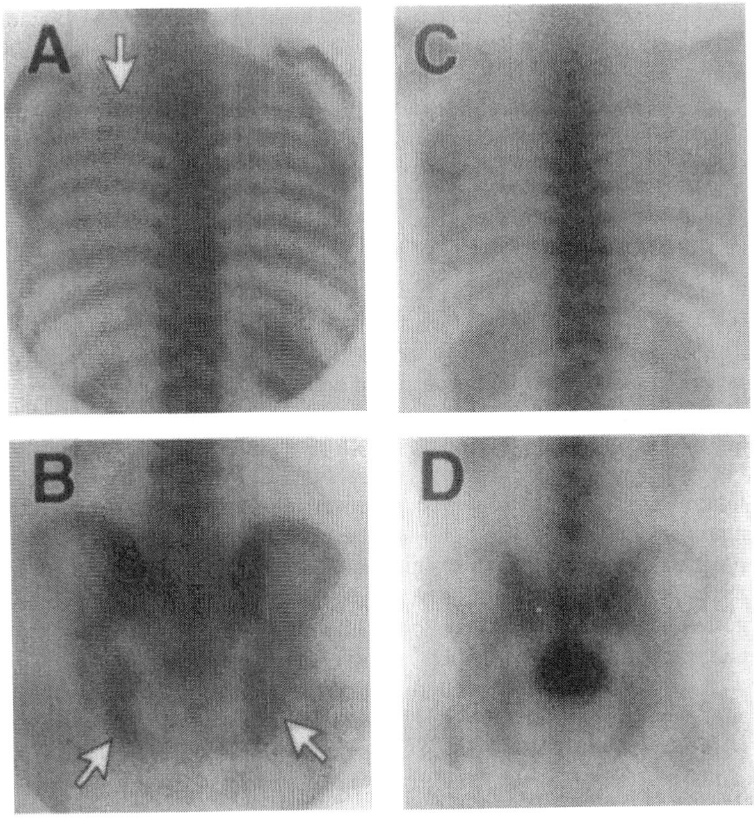

Figure 10.1. Bone scan reveals lesions of the right third rib (A) and ileum bilaterally (B) in a 73-year-old man with hormone-naive prostate cancer. After institution of coumarin, bone metastases resolved by three months and remained absent after six and a half years (C, D).

has been used as adjuvant therapy in malignant melanoma in patients with relatively lower tumor volumes and good performance status[51], in metastatic renal cell carcinomas[52] and when administered soon after inoculation of rats with sublines of the Dunning rat prostatic adenocarcinoma[46–48]. In contrast, responses are rare when coumarin has been used against advanced malignant melanoma[53,54], large volume renal cell carcinoma[55,56] and prostate cancer in patients with poor performance status[49,50]. Based upon these results and theoretical considerations, an evaluation of efficacy of coumarin, supervised by the United States Government Food and Drug Administration, was initiated in patients with stage M1 disease who have a more favorable prognosis. In addition, coumarin is being evaluated in patients who have undergone radical prostatectomy for clinically localized disease but have a high risk of recurrence.

10.6 PRELIMINARY RESULTS OF ON-GOING MULTICENTER CLINICAL TRIALS

A group of 40 patients with asymptomatic hormone-naive metastatic prostate cancer and 40 patients with androgen-independent metastatic prostate cancer who perform normally or are ambulatory with symptoms such that they are capable of light work or work of a sedentary nature (ECOG performance status less than or equal to 1 and Karnofsky index greater than or equal to 70) will receive 1 g coumarin daily and be monitored for toxicity and response in a multicenter trial. Of the 25 patients enrolled thus far, one patient discontinued drug after one month. Fourteen progressed within three months, five patients within six months, and four patients within nine months. A 77-year-old-man with androgen-independent metastatic prostate cancer remains without progression after 54 months. He underwent orchiectomy in November 1986 for Gleason sum 9 bone metastatic prostate cancer. He demonstrated recurrence June 1992 by PSA 20.9 ng/ml (Hybritech) at which time bone scan revealed metastatic disease of the right humerus, left ilium and fourth lumbar vertebra (Figure 10.2A, B). Coumarin was instituted and PSA increased to 42.7 ng/ml after 17 months and then steadily declined to a nadir of 9.9 ng/ml after 28 months, at which time bone scan revealed decreased uptake at all metastatic sites (Figure 10.2C). After 54 months, bone scan revealed no evidence of progression but PSA had risen to 53 ng/ml.

At the same three centers, patients who after radical prostatecomy have (a) microscopically positive pelvic lymph nodes, (b) seminal vesicle invasion, (c) extension of cancer to the surgical margin, or (d) detectable PSA are randomly assigned to 1 g coumarin daily or placebo. Progression is defined by the development of lesions on radionuclide bone scan, abnormal prostatic acid phosphatase levels, local recurrence proven by biopsy or PSA increases greater than or equal to 100% that exceed 10 ng/ml (Hybritech). Progression-free survival and disease-specific survival will be considered by interval analysis with an anticipated enrollment of 160 patients. A group of 42 patients treated by radical prostatectomy and who had detectable PSA post-operatively (nine patients), positive margins and/or seminal vesicles (27 patients) or microscopic pelvic lymph node metastases (six patients) have been randomized to coumarin or placebo. Two patients without progression discontinued study drug after 9 and 18 months. Eight patients progressed while on coumarin at 3, 12, 16, 18, 18, 27, 28 and 30 months. Among patients randomized to placebo, eight patients have progressed by PSA after 3, 6, 6, 9, 15, 15, 24 and 30 months. An additional patient progressed after six months but has just been crossed over to drug. The eight patients with minimum follow-up of three months have all progressed by PSA after 3, 3, 9, 9, 3, 6, 6 and 9 months, respectively. The rapidity of progression after cross-over to coumarin suggests little benefit to treatment of even small volume prostate cancer that is growing rapidly. All progressions have been detected by PSA levels and occurred in patients who entered the study for detectable PSA only (seven patients), positive margins and/or seminal vesicles (eight patients) or microscopic pelvic lymph node metastases (two patients). No patient has developed symptoms,

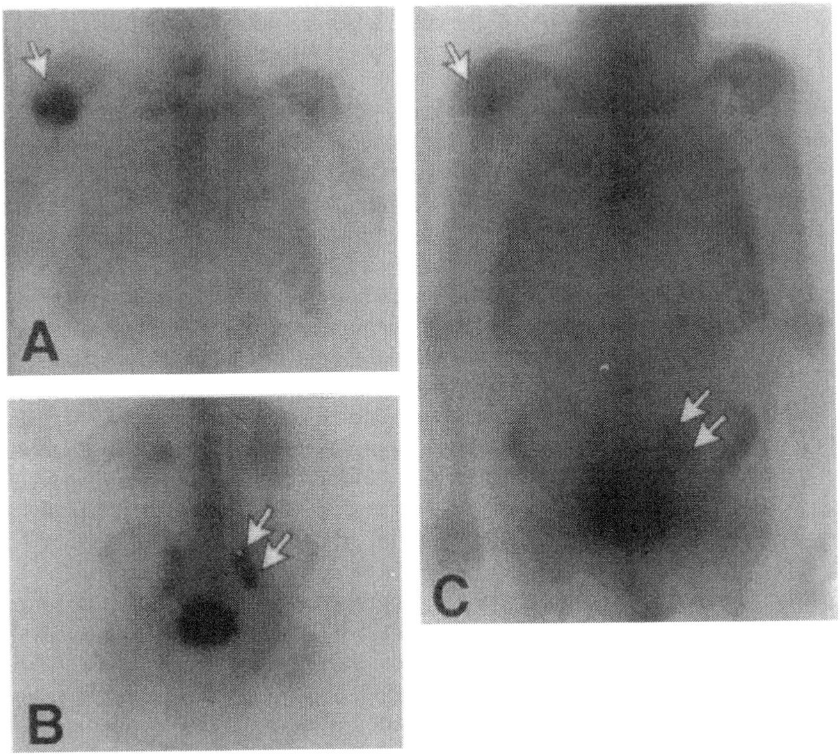

Figure 10.2. Bone scan revealed metastatic lesions of the right humerus (A), left ileum and fourth lumbar vertebra (B) in a 77-year-old man with androgen-independent metastatic prostate cancer five and a half years after orchiectomy. After institution of coumarin, PSA fell by 80%, reaching a nadir at 28 months, at which time bone scan revealed decreased uptake at all metastatic sites (C).

clinical local progression or bone metastases. Twenty-two patients have remained without progression for up to 42 months and 20 are evaluable.

10.7 CONCLUSIONS

Large-volume metastatic prostate cancer doubles at intervals of approximately 6–12 months whereas clinically localized prostate cancer may double at intervals of 2–6 years. Because prostate cancer occurs in older men at risk of death from competing causes, tumor manipulations that have even small effects upon tumor doubling times may markedly reduce disease-specific morbidity and mortality. Since no effective therapy exists for metastatic androgen-independent prostate cancer and since the use of radiation therapy and early androgen deprivation in an adjuvant fashion after radical prostatectomy is controversial, new and promising therapies should be explored in these patients.

Prostate cancer may be immune-responsive although evidence is much less convincing than in malignant melanoma and renal cell carcinoma. Coumarin has demonstrated activity against prostate cancer in an animal model and has shown rare but carefully documented activity against prostate cancer in humans. Clinical trials suggest that coumarin conforms to the results observed in other malignancies as well as therapeutic principles established with many immunotherapies. Patients with normal performance status and small tumor volumes are the most likely to respond. Continued investigation of coumarin in selected patients with prostate cancer seems warranted with the hope that even small treatment successes may translate into significant prolongation of progression-free intervals and survival with minimal or no toxicity.

REFERENCES

1 Boring CC, Squires TS, Tong T. Cancer Statistics, 1992. *CA Cancer J Clin* 1992; **42**: 19–38.

2 Wingo PA, Tong T, Bolden S. Cancer Statistics, 1995. *CA Cancer J Clin* 1995; **45**: 8–30.

3 Stamey TA, Cancer of the prostate. *Monogr Urol* 1983; **4**: 68–92.

4 Slack NH, Murphy GD. NCPC Participants. Criteria for evaluating patient responses to treatment modalities for prostatic cancer. *Urol Clin North Am* 1984; **11**: 337–42.

5 Lu-Yao GL, McLerran D, Wasson J, Wennberg JE. For the Prostate Patient Outcomes Research Team. An assessment of radical prostatectomy. Time trends, geographic variation, and outcomes. *JAMA* 1993; **269**: 2633–6.

6 Stamey TA, McNeal JE, Freiha FS, Redwine E. Morphometric and clinical studies on 68 consecutive radical prostatectomies. *J Urol* 1988; **139**: 1235–41.

7 Epstein JI, Walsh PC, Carmichael M, Brendler CB. Pathologic and clinical findings to predict tumor extent of non-palpable (stage T1c) prostate cancer. *JAMA* 1994; **271**: 368–74.

8 Catalona WJ, Richie JP, Ahman FR, Hudson MA, Scardino PT, Flanigan RC, deKernion JB, Ratliff TL, Kavousi LR, Dalkin BL, Waters WB, MacFarlane MT, Southwick PC. Comparison of digital rectal examination and serum prostate specific antigen in the early detection of prostate cancer: Results of a multicenter clinical trial of 6,630 men. *J Urol* 1994; **151**: 1283–90.

9 Grossman HB, Batata M, Hilaris B, Whitmore WF Jr. ^{125}I; implantation for carcinoma of the prostate. Further follow-up of first 100 cases. *Urology* 1982; **20**: 591–8.

10 Smith JA Jr., Haynes TH, Middleton RG. Impact of external irradiation on local symptoms and survival free of disease in patients with pelvic lymph node metastasis from adenocarcinoma of the prostate. *J Urol* 1984; **131**: 705–7.

11 Sause WT, Richards RS, Plenk HP. Prostatic carcinoma: 5-year follow-up of patients with surgically staged disease undergoing extended field radiation. *J Urol* 1986; **135**: 517–19.

12 Prout GR Jr., Heaney JA, Griffin PP, Daly JJ, Shipley WU. Nodal involvement as a prognostic indicator in patients with prostatic carcinoma. *J Urol* 1980; **124**: 226–31.

13 Kramer SA, Cline WA Jr., Farnham R, Carson CC, Cox EB, Hinshaw W, Paulson DF. Prognosis of patients with stage D1 prostatic adenocarcinoma. *J Urol* 1981; **125**: 817–21.

14 Donohue RE, Biber RJ, Mani JH, Fauver HE, Whitesel JA, Wettlaufer JN, Mohr S, Pfister RR, Scanavino D, William G, Augsburger RR. Stage D_1 adenocarcinoma of prostate. *Urology* 1984; **23**: 118–21.

15 Paulson DF, Cline WA Jr., Koefoot RB Jr., Hinshaw W, Stephani S, Uro-Oncology Research Group. Extended field radiation therapy versus delayed hormonal therapy in node positive prostatic adenocarcinoma. *J Urol* 1982; **127**: 935–7.
16 Paulson DF, Hodge GR Jr., Hinshaw W, Uro-Oncology Research Group. Radiation therapy versus delayed androgen deprivation for stage C carcinoma of the prostate. *J Urol* 1984; **134**: 901–2.
17 Golimbu M, Provet J, Al-Askari S, Morales P. Radical prostatectomy for stage D1 prostate cancer: Prognostic variables and result of treatment. *Urology* 1987; **30**: 427–35.
18 Zinke H, Utz D, Thile P, Taylor WF. Treatment options for patients with stage D1 (T0-3, N1-2, M0) adenocarcinoma of prostate. *Urology* 1987; **30**: 307–15.
19 Pontes JE, Wajsman Z, Huben RP, Wolf RM, Englander LS. Prognostic factors in localized prostatic carcinoma. *J Urol* 1985; **131**: 1137–9.
20 Epstein JI, Carmichael MJU, Pizov G, Walsh PC. Influence of capsular penetration on progression following radical prostatectomy: Study of 196 cases with long-term follow-up. *J Urol* 1993; **150**: 135–41.
21 Gibbons RP, Cole BS, Richardson RG, Correa RJ Jr., Brannen GE, Mason JT, Taylor WJ, Hafermann MD. Adjuvant radiotherapy following radical prostatectomy: results and complications. *J Urol* 1986; **135**: 65–8.
22 Lange PH, Moon TD, Narayan P, Medini E. Radiation therapy as adjuvant treatment after radical prostatectomy: patient tolerance and preliminary results. *J Urol* 1986; **136**: 45–9.
23 Anscher MS, Prosnitz LR. Postoperative radiotherapy for patients with carcinoma of the prostate undergoing radical prostatectomy with positive surgical margins, seminal vesicle involvement and/or penetration through the capsule. *J Urol* 1987; **138**: 1407–12.
24 Freiha FS, Bagshaw MA. Carcinoma of the prostate: Results of post-irradiation biopsy. *Prostate* 1984; **5**: 19–25.
25 Kabalin JN, Hodge KK, McNeal JE, Freiha FS, Stamey TA. Identification of residual cancer in the prostate following radiation therapy: Role of transrectal ultrasound guided biopsy and prostate specific antigen. *J Urol* 1989; **142**: 326–31.
26 Stamey RA, Ferrari MK, Schmid H. The value of serial prostate specific antigen determinations 5 years after radiotherapy: Steeply increasing values characterize 80% of patients. *J Urol* 1993; **150**: 1865–6.
27 Smith AA Jr., Haynes TH, Middleton RG. Impact of external irradiation on local symptoms and survival free of disease in patients with pelvic lymph node metastasis from adenocarcinoma of the prostate. *J Urol* 1984; **131**: 705–7.
28 Steinberg GD, Epstein JI, Piantadosi S, Walsh PC. Management of stage D1 adenocarcinoma of the prostate: The Johns Hopkins experience 1974 to 1987. *J Urol* 1990; **144**: 1425–32.
29 Donohue RE, Mani JH, Whitesel JA, Mohr S, Scanavino D, Augspurger RR, Biber RJ, Fauver HE, Wettlaufer JH, Pfister RR. Pelvic lymph node dissection: Guide to patient management in clinically locally confined adenocarcinoma of the prostate. *Urology* 1982; **20**: 559–65.
30 Zincke H, Utz DC, Taylor WF. Bilateral pelvic lymphadenectomy and radical prostatectomy for clinical stage C prostatic cancer: Role of adjuvant treatment for residual cancer and in disease progression. *J Urol* 1986; **135**: 1199–205.
31 Isaacs JT, Coffey DS. Adaptation versus selection as the mechanism responsible for the relapse of prostatic cancer to androgen ablation therapy as studied in the Dunning R-3327-H adenocarcinoma. *Cancer Res.* 1981; **41**: 5070–5.
32 Ellis WJ, Isaacs JT. Effectiveness of complete versus partial androgen withdrawal therapy for the treatment of prostatic cancer as studied in the Dunning R-3327 system of rat prostatic adenocarcinomas. *Cancer Res.* 1985; **45**: 6041–50.
33 Ablin RJ. Serum antibody in patients with prostatic cancer. *Br J Urol* 1976; **48**: 355–61.
34 Dmochowski L, Muruyama K, Ohsuki Y, Seman G, Bowen JM, Newton WA, Johnson DE. Virologic and immunologic studies of human prostatic carcinoma. *Cancer Chemother Rep* 1975; **59**: 17–31.
35 Robinson M, Nakhlan L, Whitaker RH. A new concept in the management of carcinoma of the prostate. *Br J Urol* 1971; **43**: 728–32.

36 Brannen GE, Gomolka DM, Coffey DS. Specificity of cell membrane antigens in prostatic cancer. *Cancer Chemother Rep* 1975; **59**: 127–38.

37 Ablin RJ. Adjuvant immunotherapy for prostatic cancer. In: Ablin RJ ed. *Prostatic Cancer.* New York: Marcel Dekker, 1981: 251–83.

38 Donovan JF, Lubaroff DM, Williams RD. Immunotherapy of prostate cancer. *Probl Urol.* 1990; **4**: 489–505.

39 Brosman S. Nonspecific immunotherapy in GU cancer. In: Crispen R ed. *Neoplasm Immunity.* Philadelphia: Franklin Institute Press, 1977: 97–107.

40 Guinan PD, John T, Baumgartner G, Sundar B, Ablin RJ. Adjuvant immunotherapy (BCG) in stage D prostate cancer. *Am J Clin Oncol* 1982; **5**: 65–8.

41 Levin I, Klein T, Kuperman O, Segal S, Shapira J, Gal R, Hart Y, Klein B. The expression of HLA class I antigen in prostate cancer in relation to tumor differentiation and patient survival. *Cancer Detect Prev* 1994; **18**: 443–5.

42 Sanda MG, Restifo NP, Walsh JC, Kawakami Y, Nelson WG, Pardoll DM, Simons JW. Molecular characterization of defective antigen processing in human prostate cancer. *J Natl Cancer Inst* 1995; **87**: 280–5.

43 Vieweg J, Rosenthal FM, Bannerji R, Heston WDW, Fair WR, Gansbacher B, Gilboa E. Immunotherapy of prostate cancer in the Dunning rat model: Use of cytokine gene modified tumor vaccines. *Cancer Res* 1994; **54**: 1760–5.

44 Moody DB, Robinson JC, Ewing CM, Lazenby AJ, Isaacs WB. Interleukin-2 transfected prostate cancer cells generate a local antitumor effect *in vivo. Prostate* 1994; **24**: 244–51.

45 Sanda MG, Ayyagari SR, Jaffee EM, Epstein JI, Clift SL, Cohen LK, Dranoff G, Pardoll DM, Mulligan RC, Simons JW. Demonstration of a rational strategy for human prostate cancer gene therapy. *J Urol* 1994; **151**: 622–8.

46 Omarbasha B, Heston WDW, Huryk R, Fair WR. Treatment of a metastatic prostate derived tumor with cimetidine and coumarin (abstract). *J Urol* 1988; **139**: 176.

47 Maucher A, Kager M, von Angerer E. Evaluation of the antitumor activity of coumarin in prostate cancer models. *J Cancer Res Clin Oncol* 1993; **119**: 150–4.

48 Omarbasha B, Fair WR, Heston WDW. Effect of coumarin on the normal rat prostate and on the R-3327 H prostatic adenocarcinoma. *Cancer Res* 1989; **49**: 3045–9.

49 Marshall ME, Butler K, Hermansen D. Treatment of hormone refractory stage D carcinoma of prostate with coumarin (1,2-benzopyrone) and cimetidine: A pilot study. *Prostate* 1990; **17**: 95–9.

50 Mohler JL, Gomella LG, Crawford ED, Glode LM, Zippe CD, Fair WR, Marshall ME. Phase II evaluation of coumarin (1,2-benzopyrone) in metastatic prostatic carcinoma. *Prostate* 1992; **120**: 123–31.

51 Thornes D, Daly L, Lynch G, Browne H, Tanner A, Keene F, O'Loughlin S, Corrigan T, Daly P, Edwards G, Breslin B, Browne HY, Shine M, Lennon F, Hanley J, McMurray N, Gaffney E. Prevention of early recurrence of high risk malignant melanoma by coumarin. *Eur J Surg Oncol* 1989; **15**: 431–5.

52 Marshall ME, Mendelsohn L, Butler K, Riley L, Cantrell J, Wiseman C, Taylor R, MacDonald JS. Treatment of metastatic renal cell carcinoma with coumarin (1,2-benzopyrone) and cimetidine: A pilot study. *J Clin Oncol* 1987; **5**: 862–6.

53 Marshall ME, Butler K, Cantrell J, Wiseman C, Mendelsohn L. Treatment of advanced malignant melanoma with coumarin and cimetidine: A pilot study. *Cancer Chemother Pharmacol* 1989; **24**: 65–6.

54 Pedersen L, Rose C, Langvad E. Combined treatment of advanced malignant melanoma with coumarin and cimetidine: A phase II study. *Cancer Immunol Immunother* 1987; **24**: 178–9.

55 Venook AP, Davenport Y, Tseng A Jr. Activity of coumarin and cimetidine in metastatic renal cell carcinoma. *J Clin Oncol* 1989; **7**: 402–3.

56 Dexeus FH, Logothetis CJ, Sella A, Fitz K, Almata R, Reuben JM, Dozier N. Phase II study of coumarin and cimetidine in patients with metastatic renal cell carcinoma. *J Clin Oncol* 1990; **8**: 325–30.

CHAPTER 11

Clinical and Biological Observations Associated with Coumarins

R.D. Thornes
Colman K. Byrnes Research Centre, Beaumont Hospital, Dublin, Ireland

11.1 INTRODUCTION

In 1820, coumarin (1,2-benzopyrone) was first isolated from Tonka bean and *Melilotus officinalis* (sweet clover) by Vogel[1]. It is present in many biological systems, where it is associated with growth control, germination and defence mechanisms. In 1973 Feuer[2] wrote an excellent review of the biological actions of coumarins (see Chapter 13). Clinically, coumarin is of proven value in the treatment of lymphoedema (see Chapter 7), its use in the treatment of burns is long overdue (see Chapter 8), and in malignant disease it appears to have two actions: one direct (see Chapter 9) and the other indirect through the cellular immune mechanism. This latter has been my interest since I first started medicine and found out about man's ability to stay healthy in the presence of so much potential pathology.

11.2 EARLY PERSONAL EXPERIENCE WITH COUMARINS

In 1948, I assisted Dr Desmond Riordan in an *in vitro* study of growth control by lactones. The two lactones that were of clinical interest were ascorbic acid and coumarin as these were already being consumed by humans as food additives. In

Coumarins: Biology, Applications and Mode of Action. Edited by R. O'Kennedy and R.D. Thornes.

February 1952, at the Royal City of Dublin Hospital (Baggot Street), three patients with advanced lung cancer were selected to receive coumarin 50 mg (50 ml of a 1 g/l solution) three times daily. Lung cancer was chosen because there was no available treatment and tumour size could be evaluated radiologically. There was a remarkable improvement in their well-being, which greatly impressed me and I felt at that time that we were treating the hosts and not just the tumour, but there was no observable effect on their tumours. This contrasts with the initial observations of Marshall and co-workers (Chapter 9), who used coumarin to treat pulmonary metastases of renal cell carcinoma and found that it resulted in rapid regression. A further nine patients with various advanced cancers were then treated but again there was only subjective response. Dr Riordan later treated more than 600 patients with advanced cancers at St Anne's Skin and Cancer Hospital, Dublin. He used oral doses of coumarin up to 250 mg daily and also injected alcoholic solutions of coumarin into the tumours but no regression of tumour was noted. Nevertheless, patients who showed subjective improvement often had tumours that remained static for long periods and it was noted that patients with multiple myeloma responded well. The advent of chemotherapy, however, displaced coumarin as a treatment and the results were never published.

In 1954, during my post-graduate training at the Boston Lying-In Hospital, vaginal thrush was a problem in many pregnant patients and when coumarin was found *in vitro* to inhibit *Candida albicans* it was sufficiently effective to initiate a study of its teratogenic effects in rabbits before using it in patients during pregnancy. Coumarin was very cheap at $100 for 2 × 50 lb sacks, and it was simply added to the drinking water in sufficient quantity to produce a saturated solution for the rabbits to drink. Due to a misunderstanding, the two sacks were placed in the water tank supplying the automatic watering system of the entire animal house. Within a few weeks other researchers using the animal facilities complained of failures in their breeding 'programs'. Following discrete investigations the sacks of coumarin were discovered, and when they were removed breeding rapidly returned to normal. This episode suggested the possible use of coumarin as a contraceptive. It was tested by a pharmaceutical company but was found to be ineffective. Many years later I heard that the company had given the coumarin by subcutaneous injection. It was probably all excreted within 2 h which was hardly enough protection for animals on '*ad lib*' sex. The possible role of coumarin as a contraceptive had still to be elucidated, but the episode does highlight the vulnerability of our water supply.

In 1961, following the discovery of the 'cancer coagulative factor'[3–5], coumarin was replaced by warfarin [3-(α-acetonylbenzyl)-4-hydroxycoumarin] in our research efforts to control cancer. Initially it was assumed that full anticoagulant cover was required to be effective, but eventually it was found that not more than a 50% increase in the prothrombin time was sufficient to prevent progression of the disease in multiple myeloma, lymphoma and myeloid leukaemia. There was no evidence that coumarin-type anticoagulants alone produced any regression of tumour but once regression was obtained with cytotoxic drugs, anticoagulants appeared to be adequate maintenance therapy[6].

An explanation of the effects of warfarin came with the use of Sumner Wood's rabbit ear chamber[7]. Warfarin, in anticoagulant doses, inhibited locomotion of V2 tumour cells in rabbits and enhanced activity of polymorphonuclear cells and macrophages. Vitamin K reversed the effect of warfarin[8]. Paradoxically, high warfarin dosage was also associated with a return of motility to the tumour cells. This was assumed to be due to increased resistance to warfarin, however, it is more likely to be due to inhibition of cellular immunity by warfarin in high dosage[9,10] because the effect on the cellular immune system is dose-dependent[11,12].

11.3 FIBRINOLYSIS

Casley-Smith and Pillar (Chapters 7 and 8) have shown that coumarin stimulates the macrophage to produce proteases which break down the proteins in lymphoedema. In patients with cancer, proteases were initially used to inhibit the cancer coagulative factor[5]. Prolonged induction of fibrinolysis by streptokinase was impossible because of antibody production and urokinase induced a gradual increase in antiplasmin activity. The protease 1 of *Aspergillus oryzae*, later called Brinase by Astra Pharmaceutical, Sweden, was found to inhibit antiplasmin[13] and was then used to treat patients with leukaemia, resulting in a marked improvement in well-being and a decrease in blast cell counts[14]. Protease 1 also induced formation of autocytotoxic antibodies that were complement-dependent in patients with acute leukaemia[15]. In mice, it was found to increase the permeability of tumour cell membranes[16]. In humans, intravenous infusion of protease 1 caused apoptosis of leukaemic cells *in vivo* but was found to be without effect *in vitro*[17]. This again suggested an indirect effect that was complement-dependent. Warfarin therapy was added to prolong the effects of protease 1 (Brinase) in leukaemia[18]. Warfarin was also tried as a maintenance therapy for other forms of recurrent cancers in an open controlled trial involving 128 patients. After 24 months, 40.6% of the warfarin patients were alive compared with 17.8% of the controls[19]. The induction of fibrinolysis by protease 1 was shown to unblock the cellular immune system in a patient with melanoma[20], and this led to further studies of the effect of induced proteolysis on cellular immunity[21].

11.4 CELLULAR IMMUNITY—UNBLOCKING

The mode of action of coumarin on the cellular immune mechanism is complex (see Chapter 5). Clinical observation suggested that coumarin acted mainly by preventing the blocking of cellular immunity in a dose-dependent manner. It also appeared that the daily dose requirement varied from day to day and from patient to patient. Clinical measurement of the day-to-day activity of the cellular immune system is not satisfactory despite all the immunological research of the past two decades. A simple method of clinical measurement, such as the prothrombin time used in dicoumarin-type anticoagulation, is needed if the correct dosage of an immune stimulator is to be given and inhibition due to overdose is to be avoided. The quantification of cell-mediated immunity in the human has hardly altered in the past 20 years. The common clinical tests still remain:

(1) Lymphocyte count (absolute count should be above $2.0 \times 10^9/l$).
(2) Skin tests for delayed hypersensitivity.
(3) T-cell count using CD3 marker or E-rosetting with sheep red cells (or CD2–CD58 linkage).
(4) Helper/suppressor cell ratio (CD4/CD8 ratio).
(5) Lymphocyte response to mitogens and cytokine release *in vitro*.

These tests are of little value for monitoring day-to-day dosage of agents that may affect cell-mediated immunity. It has been possible to use them, however, to study blockage of the cellular immune system. The well-being of the patient is remarkably improved when cellular immunity is unblocked but, although this is clinically desirable, it is of little scientific value in assessing dosage.

Skin tests for delayed hypersensitivity became positive after induction of fibrinolysis by protease 1 of *Aspergillus oryzae* (Brinase) but were negative when tested again three days later[20]. Repeated weekly infusions of Brinase combined with BCG vaccinations resulted in the skin tests for delayed hypersensitivity remaining positive[21]. The mechanisms are discussed elsewhere[22].

The concept of non-specific blocking factors in cancer was widely studied by Israel and his colleagues in France and the USA. They concentrated on 'blocking' proteins, in patients, that could be removed by repeated plasmapheresis[23]. The Hellstroms[24] also studied blocking activity and Currie and Basham used neuraminidase to remove blocking antigens from the tumour cell surface[25].

In Dublin we concentrated on blocked T lymphocytes and used an IBM/NCI continuous-flow cell separator to harvest T cells for washing before returning same to patients to produce enhanced T-cell function[26]. The blocking factors could be removed from blocked lymphocytes by simply washing in TC199. Subsequently, if normal T cells were incubated in these washings their E-rosetting capacity was markedly reduced[26]. The blocking factors were found to be sialopeptides and were present on lymphocytes of other anergic patients and not only of patients with cancer[27]. The problem of reblocking remained and repeated washings of inductions of fibrinolysis by streptokinase, urokinase or Brinase were not clinically feasible over long periods of time. To prolong the unblocking effect we tried repeated vaccination with BCG or *C. parvum*, high-dose vitamin A, levamisole or low-dose warfarin. Finally we found that coumarin in an oral dose of 25–100 mg daily gave the best results[28]. The beneficial effects, if any, of unblocking the cellular immune mechanisms in patients with cancer takes a long time to assess, so patients with chronic brucellosis, a disease with a blocked cellular immune system, were used to evaluate coumarin[29] while melanoma was the main cancer chosen to study the effects of maintaining an unblocked cellular immune system[30].

11.5 MELANOMA

Cancer therapy falls into three categories:

I Treatment of primary tumour.
II Prevention of recurrence by adjuvant therapy.
III Treatment of recurrent or advanced cancer.

In terms of overall expenditure on therapy, approximately 15% is spent in category I, 5% in category II and 80% in category III. The cellular immune system is most likely to have a major role in category II when tumour load is minimal and host resistance can still be manipulated. Immunotherapy will only be of value in an immunocompetent host.

Malignant melanoma was chosen as a suitable tumour for study as its fate appeared to be dependent upon cellular immunity and its recurrence dependent upon tumour thickness at the time of surgical excision in category I. Thus, with an accurate means of measurement the effects of adjuvant therapy could be evaluated. In addition, the prevention of recurrence is a major challenge as there is no satisfactory treatment for recurrent malignant melanoma.

At first warfarin was the coumarin used in patients with early category III disease. These patients had minimal recurrence which could be removed surgically. They were then treated with vincristine [5-(3,3-dimethyl-l-triazeno) imidazole-4-carboxamide] and BCG (Glaxo). Seventeen patients were randomized: nine received warfarin sufficient to increase their prothrombin time by not more than 50%, and eight controls received no warfarin. There were three recurrences in each group in the first year. In the second year there were four further recurrences in the control group and in the third year the last control patient had a recurrence. None of the six remaining patients on continuous warfarin therapy had recurrence[30]. In a further open trial a group of five patients with their primary lesions of melanoma excised, were anticoagulated with warfarin and another group of six similar patients were given coumarin (25 mg) daily. There was one recurrence in each group within a year[31]. A third group of five patients given coumarin (50 mg) daily after surgery had no recurrences at follow-up two years later.

In 1984 a controlled double-blind, randomised clinical trial was set up using coumarin 50 mg daily and identical placebo tablets for controls. The tablets were generously provided by Schaper & Brümmer, Salzgitter, Germany[32]. Twenty-seven patients with melanoma lesions of TNM classification stage II or stage IB with tumour (Breslow) thickness above 1.70 mm were admitted to the trial. Lesions with a thickness above 1.70 mm have a recurrence rate greater than 50% and so were ideal for the trial[33]. All patients were treated by standard wide excision surgery, checked histologically and skin grafted where necessary. Coumarin (50 mg) tablets or placebo were taken daily for two years after surgery.

In May 1987, the results when reviewed were highly significant ($P<0.01$). There were 10 recurrences in the control group ($N=14$) but only two recurrences in the treated group ($N = 13$)[32]. The trial was stopped and the remaining four patients in the control group were given 50 mg daily for two years and none has recurred to date. In the treated group there have been two further recurrences at four and six years after surgery. This raises the question of whether or not coumarin should be continued beyond two years.

This study needs to be repeated at another centre using the same trial criteria and the same dosage of coumarin (50 mg daily for two years). An unpublished open study in Australia using 300 mg of coumarin daily found no effect. This could be because the dose of coumarin required to maintain activation of the cellular immune system is critical (see above); therefore, overdose could lead to inhibition of immunity and thereby enhance the risk of recurrence.

11.6 HERPES SIMPLEX

The use of warfarin in the treatment of *Herpes simplex* infection is an example of its action on cellular immune response being dose-dependent. Warfarin was used by Berkarda in Turkey to prevent recurrence of cold sores in humans (*Herpes simplex* Type I). A maintenance dose of 2.5 mg one day per week was given following an initial loading dose of 5 mg daily for two days, then 2.5 mg for three days. At no time did the prothrombin activity fall below 40%. At six months the average number of recurrences in the untreated group was 4.2 ± 1.2 while in the warfarin (low dose) treated group it was 0.6 ± 0.6[34]. Berkarda also showed that the cellular immune system in mice is enhanced by warfarin[35] and in humans by both warfarin and coumarin[12].

In patients on high-dose warfarin (anticoagulation dose) the cellular immune system is suppressed[10] and the risk of getting a cold sore is increased by 70%. In 99 patients on full anticoagulant therapy with warfarin, 23% had more than one recurrence in a year compared to 13% of 125 controls[36].

The traditional use of the coumarin-rich Tonka bean camphor for the treatment of burns and ulceration led Sr Majella McKernan of the Medical Missionaries of Mary to devise a cream—1% coumarin in Silcock's base—for the treatment of sunburn in Irish missionaries in the tropics. During its use it was also found to be effective in the prophylaxis and treatment of *Herpes simplex* caused by ultraviolet light[37]. It became known as 'Sr. Majella's Cold Sore Cream'. It is interesting to note that ultraviolet light suppresses immunological response[38].

At the Ganta Leprosy Centre in Liberia, oral coumarin was ineffective against leprosy but was found to be useful in the treatment of burns (see Chapter 8) when applied locally as Sr Majella's Cold Sore Cream. It aided healing, reduced odour and prevented keloid formation[39].

11.7 IgA Nephropathy

The biological effects of both coumarin and warfarin appear to be macrophage-dependent[40,41], so the dosage of warfarin should be of importance in preventing progression of IgA nephropathy, a kidney disease that is the single commonest cause of end-stage renal failure requiring dialysis or renal transplant[42]. The objective of treatment is to prevent progression of disease which is characterized by a deposition of IgA and C3 in the glomerular mesangium. The mesangial cells, which are considered renal macrophages, swell and multiply within the glomerulus and compromise the blood supply, causing thrombosis and ischaemia resulting in glomerular sclerosis, tubular atropy and interstitial fibrosis[43].

In 1987, Woo Keng Thye and his colleagues in Singapore published a study of IgA nephropathy using triple therapy with cyclophosphamide, dipyridamole and low 'antithrombotic' doses of warfarin that prevented progression of the disease[44]. The five-year follow-up confirmed the results[45]. In 1990, Walker and his co-workers in Australia published their trial results with triple therapy. They, however, used full anticoagulant doses of warfarin and renal function was not preserved[46]. The different outcome in the two trials can be explained by the differing dosage of warfarin. Low-dose warfarin would be expected to stimulate macrophage activity whereas high-dose warfarin may be inhibitory.

It was decided to test this hypothesis in mice as Sato *et al.* had earlier shown that experimental IgA nephropathy could be induced in mice if the reticulo-endothelial system was blocked[47]. We used the ddY strain of mice which develop IgA nephrology with age[48]. We were fortunate to receive a present of six breeding pairs of these mice from Toshikiko Asano of the Department of Veterinary Science at the National Institute of Health in Tokyo (Director: Masaharu Naiki). Instead of warfarin we used coumarin and 7-hydroxycoumarin as the macrophage stimulant. Our experiments involving over 128 ddY mice are on-going, but it is already clear that 7-hydroxycoumarin is more effective than coumarin in preventing the accumulation of IgA in the glomeruli. This may be due to the ddY mice's inability to convert coumarin to 7-hydroxycoumarin. The ddY mice also spontaneously develop various tumours subsequent to the appearance of IgA nephropathy and so may be useful for the study of chemoprevention of cancer[48]. Tseng has already demonstrated the chemopreventive properties of coumarin in MTV-H*ras* transgenic mice[49]. In both these studies the treated mice received 200 μM coumarin in their drinking water.

11.8 FARMING

In farming, good husbandry involves the regular administration of chemical agents and vaccines to prevent both infections and infestations. Coumarin is currently being investigated for the chemopreventation of mastitis in cows and as an antihelmintic in cattle and sheep. The main problem is finding the effective dosage.

Plants appear to produce and use coumarin for chemoprevention against pests. *Eupatorium japonicum* has an ovicide and larval inhibitor in its leaves against the fruit fly (*Drosophila melanogaster*), and this was found to be coumarin. It inhibits egg hatching at a dose of 12.5 $\mu g/cm^2$ on a paper disk. The larva are inhibited at a concentration of 2.5 mg/2 g of medium[50]. The blow fly (*Lucila sericata*) eggs are also inhibited by coumarin; thus to prevent maggots (fly strike) in sheep a controlled trial comparing coumarin 1 g/l of water and cyromazine (Vetrazin, Ciba Geigy) was arranged. The results recorded 26 fly strikes in 80 control sheep (30%) sprayed with water every two weeks, 2 strikes in 52 sheep (4%) treated once in eight weeks with cryomazine and 7 strikes in 98 sheep (7%) sprayed with coumarin solution every two weeks. Coumarin is easily washed out of the sheep's wool so a binding agent would be necessary to permit a single application for comparison with commercial agents[51]. However, coumarin is inexpensive and non-toxic[52] and would make an environmentally friendly product.

In dog breeding kennels an infectious disease called fading puppy syndrome can affect brood bitches. They produce litters of pups with lymphocytopenia and these pups die slowly over several days from diverse infections. This lymphocytopenia in the pups was completely prevented if the brood bitches were treated with coumarin (25 mg) daily during pregnancy[53].

In fish farms there is an endemic protozoal disease of freshwater salmonoids called proliferative kidney disease (PKD). In a small-scale study 0.1% and 0.5% coumarin was added to the fish food. There were four groups: non-infected controls, infected controls, 0.1% coumarin and 0.5% coumarin treated groups. The presence of PK'X' parasite in tissues, external kidney swelling and mortalities in each group were monitored at 6, 12 and 16 weeks. At six weeks PK'X' parasites were observed in kidney and spleen tissue of all groups except the control non-infected group. Survival of rainbow trout treated with coumarin was higher than the untreated PKD-infected fish at 6, 12 and 16 weeks. Survival appeared highest and external kidney swelling lowest in the group treated with 0.1% coumarin compared to either the 0.5% coumarin treated group or the untreated PKD-infected group. All the PKD-infected groups developed severe fungal infection at 16 weeks, none of the positive controls survived but 70% of the negative controls, 35% of the 0.1% coumarin group and 25% of the 0.5% coumarin group survived[54]. This is another example in which the effect of the coumarin appears to be dose-dependent.

11.9 CONCLUSION

Coumarin has many different effects in nature, varying from its vitamin P-like action to its antiparasitic actions in plants. The full potential of coumarin alone or as adjuvant therapy for immunostimulation by proteolytic enzymes, interferons or interleukins, awaits a clinical test that can measure overall function of the cellular immune system from day to day. Then it will be possible to give the correct dosage over long periods to prevent disease.

REFERENCES

1 Vogel A. Dissertation on benzo-acid from the Tonka bean and from the Melilotus (sweet clover) [German]. *Annalen Der Physik* 1820; **64**: 161–6.

2 Feuer G. The metabolism and biological actions of coumarin. *Progress in Medicinal Chemistry* 1974; **10**: 85–158.

3 O'Meara RAQ, Thornes RD. Some properties of the cancer coagulative factor. *Irish J Med Sci* 1961; **423**: 106–12.

4 Thornes RD. Fibrin and fibrinolysis in cancer. PhD Thesis, University of Dublin, Trinity College (1961).

5 Thornes RD. Host factors in neoplasia. The Graves Lecture 1966. *Irish J Med Sci* 1966; **487**: 265–76.

6 Thornes RD. Anticoagulant therapy in patients with cancer. *J Irish Med Assoc* 1969; **62**: 426–9.

7 Wood S Jr. Cinephotomicroscopy of living tissue in medical research and teaching. Insertion and use of a rabbit ear chamber for time lapse and standard cinephotomicroscopy in the living animal. *J Soc Motion Picture and Television Engineers* 1965; **74**: 3–42.

8 Thornes RD, Edlow DW, Wood S Jr. Inhibition of locomotion of cancer cells *in vivo* by anticoagulant therapy. 1. Effect of sodium warfarin on V2 cancer cells, granulocytes, lymphocytes and macrophages in rabbits. *The Johns Hopkins Medical Journal* 1968; **123**: 305–16.

9 Stuhlmeier K, Theyer G, Baumgartner G, Zlabinger GL. Synergistic effect of coumarin (1,2-benzopyrone) and endotoxin in the induction of human interleukin-1. *Clin exp immuno* 1991; **84**: 317–23.

10 Cohen S, Benacerraf B, McCluskey RT, Ovary Z. Effect of anticoagulants on delayed hypersensitivity reactions. *J Immunology* 1967; **98** (2): 351–8.

11 Triozzi PL, Vicstein F, Ailabouni I, Fertel R, Rinehart JJ. Effects of coumarins in Interleukin 2 (IL2) activity. *Proc Amer Assoc Cancer Res* 1990; **31**: 277.

12 Berkarda B, Bouffard-Eyuboglu H, Derman U. The effect of coumarin derivatives on the immunological system of man. *Agents & Actions* 1983; **13**: 50–2.

13 O'Brien ET, Thornes RD, O'Brien D, Hogan B. Inhibition of antiplasmin and fibrinolytic effect of protease in patients with cancer. *Lancet* 1968; **1**: 173–6.

14 Thornes RD. Inhibition of antiplasmin and effects of protease in patients with leukaemia. *Lancet* 1968; **2**: 1220–3.

15 Thornes RD, Devlin JG, Lim Hoe Kee J, MacDonell JD. Complement dependent cytotoxic antibodies in patients with acute leukaemia following fibrinolytic therapy. *Irish J Med Sci* 1970; **3**: 107–14.

16 Smyth H, Flahavan E, Thornes RD. The effect of protease 1 of *Aspergillus oryzae* (Brinase) on membrane permeability and growth of Landschütz ascites tumour cells. *Int J Cancer* 1971; **7**: 476–82.

17 Thornes RD. Fibrinolytic therapy of leukaemia. *J Royal Coll Surgeons in Ireland* 1971; **6**: 123–8.

18 Thornes RD, Deasy PF, Carroll R, Reen DJ, MacDonell JD. The use of the proteolytic enzyme Brinase to produce autocytotoxicity in patients with acute leukaemia and its possible role in immunotherapy. *Cancer Research* 1972; **32**: 280–4.

19 Thornes RD. Warfarin as maintenance therapy for cancer. *J Irish Coll Phys Surg* 1972; **2**: 41–2.

20 Thornes RD, Smyth H, Browne O, Holland PDJ. BCG plus protease 1 in malignant melanoma. *Lancet* 1973; **1**: 1386.

21 Thornes RD, Smyth H, Browne O, O'Gorman M, Reen DJ, Farrell D, Holland PDJ. The effects of proteolysis on the human immune mechanism in cancer. *J Med* 1974; **5**: 92–7.

22 Thornes RD. Adjuvant therapy of cancer via the cellular immune mechanism or fibrin by induceed fibrinolysis and oral anticoagulants. *Cancer* 1975; **35**: 91–7.

23 Israel L, Edelstein R, Mannoni P, Radot E, Greenspan EM. Plasmapheresis in patients with disseminated cancer: clinical results and correlations with changes in serum protein. *Cancer* 1977; **40**: 3146–54.

24 Hellstrom KE, Hellstrom I. Lymphocyte mediated cytotoxicity and blocking serum activity to tumor antigens. In: Dixon FJ, Kinkel HG (eds). *Advances in Immunology*. New York: Academic Press, 1974: 249–77.

25 Currie GA, Basham C. Serum mediated inhibition of the immunological reactions of the patient to his own tumour: a possible role for circulating antigen. *Brit J Cancer* 1972; **26**: 427–32.
26 Browne O, Bell J, Holland PDJ, Thornes RD. Plasmapheresis and immunostimulation. *Lancet* 1976; **2**, 96.
27 O'Kennedy R, Smyth H, Thornes RD, Corrigan A. Increased sialic acid in 'blocked' lymphocytes and correlation of spontaneous release with clinical state. *Europ J Cancer* 1980; **16**: 1163–9.
28 Thornes RD. Acquired immune suppression in chronic brucellosis. *Irish Med J* 1983; **26**: 225.
29 Thornes RD. The anergy of chronic brucellosis. *J Irish Med Assoc* 1979; **70**: 480–3.
30 Thornes RD. Prevention of recurrence of malignant melanoma by coumarin (1,2 benzopyrone). *J Irish Coll Phys Surg* 1993; **22** (Suppl): 17–19.
31 Thornes RD. Coumarins, melanoma and cellular immunity. In: McBrien OCH, Slator TF (eds). *Protective Agents in Cancer*. London: Academic Press, 1983; 43–56.
32 Thornes RD, Daly L, Lynch G, Browne H, Tanner A, Keane F, O'Loughlin S, Corrigan T, Daly P, Edwards G, Breslin B, Browne Hy, Shine M, Lennon F, Hanley J, McMurray N, Gaffney E. Irish Melanoma Group. Prevention of early recurrence of high risk malignant melanoma by coumarin. *Eur J Surg Oncol* 1989; **15**: 431–5.
33 Day CL, Lew RA, Mihm MC, Harris MN, Kopf AW, Sober AJ, Fitzpatrick TB. The natural break points for primary tumour thickness in clinical stage I melanoma. *N Eng Med J* 1981; **305**: 1155.
34 Berkarda, B. Preliminary report on warfarin for the treatment of *Herpes simplex*. *J Irish Coll Phys Surg* 1993; **22** (Suppl): 56.
35 Berkarda B, Marrack P, Kappler JW, Bakeemeier RF. Effects of warfarin on the immune response in mice. *Arzneimitt Forsch/Drug Res* 1978; **28**: 1407–11.
36 Thornes BS, Graham IM. Does warfarin modify cellular immune status to influence the occurrence of cold sores? A case control study. Student Project, Adelaide Hospital, Dublin 1993.
37 Thornes RD. Prevention of ultraviolet-light-induced *Herpes labialis*. *Lancet* 1992; **339**: 133.
38 Kripke ML. Immunological unresponsiveness induced by ultra violet radiation. *Immunol Rev* 1984; **80**: 87–102.
39 Chambers M, Thornes RD, McKernan M. Necrosis of skin by coumarin. *BMJ* 1989; **298**: 755.
40 Pillar NB. The ineffectiveness of coumarin treatment on thermal oedema of macrophage-free rats. *Br J Exp Pathol* 1976; **57**: 170–4.
41 Maat B. Selective macrophage inhibition abolishes warfarin-induced reduction of metastases. *Br J Cancer* 1981; **41**: 313–16.
42 Emancipator SN, Lamm ME. Biology of disease. IgA nephropathy: Pathogenesis of the most common form of glomerulonephritis. *Lab Invest* 1989; **60**: 168–83.
43 Browne O, Doyle GD, Campbell E. An immunological study of IgA nephropathy and a report of its incidence in Ireland. *Irish J Med Sci* 1985; **154**: 461–5.
44 Woo KT, Edmonson RPS, Yap HK,Wu AYT, Chiang GSC, Lee EJC, Pree HS, Lim CH. Effects of triple therapy on the progression of mesangial proliferative glomerulonephritis. *Clin Nephrol* 1987; **27**: 56–64.
45 Woo KT, Lee GSL, Lau YK, Chiang GSC, Lim CH. Effects of triple therapy in IgA nephritis: a follow-up study five years later. *Clin Nephrol* 1991; **36**: 60–6.
46 Walker RG, Yu SM, Owen JE, Kincaid-Smith P. The treatment of mesangial IgA nephropathy with cyclophosphamide, dipyridamole and warfarin. A two year prospective trial. *Clin Nephrol* 1990; **34**: 103–7.
47 Sato M, Ideura T, Koskiowa S. Experimental IgA nephropathy in mice. *Lab Invest* 1986; **54**: 377–385.
48 Imai H, Nakamoto Y, Askaura K, Miki K, Yasuda T, Miura AB. Spontaneous glomerular IgA deposition in ddY mice: An animal model for IgA nephritis. *Kidney Int* 1985; **27**: 756–61.
49 Tseng A. Coumarin prevents tumourigenesis in MTV-H-ras transgenic mice. *J Irish Coll Phys Surg* 1993; **22** (Suppl): 49–50.
50 Nakajima S, Kawazu K. Coumarin and euponin, two inhibitors for insect development from the leaves of *Eupatorium japonicum*. *Agric Biol Chem* 1980; **44**: 2893–9.

51 Thornes RD, Wall PG. Control of blow fly strike in sheep by coumarin. *Veterinary Record* 1991; **129**: 496.

52 Cox D, O'Kennedy R, Thornes RD. The rarity of liver toxicity in patients treated with coumarin (1,2-benzopyrone). *Human Toxicology* 1989; **8**: 501–6.

53 Wall PG, Thornes RD, Sheehan MV. The treatment of suspected canine immune deficiency syndrome with coumarin (1,2-benzopyrone). *J Cancer Res Clin Oncol* 1994; **120** (Suppl): 39–42.

54 O'Flynn J, Hanjavanit C. Mulcahy MF. Effect of coumarin on an experimental infection of rainbow trout with proliferative kidney disease. Aquaculture Development Centre, Zoology Department, University College, Cork, Ireland. (In press)

CHAPTER 12

Analysis of Coumarins

Declan P. Bogan[1], Gary J. Keating[1], Heiko Reinartz[1], Ciaran F. Duffy[1], Malcolm R. Smyth[2], Richard O'Kennedy[1] and R. Douglas Thornes[3]

[1]*School of Biological Sciences, Dublin City University, Dublin, Ireland*
[2]*School of Chemical Sciences, Dublin City University, Dublin, Ireland*
[3]*Colman K. Byrnes Research Centre, Beaumont Hospital, Dublin, Ireland*

12.1 INTRODUCTION

The ubiquitous nature of coumarin and its many derivatives in industry, medicine and in the environment have led many investigators to search for methods of analysis of coumarin and its derivatives[1–3]. The methods thus far developed cover a wide range of analytical techniques, encompassing more traditional methods, e.g. paper chromatography, or colorimetric approaches, and currently available general techniques, e.g. high-performance liquid chromatography, capillary electrophoresis, and immunochemical and some specific methods, e.g. centrifugal partition chromatography. Coumarin and its derivatives are determined to assess their clinical efficacy, their effects on foodstuffs, and their uses in industry. The methods need to be applicable to a variety of matrices, to a range of concentrations, and be sensitive, reproducible, reliable and accurate. In this chapter a broad range of relevant analytical methods are critically examined in relation to their application to the analysis of coumarin, and some of its derivatives, with special reference to 7-hydroxycoumarin.

12.2 HISTORICAL BACKGROUND OF COUMARIN ANALYSIS

The presence of coumarin in plant extracts or as a flavouring component was of interest to the early coumarin analysts. Paper chromatography, thin-layer

Coumarins: Biology, Applications and Mode of Action. Edited by R. O'Kennedy and R.D. Thornes.

chromatography, colorimetric assays and polarography were all used. These more traditional analytical methods were applied to the analysis of chocolate[4], sweet clover[5], flavourings[6–9] and plant extracts[10–13]. The majority of these techniques were applied primarily for the qualitative determination of coumarin and/or its derivatives. The concentration of coumarin and derivatives was determined by colorimetric assays[9] or by using the ultraviolet (UV) absorption spectra of the derivative of interest after chromatographic separation[8]. Coumarin was treated with *p*-aminophenol to selectively determine it in the presence of vanillin[8].

Paper chromatography was also used for the separation of coumarins. Butanol : acetone : water[14], acetic acid : butylene glycol : water[15], or ammonia : butanol : water[6] are some examples of mobile phases used for the separation of coumarin and its derivatives in organic flavourings or in plants. After separation of coumarin from other endogenous species present in the matrix of interest it is possible to visualise coumarin after treatment with alkali or after photocatalysis. Coumarin can be visualised by treating the chromatogram, after separation, with alkali[6,7,15,16], e.g. KOH or Na_2CO_3. The coumarinic salt produced is fluorescent and the intensity can be used to determine coumarin concentration. Photocatalysis was another technique[17] utilised for coumarin and 7-hydroxycoumarin detection after paper chromatographic separation. Here the chromatogram was treated with alkali, irradiated to form a *trans*-isomer that fluoresces blue for 7-hydroxycoumarin and yellow-green for coumarin under UV light. The limit of detection of the method was 0.005 μg.

Derivatisation of coumarin was also investigated for use in qualitative and quantitative analysis[18-20]. The analysis was based on the formation of acidic products, amides of ethylenediamine, or coloured products. After coumarin was dissolved in ethylenediamine, a rapid visual indicator or the potentiometric response of the derivative was utilised for its determination[18]. The coloured diazocoupled product[19] formed between diazotised sulfanilic acid and coumarin[19], had a linear detection range between 0.1 and 0.7 g/l. The sodium salt of coumarinic acid (formed by reaction of coumarin and alkali) was reacted with 2,6-dichloroquinone-4-chloroimide to form a blue indophenol dye[20]. This method had a linear detection range between 2 and 10 μg/ml.

The emergence of more sophisticated analytical technology in the 1960s and 1970s led to its application for the determination of coumarin and its derivatives. Gas chromatography was applied to the determination of coumarin as a contaminant of flavourings[21,22]. Limits of detection were in the order of 0.2 μg, or 0.01% of sample with an electron capture detector.

Today, high-performance liquid chromatography, gas chromatography, capillary electrophoresis and biosensors are routinely used for the analysis of coumarin and its analogues. Method development, data handling and the applicability of these techniques are made more straightforward by the development of computer controlled systems. This chapter outlines the various methods currently available for the qualitative and/or quantitative determination of coumarin and some of its analogues from a variety of sources.

12.3 CHROMATOGRAPHIC ANALYSIS OF COUMARINS

Table 12.1 gives a summary of some of the chromatographic methods available for the determination of coumarin and 7-hydroxycoumarin in a variety of matrices.

12.3.1 Thin-Layer Chromatography

The principle of thin-layer chromatography (TLC) is the separation of compounds on a coated solid surface with a liquid carrier phase. Typically a silica-coated

Table 12.1. Summary of the chromatographic methods of analysis of coumarins.

Method and application	Mode of detection	Limit of detection	References
Thin-layer chromatography			
In vivo 7-OHC	Colorimetric		22
metabolism	Fluorescence	1 ng/ml	27
	Fluorescence	5 ng	23,24
Vanilla flavouring	UV absorbance	500 ng	
	Radiochemical	(Coumarin)	26
High-performance liquid chromatography			
In vivo coumarin	UV absorbance	500 ng/ml	32
metabolism		(7-OHC)	33
		0.3 ng/ml, 50 ng/ml (7-OHC)	34,53
		0.6 ng/ml, 200 ng/ml (7-OHCG)	34,53
	UV absorbance		46
In vitro metabolism:	Radiochemical		37–40,48,49
Coumarin			
Coumarin in:	UV absorbance	0.6 ng/ml	50,51
Foodstuffs			
Plant extracts	UV absorbance		
Gas chromatography			
Coumarin in:	Mass spec.	7.7 ng/ml	54
Flavours/fragrances	FTIR		55
Tobacco	FID/mass spec.	0.05 mg/g	57
Capillary electrophoresis			
In vivo metabolism of coumarin	UV absorbance	1 μg/ml	62–63
In vitro metabolism of coumarin and 7-hydroxycoumarin	UV absorbance	1 μg/ml	64,65 66

7-OHC, 7-hydroxycoumarin; 7-OHCG, 7-hydroxycoumarin-glucuoronide.

stationary phase is used with an organic mobile phase. The compound is applied to the solid surface and interaction with the carrier phase, in which the components migrate along the solid surface, enables separation. A range of solid and mobile phases have been assessed for their applicability to the separation of coumarin and its derivatives. Typically, TLC was used as a purely qualitative method but it is possible to use TLC in quantitative analysis.

Metabolism of Umbelliferone and Herniarin

The metabolism of 7-hydroxycoumarin (umbelliferone) and 7-methoxycoumarin (herniarin) was followed by TLC[23]. Separation of the principal metabolites was achieved with benzene–glacial acetic acid–water (6 : 7 : 3, v/v/v), and with potassium chloride–glacial acetic acid (100 : 1 v/v), on thin-layer cellulose plates (Sigmacell type 19, and Macherey, Nagel & Co. MN 300 plates). After separation, the spots on the TLC plates were observed under UV light (fluorescence at 254 nm) or by spraying with fast blue B salt. The different compounds gave a variety of different colours, e.g. 7-hydroxycoumarin (purplish-white), 7-methoxycoumarin (pinkish-white), 6,7-dihydroxycoumarin (greenish-yellow), and 3,4-dihydro-7-hydroxycoumarin (reddish-violet). The method was used for studying the metabolism of 7-hydroxycoumarin in rats after oral and intraperitoneal administration.

TLC of Vanilla Flavourings

High-performance silica gel TLC was used to qualitatively and quantitatively determine coumarin in real and artificial vanilla flavourings[24]. Samples and standards were spotted on to Whatman LHPKD pre-adsorbent silica gel plates. Separation was carried out with toluene–methanol (97 : 3, v/v) and visualised under UV light by spraying with NaOH. Coumarin was detected as a fluorescent or coloured zone and quantified by scanning with a densitometer. The limit of detection was 5 ng for fluorescence and 500 ng for absorbance. Belay and Poole[25] determined vanillin and the adulterant coumarin in a variety of natural vanilla beans by TLC. They employed an automated multiple development (AMD) system which optimises the separation selectivity throughout the chromatogram for mixtures of a wide polarity range. The whole process is automated which ensures a high degree of reproducibility. Separation was carried out with a stepwise mobile phase change between chloroform, ethyl acetate, 1-propanol, acetic acid and hexane. They were able[25] to separate and identify coumarin from other vanilla extracts, e.g. vanillin, ethyl vanillin, vanillyl alcohol, vanillic acid, 4-hydroxybenzyl alcohol, 3,4-dihydroxybenzaldehyde, 4-hydroxybenzoic acid and piperonal.

A chloroform–ethyl acetate–propan-1-ol (96 : 2 : 2, v/v/v) mobile phase was also used for the separation of vanilla extracts[26] by TLC. This method allowed the determination of the authenticity of natural vanilla extracts as opposed to synthetic flavour compounds which can only be found in counterfeit vanilla flavourings (coumarin is considered as an adulterant in counterfeit vanilla flavouring). Whatman

HP-K high-performance silica gel plates were used for the separation of coumarin from the other species present in the flavourings. Plates were analysed at 280 nm or as 2,4-dinitrophenylhydrazone derivatives at 500 nm with a scanning densitometer.

TLC Analysis of 7-Hydroxycoumarin in Urine

TLC was also used for the analysis of 7-hydroxycoumarin after *in vivo* metabolism of coumarin[27]. Sixty-four patients were treated with 2 mg of coumarin. Samples were treated with β-glucuronidase, and extracted into chloroform. Separation was carried out on silica plates (pre-coated high-performance TLC glass-backed silica gel 60 F_{254}) with a mobile phase of chloroform–water–ethyl acetate–acetic acid (24 : 12 : 6 : 0.5, v/v). Quantitative results were obtained by fluorescence densitometry. The chromatographic lanes were scanned at a wavelength of 313 nm using a mercury lamp in the reflection mode in the presence of a K400 secondary filter. Peak heights were compared to those of 7-hydroxycoumarin standards. The limit of detection was 1 ng/ml with a linear range of 5–4000 ng/ml.

12.3.2 High-Performance Liquid Chromatography

High-performance liquid chromatography (HPLC) is a technique whereby molecules may be separated by exploiting their physical and/or chemical differences[28-30]. The separation by HPLC is, typically, facilitated by the degree of hydrophobicity or polarity of an analyte and its partitioning between two phases. The analyte is typically present in the mobile phase, and it is allowed to interact with a stationary phase. The degree of interaction will be unique for every species. Separation of several species in the same solution is made possible by exploiting this interaction by encouraging a varying degree of interaction, between the phases, of the species present. HPLC has been applied to the analysis of coumarin in flavourings, in plant extracts and in *in vivo* and *in vitro* metabolic studies[31–53]. Separation has been achieved on a variety of stationary phases, e.g. C_{18} and C_8, and with many different mobile phases, e.g. based on methanol or acetonitrile. Detection systems used here included UV detection, radiochemical detection, mass spectral detection, fluorescence detection and electrochemical detection. An excellent review on the determination of many coumarin derivatives by HPLC between 1983 and 1990 has been produced by Shkarenda *et al.*[52]. The conditions for the separation of complex mixtures of coumarins and the use of coumarins in the HPLC analysis of natural and synthetic compounds are discussed. The coumarin derivatives include furocoumarins, hydroxycoumarins, pyranocoumarins and isocoumarins. Detection wavelengths, column selection, mobile phases, sources and the presence of an internal standard are all included.

Metabolism Studies

Hepatic microsomes play a major role in the metabolism of coumarin to a variety of metabolites. The metabolic pathway is dependent on the species[1,31]. HPLC has been adapted to determine the many metabolites for studies both on the *in vivo* and the *in vitro* metabolism of coumarin (Figure 12.1).

Metabolism in vivo. Moran *et al.*[32], and Egan and O'Kennedy[33] both described methods for the determination of 7-hydroxycoumarin in urine after the *in vivo* metabolism of coumarin in humans. Initial sample preparation steps and long

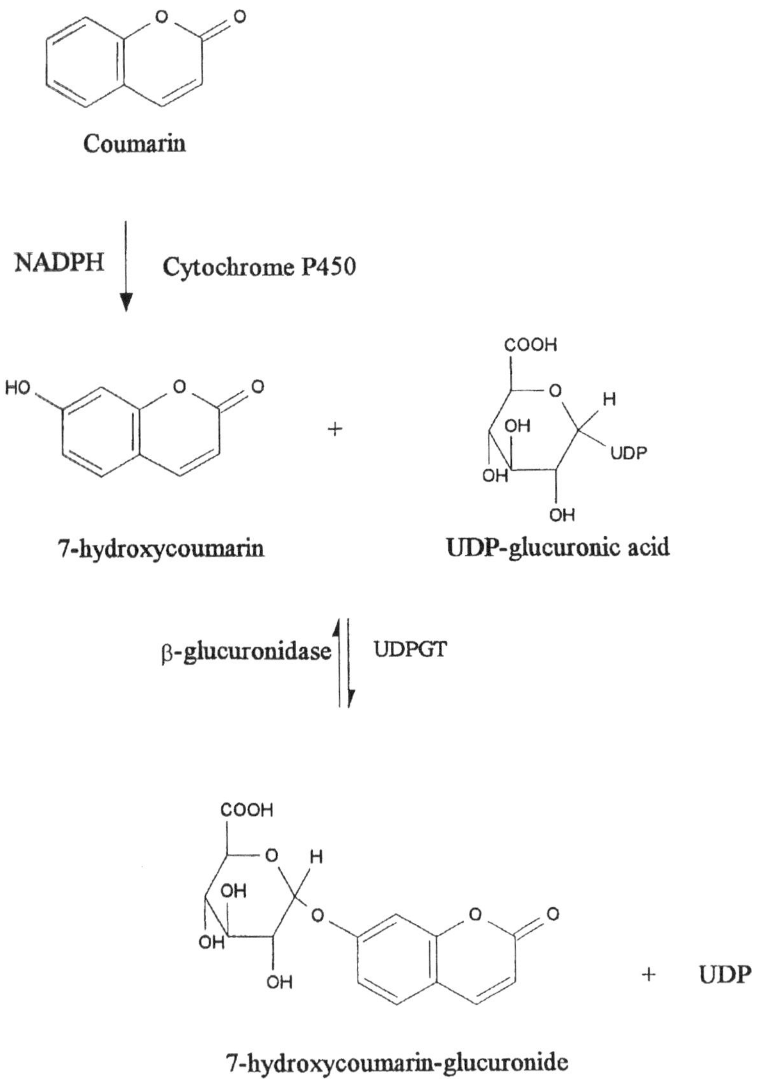

Figure 12.1. The reaction scheme for an *in vitro* or *in vivo* pathway of coumarin metabolism to 7-hydroxycoumarin and 7-hydroxycoumarin-glucuronide.

incubation steps made the method of Moran *et al.*[32] very time consuming. Large samples of urine were required, and large amounts of β-glucuronidase were also needed to liberate the glucuronide conjugate from 7-hydroxycoumarin. The method of Egan and O'Kennedy[33] made major improvements on this method. It shortened the total time of preparation by at least 16 h. The amount of β-glucuronidase was reduced from 20 000 units/ml to 5000 units/ml and the incubation time was reduced from 16 h for Moran *et al.* to 0.5 h for Egan and O'Kennedy[33], respectively. Direct extraction into diethyl ether[33] rather than freeze drying[32] was the preferred method of sample preparation before the application of the sample on to the HPLC. Direct extraction of 7-hydroxycoumarin will remove endogenous species that may cause fouling of the HPLC equipment or they may co-elute with 7-hydroxycoumarin and cause interference. Samples were separated with mobile phases based on a methanol : water : acetic acid [300 : 700 : 2 (v/v/v) for Egan and O'Kennedy, and 200 : 300 : 1 (v/v/v) for Moran *et al.*] mix. The limit of detection of the method of Egan and O'Kennedy was 500 ng/ml with a linear detection range between 0 and 100 μg/ml.

Sharifi *et al.*[34] and Bogan and O'Kennedy[53] both developed methods that avoided the deconjugation step entirely for the determination of coumarin and some of its metabolites (e.g. 7-hydroxycoumarin, 7-hydroxycoumarin-glucuronide, 3-hydroxycoumarin) in urine. Extraction of the glucuronide conjugate from rabbit urine, giving a pure standard to work with, allowed the study of samples containing 7-hydroxycoumarin-glucuronide without having to deconjugate. Sharifi *et al.*[34] described the determination of the coumarins in urine and in plasma. The urine was prepared by altering its pH to 4.5 and then diluting the sample with water and methanol. Separation was achieved with a methanol : water : tetrahydrofuran : acetic acid (45 : 40 : 10 : 5, v/v/v/v) mobile phase. Sharifi *et al.*[34] used a C_8 reverse-phase column to separate the individual components of interest whereas Moran *et al.*[32], Egan and O'Kennedy[33] and Bogan and O'Kennedy[53] used C_{18} reverse-phase columns.

The limits of detection for the method of Sharifi *et al.*[34] were far superior to those of Moran *et al.*[32] or Egan and O'Kennedy[33]. Sharifi *et al.*[34] were able to detect down to 0.3 ng/ml with a linear range between 0.6 and 26 ng/ml (250–10 000 nmol/ml) for 7-hydroxycoumarin-glucuronide, and similarly for coumarin and 7-hydroxycoumarin. The limit of detection of the method of Bogan and O'Kennedy was 50 ng/ml for 7-hydroxycoumarin and 200 ng/ml for coumarin and 7-hydroxycoumarin-glucuronide. The method of Bogan and O'Kennedy[53] was applied to the determination of coumarin, 7-hydroxycoumarin and 7-hydroxycoumarin-glucuronide in plasma, urine and serum after the *in vivo* metabolism of coumarin and 7-hydroxycoumarin. Separation was based on a gradient elution with UV detection at 320 nm (Figure 12.2).

Lamiable *et al.*[35] used a different mobile phase (acetonitrile : phosphate buffer), extraction into hexane, and reconstitution into hydrochloric acid for the determination of coumarin in plasma. They used a C_8 reverse-phase column. All the above methods used UV detection at 280 nm or 320 nm.

Metabolism in vitro. High-performance liquid chromatography has also been used to follow the *in vitro* metabolism of coumarin by hepatic microsomes. Fentem *et al.*[36]

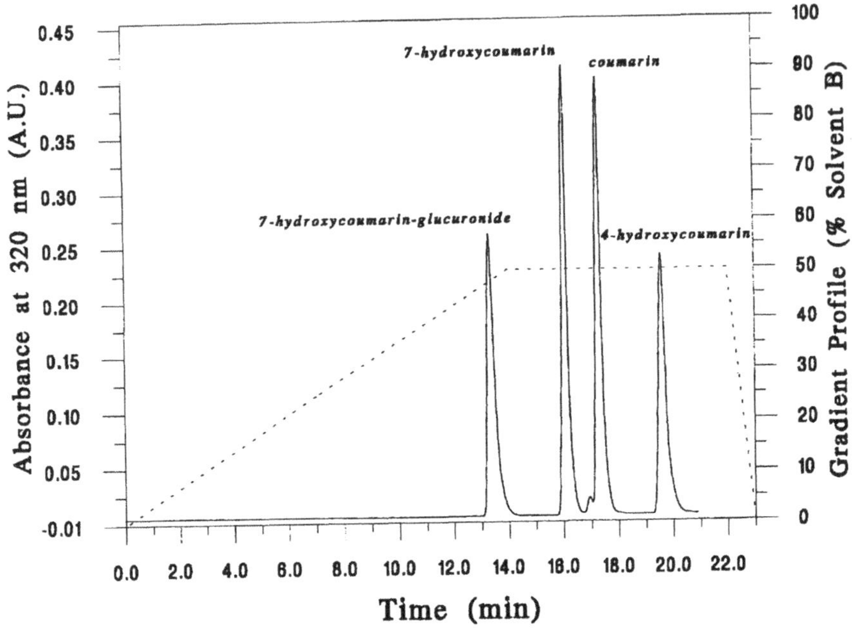

Figure 12.2. The separation of 7-hydroxycoumarin-glucuronide, 7-hydroxycoumarin, coumarin and 4-hydroxycoumarin by HPLC. The sample was prepared in water and analysed by reverse-phase HPLC on a C_{18} column under gradient elution conditions and detection at 320 nm[53]. The figure shows an overlay of the gradient mobile-phase profile and the elution order of the coumarins present. (Reproduced from *J Pharm Biomed Anal*, 1996; **14**: 1585–90.)

showed the complete separation of a standard mixture of the majority of coumarin metabolites. Separation was carried out on a 25 cm × 4.6 mm Spherisorb 5 ODS-2 column using a linear gradient elution of formic acid : water : methanol, and detection at 280 nm. *O*-Hydroxyphenylacetaldehyde was found to be a major metabolite of coumarin formed by rat, human, and gerbil liver microsomes[37]. This was determined by separation as above[36] but with mass spectral analysis of the metabolite isolated from the rat liver microsomal incubation. Cytochrome P450 2A6 activity was assessed for coumarin 7-hydroxylation, as well as for the production of other metabolites.

Metabolism of radiochemical-labelled [3-^{14}C] coumarin by hepatic microsomes from various species was also studied by HPLC[38–40]. The metabolites were determined by comparison of retention times with known standards. The concentration of metabolite was quantified by measurement of the radioactivity present in the appropriate HPLC fractions. At specific timed intervals the metabolic reaction was terminated by the addition of ice-cold methanol, followed by centrifugation. It was possible to monitor the majority of metabolites produced over time to give a pharmacokinetic profile of coumarin metabolism *in vitro*.

Evans and Relling[41] determined 7-ethoxycoumarin *O*-deethylase activity by analysing 7-hydroxycoumarin using fluorescence detection (excitation wavelength at 360 nm and an emission wavelength of 470 nm). Separation was carried out on a μBondapak phenyl column (300 mm × 3.9 mm i.d.). Separation was achieved with an isocratic mobile phase of methanol and 0.05 M acetate buffer (55 : 45, v/v; pH 4.7). The limit of detection was 70 ng/ml. 7-Hydroxycoumarin was extracted into chloroform and this was evaporated to dryness followed by reconstitution into methanol–water (70 : 30, v/v) before HPLC separation and analysis.

The *in vitro* metabolism of 7-hydroxycoumarin to 7-hydroxycoumarin-glucuronide[42] by uridine diphosphate (UDP) glucuronyl transferase (UDPGT) and the cofactor UDP-glucuronic acid was monitored with HPLC. A mixture including 7-hydroxycoumarin, UDP-glucuronic acid, and a crude protein UDPGT preparation was prepared in Tris–HCl buffer (pH 7.4). Aliquots were removed at specific time intervals, and added to trichloroacetic acid to stop the reaction. The samples were then analysed on a C_{18} column and separated with a gradient elution from 100% (A) water : methanol : acetic acid (95 : 5 : 0.2, v/v/v) to 50% (A) : 50% (B), methanol over 14 min. The eluent was monitored at 320 nm. The method allowed the separation (Figure 12.2) of coumarin, 7-hydroxycoumarin, 7-hydroxycoumarin-glucuronide, and an internal standard (4-hydroxycoumarin). It was possible to monitor the increase in 7-hydroxycoumarin-glucuronide over time, and the decrease in 7-hydroxycoumarin content. The limit of detection for 7-hydroxycoumarin was 0.5 μg/ml and the linear range was 0–100 μg/ml. Duffy *et al.*[43] adapted the method of Killard *et al.*[42] for the determination by HPLC of the UDP-glucuronyl transferase activity for 7-hydroxycoumarin in rabbit tissue samples. Crude tissue preparations of liver, kidney, bladder, large intestine, lung, spleen, heart and fat were prepared in Tris–HCl, pH 7.4, and used for the preparation of 7-hydroxycoumarin-glucuronide as above. Separation was carried out with gradient elution on a Microbonda Pak C_{18} column. The liver showed the highest enzyme activity followed by the kidney, bladder and large intestine. No activity was observed in the lung, spleen, fat or heart.

Hiller and Cole[44] and Walsh *et al.*[45] both describe HPLC methods for the separation and determination of 7-ethoxycoumarin metabolites in liver slice incubates. The assay of Hiller and Cole[44] is based on short-column gradient elution which permits baseline resolution of the parent drug and three metabolites. The total assay time is 10 min. Separation of 7-ethoxycoumarin, 7-hydroxycoumarin, 7-hydroxycoumarin-glucuronide and 7-hydroxycoumarin-sulphate is described. Separation is carried out on a 3.3 cm × 4.6 mm (i.d.), 3 μm particle size, LC-8 HPLC column. The mobile phases were A: 0.25% acetic acid (900 ml), tetrabutyl ammonium phosphate (1.7 g) and acetonitrile (100 ml) and the pH was adjusted to 4.5 with NaOH and B: a variation on the mixture of solvent A, i.e. 0.25% acetic acid (500 ml), tetrabutyl ammonium phosphate (1.9 g) and acetonitrile (500 ml). The 1.5 ml/min gradient was as follows: 0–2 min 85% A; 2–7 min 85% A–27% A; 7.1 min 27% A–85% A; 10 min—next sample. Walsh *et al.*[45] used a similar method for the determination of the same metabolites, as above[44] of 7-ethoxycoumarin *in vitro*. Separation was carried out by ion-pairing HPLC on a 250 mm × 4.6 mm C_8 column using a 1 ml/min gradient elution. The limit of quantitation for 7-ethoxycoumarin and its metabolites was *ca.* 0.025 nmol/ml. They[45] describe the chemical preparation of standards of 7-hydroxycoumarin-glucuronide and 7-hydroxycoumarin-sulphate. The structures and characterisation of the metabolites are included. The activity of the liver slices for 7-ethoxycoumarin metabolism in Krebs–Henseleit buffer results in the formation of these metabolites and they are readily quantified by HPLC with UV detection at 320 nm. The activity was *ca.* 5 nmol of glucuronide produced per incubation ($N = 2$) after 4 h.

Coumarin in Foodstuffs

Coumarin is the most common adulterant of vanilla extract[46]. The method of Thompson and Hoffman[46] utilised a methanol : water : acetic acid mobile phase (40 : 60 : 0.1 v/v/v) and a C_{18} column, with UV detection for the determination of coumarin. The method had a limit of detection of 0.6 ng/ml (from 3 × signal-to-noise ratio). It was possible to separate the coumarin from vanillin as well as other endogenous species in bean extracts. Thus, it was possible to use HPLC in the assessment of quality. Phenolics and flavonoids are common substituents of plant-derived beverages[47] playing a role in the stability, coloration, nutritional value, etc. Many coumarin derivatives, including 7-hydroxycoumarin, were separated by HPLC with coulometric array detection. Limits of detection were in the order of 1 ng/ml. Electrochemical detection provides selectivity and sensitivity to the study of trace amounts of these compounds in such matrices. Mazza[48] also describes the separation and determination of coumarin in alcoholic beverages. Separation was carried out on a LiChrosorb RP-18 column with a 1:1 water : methanol mobile phase.

Plant Extract Analysis

Many of the reported analyses of coumarin and its derivatives involved some form of sample preparation. The method might require organic solvent extraction[32–34,49,50], precipitation of any protein present[37], or filtration[51]. Nykolov *et al.*[50] determined numerous hydroxycoumarins present in the bark of *Fraxinus ornus*, which is a major source of esculin, an anti-inflammatory and vitamin-P-like agent. Climatic conditions and the stage of plant growth lead to great variations in hydroxycoumarin content. The samples were extracted into ethanol and analysed on a C_8 column, with gradient elution of the components (mobile phases were water : phosphoric acid, pH 3.0, and methanol) and UV detection at 210 nm. Preparative HPLC was used to purify the individual components. Fractions were taken and analysed separately by thin-layer chromatography, and by mass spectrometry to characterise each of the hydroxycoumarins. The determination of coumarin in cinnamon and cassia[51] was achieved by separation on a LiChrosorb RP-8 reversed-phase column with a water–

methanol–acetonitrile–tetrahydrofuran (60 : 12 : 20 : 8, v/v/v/v) mobile phase. Vanillin was used as an internal standard.

12.3.3 Gas Chromatography

Gas chromatography (GC) is a technique that allows the separation of thermally stable and volatile organic and inorganic compounds[28]. The early work on coumarin analysis[22] centred on the separation of coumarin as an adulterant in flavourings. More recent work has focused on coumarin as a flavouring[54,55], as a fragrance[54,56], in tobacco and in vegetables[57]. Separations were typically carried out on fused-silica capillaries with helium as the carrier gas. The temperature gradients for the different methods varied from 8 °C/min or 10 °C/min up to a final temperature of 300 °C[57,58].

Flavours and Fragrances

Supercritical fluid extraction (SFE) was used as the preparation step to determine coumarin in some natural flavour and fragrance compounds[54]. The samples were extracted into 300 atm of supercritical carbon dioxide at 45 °C. The extraction chamber outlet restrictor capillary was connected to the standard on-column injection port of the gas chromatogram, allowing for an on-line purification and separation. This reduces sample loss due to sample transfer after extraction. The sample is then separated and detected by either mass spectrometry or flame ionisation (FID). The method was used for the determination of coumarin in cinnamon and in rosemary and for the analysis of spices, chewing gum, spruce needles and cedar wood. Separation was carried out on the GC with a gradient temperature program from 70 °C to 320 °C at 8 °C/min and 1.14 ± 12% mg of coumarin was detected per gram of rosemary.

The efficacy of using GC was also assessed as a tool for the accurate identification of coumarin and other flavourings[55]. In the report by Grundschober coumarin is separated by GC with mass spectral (MS) and Fourier transform infrared (FTIR) detection to distinguish coumarin from other flavourings present in the species of interest. Infrared bands and their intensity and mass spectral coumarin ion relative abundances are given. The relative abundances for coumarin ion profile after mass spectral analysis are given in Table 12.2, and Table 12.3 is a list of the infrared bands observed for coumarin as analysed by Fourier transform infrared analysis.

The effective blood concentration of coumarin in mice, after the inhalation of coumarin, is described by Jirovetz *et al.*[56]. Sandalwood oil was added into the cages of mice, and blood samples were taken and assessed for coumarin content. The levels of coumarin found in the blood were 7.7 ng/ml (± 2 ng/ml), after the addition of coumarin (3.0 g) into the mouse cage. Samples were separated and detected by a variety of methods, including FTIR, MS and atom emission detection.

Table 12.2. Mass spectral analysis data for relative abundances of coumarin ions.

m/z	Range of relative abundance
147	7–12
146	66–100
119	8–10
118	82–100
90	28–41
89	30–40
64	8–12
63	21–32
62	8–12
59	5–15
51	8–15
44	7–11
43	5–14
39	11–16

(Modified table from Lebens Z. *Unters Forsch*, 1991; **192**(6): 530–4.)

Tobacco and Vegetable Juice Analysis

Thin-layer chromatography and high-performance liquid chromatography were used to purify coumarin from tobacco samples before their analysis by gas chromatography[57]. A chloroform–hexane (25 : 75, v/v) solution was used to extract the coumarin from the various tobacco leaf samples chosen. The HPLC mobile phase was based on a chloroform–hexane gradient [6 : 94 (v/v) at T_0, 40 : 60 at T_{18}, and at T_{24} 6 : 94]. Coumarin was separated on a silica gel TLC plate, with a pentane–ethyl acetate (70 : 30, v/v) mobile phase. The coumarin fraction from the HPLC and the coumarin TLC spot were then analysed by GC-MS. The limit of detection of the method was

Table 12.3. The infrared bands observed for coumarin as analysed by Fourier transform infrared analysis.

Infrared bands	Weak/medium
825	w
877	w
919	w
1096	m
1173	w
1265	w
1451	w
1608	w
1778	w
3079	w

(Modified table from Lebens Z. *Unters Forsch*, 1991; **192**(6): 530–4.)

0.05 mg/g tobacco. The identity and purity of coumarin were confirmed by nuclear magnetic resonance (NMR), and mass spectrometry (MS). Radiolabelled [^{13}C]coumarin was used as an internal standard for mass spectral determination of coumarin.

Weinberg *et al.*[58] describe the analysis of coumarins in vegetable juices. The coumarins present in vegetable juices include umbelliferone, psoralen and bergapten. The coumarins were extracted into methylene chloride, and analysed by GC-MS.

12.3.4 Capillary Electrophoresis Analysis

There has only been a limited amount of work reported on the analysis of coumarin and its derivatives by capillary electrophoresis. This powerful analytical technique is a relatively new addition to the tools available to the analyst and it is gaining in widescale acceptance for the determination of many species in a multitude of areas[59-61]. Several researchers[62-69] have applied this tool to the separation of 7-hydroxycoumarin, 7-hydroxycoumarin-glucuronide, and warfarin in a variety of matrices and in *in vitro* and *in vivo* studies.

Determination of Free and Total 7-Hydroxycoumarin in Urine and Serum

Separation of the species is normally carried out on an untreated silica capillary with ultraviolet detection at either 210 nm or 320 nm. Separation is typically carried out in a phosphate buffer-based solution as electrolyte. The separation of 7-hydroxycoumarin from coumarin[62], and endogenous species in urine or serum can be accomplished in less than a minute (Figure 12.3). Samples were prepared by extraction of the analyte into diethyl ether, and reconstitution into electrolyte buffer. Separation was carried out in 25 mM phosphate buffer, pH 7.5 on a 27 cm capillary. The analyte was determined at 210 nm, and the method was found to have a linear range between 0 and 50 µg/ml with a limit of detection of 1 µg/ml. Warfarin was used as an internal standard for the determination of free 7-hydroxycoumarin in the urine and serum. The analysis of total 7-hydroxycoumarin requires the deconjugation of the 7-hydroxycoumarin from the glucuronide form of the drug. The addition of β-glucuronidase to the serum or urine samples appears to cause an interaction between warfarin and endogenous species present in the biological fluid. Endogenous species comigrate with the warfarin and thus it cannot be used for the determination of total 7-hydroxycoumarin present in urine or serum. The method was found to have very good accuracy and precision with percentage relative standard deviations between 0.8 and 7.3%.

The Direct Determination of 7-Hydroxycoumarin and 7-Hydroxycoumarin-glucuronide in Urine

Separation of 7-hydroxycoumarin and 7-hydroxycoumarin-glucuronide in urine, without sample clean-up[63], was carried out on a 47 cm untreated silica capillary with a phosphate buffer/deoxycholic acid/acetonitrile electrolyte solution (Figure 12.4). Detection was at 320 nm. At 214 nm it was possible to observe the

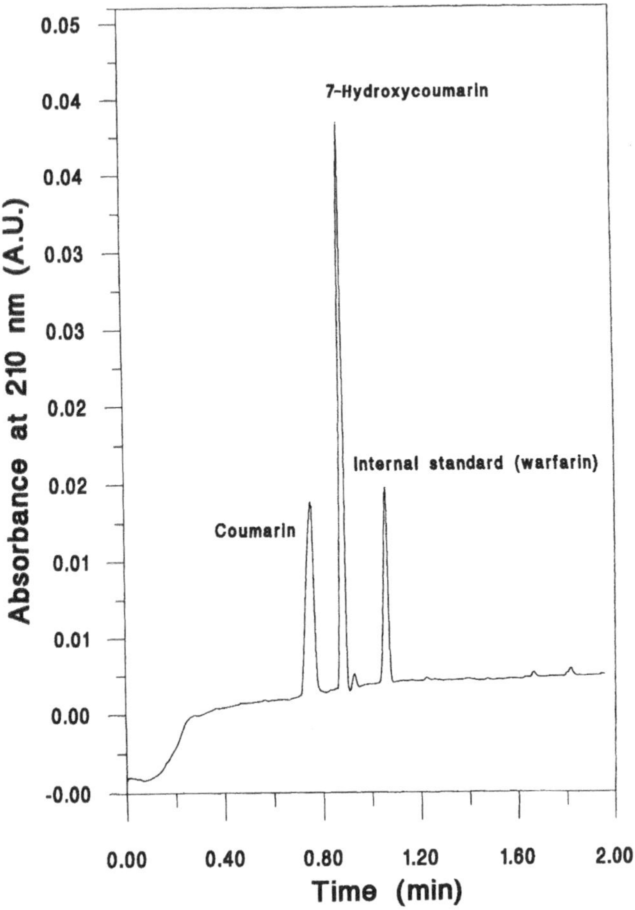

Figure 12.3. The separation of coumarin, 7-hydroxycoumarin, contaminant of 7-hydroxycoumarin standard, and warfarin by capillary electrophoresis[62]. The compounds were separated on a 27 cm × 50 μm untreated silica capillary in 25 mM phosphate buffer (pH 7.4) at 20 kV with detection at 210 nm. (Modified figure from *J Chromatogr B*, 1995; **663**: 371–8.)

endogenous species present in urine that co-migrate with both analytes. To overcome this, at the expense of sensitivity, detection at 320 nm was chosen. At this wavelength 7-hydroxycoumarin could be selectively analysed and 7-hydroxycoumarin-glucuronide, as the interfering species, did not have any absorbance. The linear range of the assay is 0–100 μg/ml, with a limit of detection of 2 μg/ml. The separation of 7-hydroxycoumarin and 7-hydroxycoumarin-glucuronide is achieved in less than 7.5 min. In comparison to the method of Bogan *et al.*[62], there is minimal sample clean-up. There is no analyte loss during deconjugation, extraction, evaporation or during reconstitution. It is possible to directly determine both major human metabolites of coumarin after administration *in vivo*. Percentage relative standard deviations were in the order of 1% to 10%.

in vitro Metabolism Studies

Applications of capillary electrophoresis to the *in vitro* study of coumarin[64,65] and 7-hydroxycoumarin[66] metabolism have also been reported. Liver microsomal

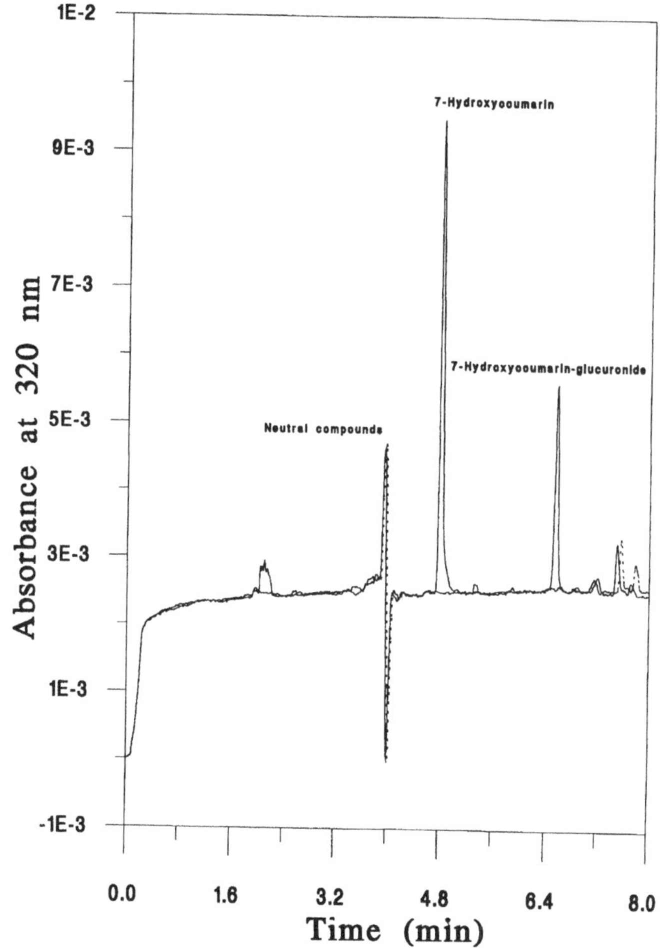

Figure 12.4. Overlay of two electropherograms (0 μg/ml and 50 μg/ml standards) of 7-hydroxycoumarin and 7-hydroxycoumarin-glucuronide prepared in urine and analysed directly by capillary electrophoresis[63]. Separation was carried out on a 47 cm × 50 μm untreated silica capillary with a 90% of 100 mM phosphate: 11 mM deoxycholic acid: 10% acetonitrile buffer, at 25 kV with detection at 320 nm. (Reproduced from *Analyst* 1996; **121**: 243–7 by permission of the Royal Society of Chemistry.)

preparations were prepared from a variety of sources (human, rabbit, bovine, porcine and murine). A reaction mixture was prepared in phosphate buffer and it included an NADPH regeneration system, coumarin, and a microsomal preparation (Figure 12.5). The reaction commenced with the addition of the microsomes, and at specific timed intervals samples were removed and, without any sample preparation, analysed by capillary electrophoresis. It was possible to monitor the production of 7-hydroxy-coumarin over time, and to compare the ability of the microsomes to metabolise coumarin. Thus, cytochrome P450 2A6 activity in the different species[64] could be

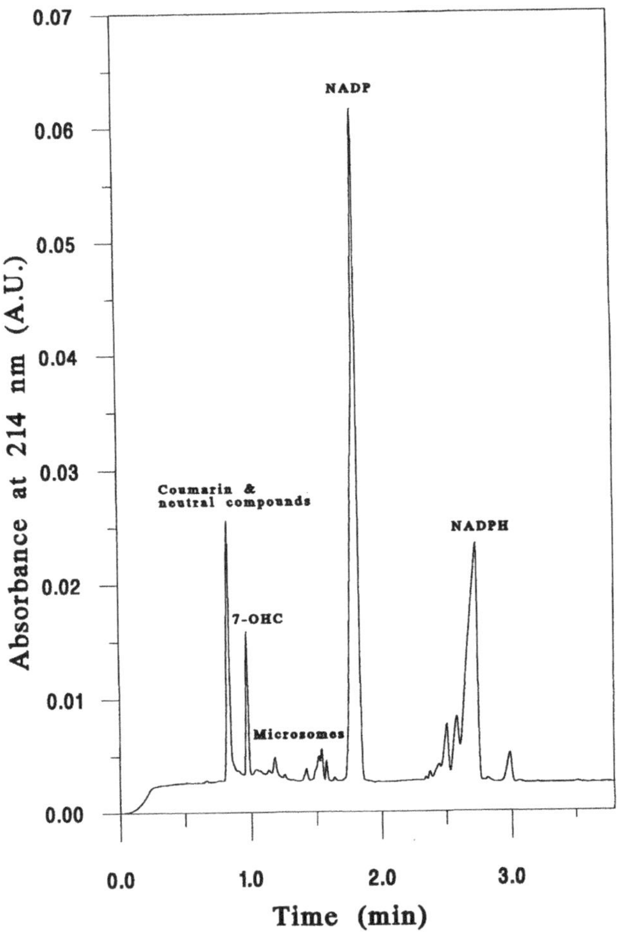

Figure 12.5 The capillary electrophoresis separation of the coumarin metabolic reaction mixture[64,65]. The electropherogram shows the separation of coumarin, 7-hydroxycoumarin, microsomes, NADP and NADPH. Separation was carried out as in Figure 12.3, with detection at 214 nm. (Reproduced from *Xenobiotica*, 1996; **26**(4): 437–45.)

assessed as well as inter-individual differences within a species[65]. The limit of detection was 1 μg/ml, with a linear detection range 0–50 μg/ml. The study showed that human liver microsomal preparations have the highest 7-hydroxycoumarin production, with rat, pig and dog having undetectable levels.

Capillary electrophoresis was utilised to monitor the *in vitro* production of the principal phase II coumarin metabolite, 7-hydroxycoumarin-glucuronide[66] (Figure 12.6). A crude preparation of bovine liver uridine diphosphate (UDP) glucuronyl

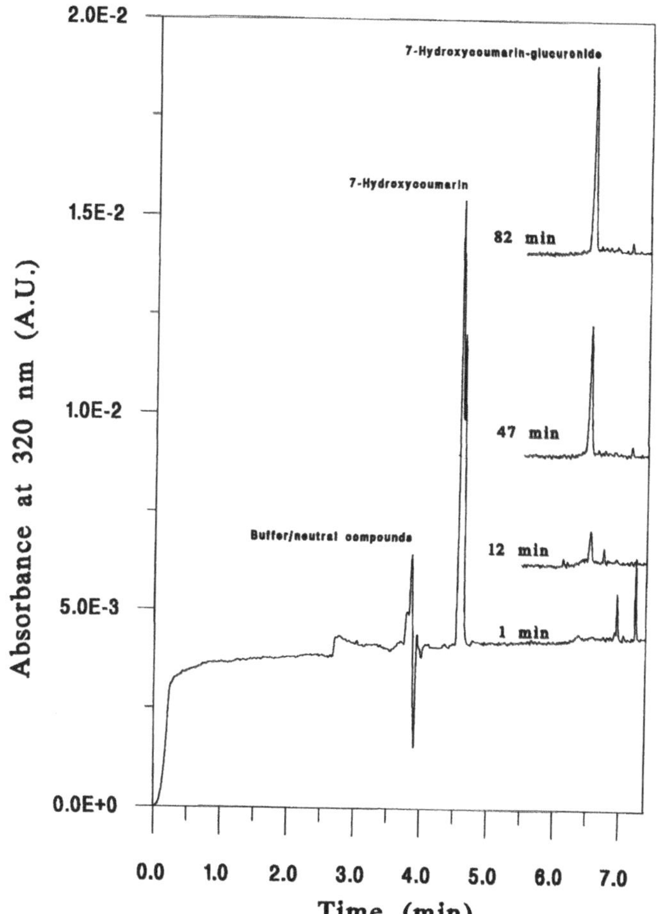

Figure 12.6. Overlay of the electropherograms of samples taken at different times after the glucuronidation of 7-hydroxycoumarin (shown in offset)[66]. It illustrates the increase in 7-hydroxycoumarin-glucuronide content over time as it is formed by the metabolism of 7-hydroxycoumarin by uridine diphosphate glucuronyl transferase. Separation was carried out as in Figure 12.4. (Reproduced from *J Capillar Elec*, 1995; **5**: 241–5.)

transferase was prepared and used for the glucuronidation of 7-hydroxycoumarin (Figure 12.1). The UDP-glucuronyl transferase glucuronidates the 7-hydroxycoumarin through the -OH moiety. It is rendered more water soluble, thus making it more readily excretable in the urine. Samples are removed from the reaction mixture incubating at 37 °C and, similar to the phase I study, no sample preparation is necessary before capillary electrophoresis analysis. Separation was carried out on a 47 cm capillary with a phosphate buffer : deoxycholic acid : acetonitrile electrolyte solution, as in the method of Bogan *et al.*[63]. The method allowed the detection of 7-hydroxycoumarin-glucuronide from 2 to 100 μg/ml. The rate of reaction for the first 70 min was 3.1 ± 0.13 nmol of 7-hydroxycoumarin-glucuronide produced per min per mg of protein. The method was fast and very reliable. HPLC was also used to determine the activity of the liver enzyme preparation for 7-hydroxycoumarin-glucuronidation[42,43]. There was no statistical difference between the results obtained by these methods.

Using 7-Hydroxycoumarin to Determine Separation Efficiencies

The electro-osmotic velocity occurring in capillaries made from polyfluorocarbon, polyethylene or polyvinylchloride was determined by the use of electrically neutral marker substances in dependence on the pH in aqueous solutions[67]. The fluorescent characteristic of 7-hydroxycoumarin was exploited. 7-Hydroxycoumarin is determined as it is neutral in the acidic pH range up to 5.8. The pH and composition of electrolyte solutions are changed and the effective influence of the coating at the capillary surface assessed.

Separation of Coumarins in Plant Extracts

Seven coumarins were separated by micellar electrokinetic capillary chromatography (MECC)[68]. Coumarin, and several of its hydroxy- and methyl- derivatives were separated by a phosphate-borate (pH 7.0) buffer with sodium dodecyl sulphate (SDS, anionic surfactant) as electrolyte. The eluent was monitored at 195 nm. The samples were analysed on a 70 cm untreated capillary at an applied voltage of 18 kV. These neutral coumarins separate based upon the differential partitioning between the SDS micelle hydrophobic interior and the aqueous mobile phase.

Warfarin Determination in Plasma

3-(α-Acetonylbenzyl)-4-hydroxycoumarin (warfarin) is widely used as an anticoagulant. As a racemic compound it was necessary to develop the separation anticoagulant of the two forms of warfarin based on chiral capillary electrophoresis[69]. A methylated *β*-cyclodextrin-containing electrolyte was used to successfully resolve the two isomers. Plasma samples were analysed on a 72 cm fused silica capillary. The limit of detection was 0.65 μM of each enantiomer. The optimisation of the separation is described. Voltage, electrolyte ionic strength and content have a dramatic effect on separation.

12.4 SPECTROSCOPY

The absorbance, fluorescence and luminescence of any compound are dependent on the interaction of electromagnetic waves and matter[70]. Interactions are dependent on the molecular structure of the compound and are unique for all compounds. Coumarin and its derivatives all have their own unique spectroscopic properties that can be exploited for their determination. Spectrofluorimetry and luminescence have been used to determine 7-hydroxycoumarin directly or coumarin indirectly after derivatisation. Table 12.4 is a summary of the spectroscopic methods of analysis and their applications.

12.4.1 Spectrofluorimetry

Pharmacokinetic and Biopharmaceutical Studies

Tan *et al.*[71] described a method for the quantitative analysis of coumarin and 7-hydroxycoumarin, and mixtures thereof, in whole blood. Coumarin and 7-hydroxycoumarin were selectively isolated from blood by solvent extraction and analysis was based on the measurement of fluorescence of the coumarin fluorophore or 7-hydroxycoumarin (which is fluorescent). The coumarin is extracted into ether and evaporated to almost dryness with a constant nitrogen flow at low heat. One millilitre of methanol is added and the evaporation of the ether is completed. Then 5.0 ml of 2 N methanolic KOH is added to the solution and the coumarin exposed to UV light under alkali conditions. The fluorescence is determined at excitation and emission wavelengths of 361 nm and 491 nm, respectively. The linear range for coumarin determination is 0.02–0.2 μg/ml.

7-Hydroxycoumarin, as well as coumarin, is extracted into the ether. A second extraction step is then used to isolate 7-hydroxycoumarin before fluorescence detection. Glycine buffer (pH 10.0) is added to the ether layer and centrifuged. The glycine layer is removed and fluorescence is determined at activation and emission wavelengths of 370 nm and 450 nm, respectively. The linear range for 7-hydroxycoumarin was 1–10 ng/ml. For the analysis of a mixture of coumarin and

Table 12.4. Summary of the spectroscopic methods of analysis of coumarins.

Method and application	Mode of detection	References
Spectrofluorimetry:		
Pharmacokinetics of coumarin metabolism	Fluorescence of 7-OHC and derivatised coumarin	71–74
Glucuronidation of 7-OHC	Fluorescence of 7-OHC	75
In vivo metabolism of coumarin	Fluorescence of 7-OHC	76,77,79,80
	Microplate assay	78
	TLC separation	27

7-OHC, 7-hydroxycoumarin.

7-hydroxycoumarin the procedure is as just described with the ether layer used for coumarin determination and 7-hydroxycoumarin determined by its fluorescence in glycine buffer. The method was shown to be sensitive, and reproducible with little interference from other constituents in blood and there was no need to precipitate out the proteins in the blood. Other umbelliferone conjugates or other hydroxycoumarins do not interfere.

The spectrofluorimetric method of Tan *et al.*[71] was used to determine the pharmacokinetics of coumarin after oral and intravenous administration[72] in humans. Levels of 7-hydroxycoumarin and 7-hydroxycoumarin-glucuronide were determined by spectrofluorimetry. Samples were analysed after extraction for free 7-hydroxycoumarin and after treatment with β-glucuronidase to liberate 7-hydroxycoumarin from the glucuronide conjugate for total 7-hydroxycoumarin content determination.

Coumarin must be derivatised and irradiated with UV light before it can be analysed by fluorimetry[73]. Coumarin was first extracted into anhydrous ether, then evaporated to dryness, and finally reconstituted into methanol. An alkali solution of 2 N KOH was added to the reconstituted sample and irradiated with UV light. After irradiation it was analysed by fluorimetry with an excitation wavelength of 361 nm and emission wavelength of 491 nm. 7-Hydroxycoumarin can be analysed after clean-up without the need for derivatisation. The excitation and emission wavelengths for 7-hydroxycoumarin are 370 nm and 450 nm, respectively. This procedure was utilised for determining binding of coumarin and 7-hydroxycoumarin to proteins (in human serum) and to erythrocytes and for assessing their solubility in water, phosphate buffer and human serum.

Coumarin 7-hydroxylation activity in the liver microsomes of eight different species was investigated by Kaipainen *et al.*[74]. An incubation solution containing coumarin, NADPH, $MgCl_2$, and the microsomal preparation was prepared. At predefined times the reaction was stopped by precipitating the protein etc. with trichloroacetic acid. NaOH-glycine buffer (pH 10.3) was added to the solution and fluorescence was determined at excitation and emission wavelengths of 390 nm and 440 nm, respectively. Antibodies were raised, in mice, against cytochrome P450, with high activity against coumarin 7-hydroxylation. The microsomes were pre-incubated with the antibody, and inhibition of coumarin 7-hydroxylase activity was assessed in the microsomes from the different species. There was almost 100% inhibition of coumarin 7-hydroxylation in some species and little, if any, in the rat or rabbit microsomes.

Glucuronidation Studies

The major site of glucuronidation (a detoxification pathway that makes compounds more hydrophilic and more readily excretable in urine) is the liver. The majority of coumarin administered in humans is excreted as 7-hydroxycoumarin-glucuronide[33,53,63]. A novel method was developed that allowed the monitoring of the decrease in 7-hydroxycoumarin content[75]. A fluorimeter is connected via optical fibres to microlight guides that were placed on the surface of liver tissue samples. It was possible to monitor the fluorescence of 7-hydroxycoumarin. As the 7-hydroxycoumarin was converted to 7-hydroxycoumarin-glucuronide the decrease in fluorescence could be observed[75]. One strand of the microlight guide is connected to a 100 W mercury arc lamp and the other strand to a photomultiplier. Excitation and emission wavelengths were at 366 nm and 450 nm, respectively. Different parameters were changed to assess the differences in glucuronidation in the different liver regions. Glucuronyl transferase activity was determined by measuring 7-hydroxycoumarin-glucuronide spectrofluorimetrically after treating it with β-glucuronidase in a fluorimeter. Maximal rates of glucuronidation in periportal and pericentral regions of the liver lobule were 9.6 and 35 μmol/g/h, respectively.

Urine Analysis

Several investigators have used the fluorescence characteristics of 7-hydroxycoumarin for its determination in urine[76-80]. A spectrofluorimetric assay was used to assess inter-individual variability in coumarin 7-hydroxylation by Rautio *et al.*[76] and Iscan *et al.*[80]. Samples of urine were treated with β-glucuronidase before extraction into chloroform and analysis. Results were compared to analysis of the same samples by a HPLC method[76]. Samples were analysed on an Interchrom Nucleosil C_{18} column with UV detection at 313 nm. Samples were separated by gradient elution [buffer A: 20% of a 1.5% acetic acid (pH 4.85, adjusted with NH_4OH) and 80% buffer B; buffer B: 50% buffer A + 50% acetonitrile]. Different doses of coumarin were used in a preliminary experiment to study the relationship between the dose and the 7-hydroxycoumarin excretion[76]. However, both research groups administered a single 5 mg dose of coumarin for the inter-individual variability study.

In the two studies 110[76] and 100[80] people, respectively, were administered coumarin and the production of total 7-hydroxycoumarin was determined. There was a large inter-individual variability in the metabolism of coumarin to 7-hydroxycoumarin. Up to 80% of the 7-hydroxycoumarin formed was excreted in the first 2 h, and 59.8% (± 21.5%) of the administered coumarin was derivatised to 7-hydroxycoumarin or 7-hydroxycoumarin-glucuronide[80]. Coumarin 7-hydroxylation was found to be lower in males and in cigarette smokers[80], however, Rautio *et al.*[76] found no statistical difference between male and female coumarin 7-hydroxylase activity.

Merkel *et al.* also investigated 7-hydroxycoumarin production by analysing urinary excretion of the coumarin metabolites[79]. Samples were prepared by a similar method to that of Rautio *et al.*[76]. Merkel *et al.*[79] report that grapefruit juice flavonoids inhibit cytochrome P450 2A-dependent metabolic pathways. They report that coumarin 7-hydroxylation was significantly decreased with 7-hydroxycoumarin excretion delayed by up to 6 h. Thirteen volunteers were assessed for the effect of the grapefruit juice on their production of 7-hydroxycoumarin *in vivo*.

Egan and O'Kennedy[78] developed a spectrofluorimetric microassay for the determination of 7-hydroxycoumarin in urine and plasma. The samples were initially analysed without sample clean-up. The samples were then treated with *β*-glucuronidase, extracted into diethyl ether, evaporated to dryness and reconstituted into methanol. The linear range for detection of 7-hydroxycoumarin, before or after extraction, in urine and plasma was 0.5–10 μg/ml and 10–100 μg/ml, respectively, and the limit of quantification was 0.5 μg/ml. The excitation wavelength chosen was 370 nm and the emission wavelength was monitored at 450 nm. The assay was carried out in 96-well Dynatech microtitre plates.

12.4.2 Luminescence

Coumarin is added as a marker compound to separate rebated and unrebated kerosene for the calculation of customs excise duty[81]. A quick method of analysis of the fuel oils is to convert the coumarin present into the potassium salt of *cis*-coumaric acid, and, furthermore, into the *trans* isomer by irradiation with UV light at 365 nm. This isomer is fluorescent and coumarin can be qualitatively determined. Coumarin is quantified by measuring the fluorescence obtained by excitation at 365 nm and emission at 500 nm. Total luminescence spectra of the fuel samples are obtained by examining all the fluorescence spectra across a range of wavelengths (300–600 nm).

12.5 ANTIBODY-BASED IMMUNOANALYTICAL TECHNIQUES

The antibody–antigen interaction can be used as an analytical tool for selective and sensitive determinations of drugs, hormones, etc.[82,83]. The production of antibodies, their functions and their applications are reviewed by Killard *et al.*[83]. The specificity of an antibody determines its suitability as a potential tool for sensitive, accurate analysis of coumarin and its derivatives. The use of ELISA, biospecific interaction analysis and electroanalytical methods are described in this section. Table 12.5 summarises the immunoanalytical methods used for analysis of coumarin and 7-hydroxycoumarin.

12.5.1 Enzyme-linked Immunosorbent Assay (ELISA)

An enzyme-linked immunosorbent assay (ELISA) is an antibody-based method for the detection of drugs in body fluids. A large variation of assay designs have been used for the detection of antibodies and antigens[83-85]. Immunoassays can be divided into heterogeneous and homogeneous assays. In a heterogeneous assay, a separation step is required where the free antibody and antigen is separated from the complexed solid phase bound antibody–antigen. The quantifiable signal from the complex does not differ from the free entity and therefore without this separation, interference would occur. In a homogeneous assay, this separation step is not required. The binding of antigen and antibody results in a change in the activity of the label and therefore this activity can be detected without the problems of interference. A heterogeneous immunoassay will have the following steps common to all variations:

Table 12.5. Summary of the immunoanalytical methods of analysis of coumarins.

Method and application	Mode of detection	Limit of detection	References
ELISA:			
Analysis of coumarin or 7-hydroxycoumarin—antigen-inhibition ELISA with anti-coumarin/7-OHC antibody	OPD as substrate with detection at 405 nm	0.5 μg/ml	86,87
	*p*NPP as substrate with detection at 414 nm	40 ng/ml	91
7-Hydroxycoumarin–alkaline-phosphatase bispecific antibody assay	*p*NPP as substrate with detection at 414 nm	6 ng/ml	91
Electroanalytical analysis:			
Direct analysis of 7-OHC	Electrochemistry of 7-OHC	10 μM	92
Indirect analysis of 7-OHC	Electrochemistry of HRP	24 μM	93
Biospecific interaction analysis:			
7-Hydroxycoumarin assay	Surface plasmon resonance	0.5 μg/ml	96

(1) Immobilisation of antigen or antibody on a solid surface (usually a microtiter plate).
(2) Incubation with a sample containing the corresponding antigen or antibody.
(3) Detection of specifically bound antibody or antigen by binding an enzyme-linked antibody or antigen, using a colorimetric reaction employing the enzyme.

The method most frequently used for the quantification of haptens (e.g. drugs, hormones, etc.) is the antigen-inhibition ELISA[86].

Determination of Coumarin and 7-Hydroxycoumarin by Antigen-inhibition ELISA

Antigen-inhibition ELISA has been used to detect both coumarin and 7-hydroxy-coumarin in urine[86-88] (Figure 12.7). For the detection of coumarin[86], a coumarin–thyroglobulin conjugate was immobilised on an ELISA microtitre plate. Coumarin standards and unknowns ($N=4$) were incubated at 37 °C for 2 h with purified polyclonal anti-bovine serum albumin–coumarin IgG antibodies raised in rabbits. A competition occurs between the free coumarin in solution and the coumarin coated on the plate surface for the antibody. Unbound antibody and free coumarin–antibody complexes are washed off leaving bound coumarin-antibody complexes behind on the plate. Bound antibody was detected using HRP-labelled secondary goat anti-rabbit IgG, with *o*-phenylene diamine (OPD) as the substrate. The linear range of the assay is from 10 to 100 μg/ml. A similar protocol is followed for the determination of 7-hydroxycoumarin by ELISA[87]. The antibody was raised against BSA–

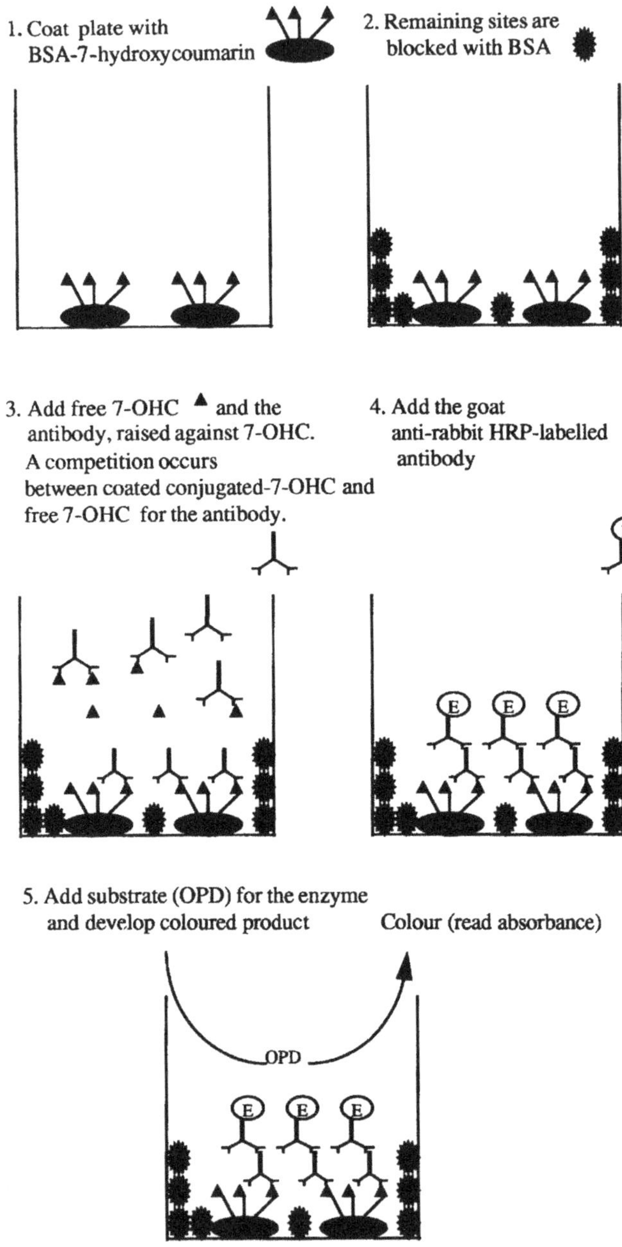

Figure 12.7. The principal steps in a competitive ELISA for the determination of 7-hydroxycoumarin and coumarin in urine, by the use of anti-7-hydroxycoumarin or anti-coumarin polyclonal antibodies raised in rabbits[91]. The steps in the ELISA are described in the text.

7-hydroxycoumarin and screened with ovalbumin-7-hydroxycoumarin. The linear range of this assay is from 5 to 100 μg/ml with a limit of detection of 0.5 μg/ml.

A similar system, employing a BSA–7-hydroxycoumarin conjugate immobilised at the microtitre plate surface, an anti-7-hydroxycoumarin IgG raised in rabbits, and alkaline phosphatase-labelled goat anti-rabbit IgG as detection antibody with *para*-nitrophenol phosphate (*p*NPP) as substrate, was found to be more sensitive[88]. The linear range was from 50 ng/ml to 5 μg/ml, with a limit of detection of 40 ng/ml.

Antigen-inhibition ELISA Employing a Bispecific Antibody

Bispecific antibodies are antibodies that, in contrast to natural antibodies, can bind to two different antigens simultaneously. Bispecific antibodies have been produced chemically by cross-linking whole antibodies (bispecific multivalent antibodies, bsMab) or fragments (Fab^1- $bs(Fab^1)_2$) biologically using hybridoma technology or through genetic engineering. Their production, potential, and advantages in enzyme immunoassays are reviewed by Nolan and O'Kennedy[89].

bsMabs can be rapidly synthesised by the heterobifunctional cross-linkage of two parental antibodies[90]. Such a bispecific antibody was produced from an anti-thyroglobulin–7-hydroxycoumarin and an anti-alkaline phosphatase antibody, using the method of Cook and Wood[90]. The product was used to develop a one-step antigen inhibition ELISA for the detection of 7-hydroxycoumarin in urine[91]. The plate was coated with BSA–7-hydroxycoumarin. Then 25 μl of 7-hydroxycoumarin standard or unknown, 50 μl of bsMab at its working dilution and 25 μl of alkaline phosphatase at a well concentration of 100 μg/ml, were added and incubated for 90 min at 37 °C. Colour was developed with *p*NPP as substrate and absorbance at 414 nm.

Free 7-hydroxycoumarin was determined in urine using this method, and had a linear range between 20 ng/ml and 4 μg/ml. Total 7-hydroxycoumarin determination required the deconjugation of 7-hydroxycoumarin–glucuronide by the addition of β-glucuronidase. All samples were diluted 1:10 in 60 mM PBS–Tween 20 and the samples were analysed as above. The linear range for the determination of total 7-hydroxycoumarin was from 100 ng/ml and 1 μg/ml with a limit of detection of 30 ng/ml. Accuracy and precision was determined by spiking 7-hydroxycoumarin and 7-hydroxycoumarin-glucuronide into control urine. Samples were also analysed after sample clean-up. Samples were extracted into diethyl ether, evaporated off and reconstituted into PBS–Tween 20. The limit of detection was 6 ng/ml with a linear range between 6 ng/ml and 800 ng/ml. Samples were analysed by this method and by HPLC[33] and capillary electrophoresis (CE)[62]. There was no statistical difference between the different methods.

12.5.2 Electroanalytical Analysis

Dempsey *et al.*[92], Deasy *et al.*[93] and Lu *et al.*[94] have investigated the use of anti-7-hydroxycoumarin antibodies for the determination of 7-hydroxycoumarin. A biosensor based on the immobilisation of anti-7-hydroxycoumarin antibodies at the surface of a glassy carbon electrode behind a permeable cellulose dialysis membrane was developed by Dempsey *et al.*[92]. The electrochemical behaviour of 7-hydroxycoumarin at a bare glassy carbon electrode is documented[95]. However, when the antibody is present at the electrode surface there is a 90% decrease in 7-hydroxycoumarin electrochemical response. The binding of 7-hydroxycoumarin to the antibody is believed to occur at its electroactive site which would explain the decrease in the response at the electrode surface. Anodic detection of the oxidised product was achieved at 0.6 V using DC amperometry. The limit of detection was 10 μM. Antibody–antigen kinetics were also investigated. A first-order reaction was observed with $k = 0.0329$ mA/min.

An enzyme-linked electrochemical competitive immunoassay was developed by Deasy *et al.*[93] (Figure 12.8). The anti-7-hydroxycoumarin antibodies were labelled with horseradish peroxidase (HRP) and the electrochemical character of HRP utilised for the indirect determination of 7-hydroxycoumarin. The glassy carbon surface was coated with thyroglobulin–7-hydroxycoumarin. Different concentrations of 7-hydroxycoumarin were added together with antibodies raised, in rabbits, against BSA–7-hydroxycoumarin. A competition, similar to the competitive ELISA above, occurs between free 7-hydroxycoumarin and 7-hydroxycoumarin at the electrode surface. After antibody–antigen interaction all unbound solutions are removed by washing, leaving labelled antibody at the surface. There should only be anti-7-hydroxycoumarin antibodies labelled with HRP remaining at the surface. The electrochemical activity of the HRP on the labelled antibody at the electrode surface is then determined. The current is measured after the addition of 1 mM hydrogen peroxide (enzyme substrate), in the presence of 2 mM hydroquinone. The current output was inversely proportional to the concentration of free 7-hydroxycoumarin in solution. The limit of detection was 24 μM and 7-hydroxycoumarin concentrations could be determined up to 1 mM.

Lu *et al.*[94] developed an enzyme immunosensor for 7-hydroxycoumarin based on electrochemical detection that can be regenerated. BSA–7-hydroxycoumarin is chemically immobilised at the surface of a glassy carbon electrode. A competition is set up between free 7-hydroxycoumarin in solution and BSA–7-hydroxycoumarin coated at the electrode surface for binding of an HRP-labelled anti-7-hydroxy coumarin antibody. The reduction of hydrogen peroxide in the presence of potassium ferrocyanide (a mediator in the reaction) was used for the determination of 7-hydroxycoumarin. After binding of the antibody to the BSA–7-hydroxycoumarin conjugate the excess antibody and free 7-hydroxycoumarin were removed and the reduction of hydrogen peroxide is determined by DC voltammetry at a potential at 0.0 V versus Ag/AgCl. The surface of the electrode was regenerated by immersing the electrode in 10 mM HCl to dissociate the antibody–hapten complex. The regenerated electrode was then washed with phosphate-buffered saline, ready for the next analysis.

12.5.3 Biospecific Interaction Analysis

A competitive immunoassay for the analysis of 7-hydroxycoumarin in biological samples has been developed, based on a commercial biosensor system—the

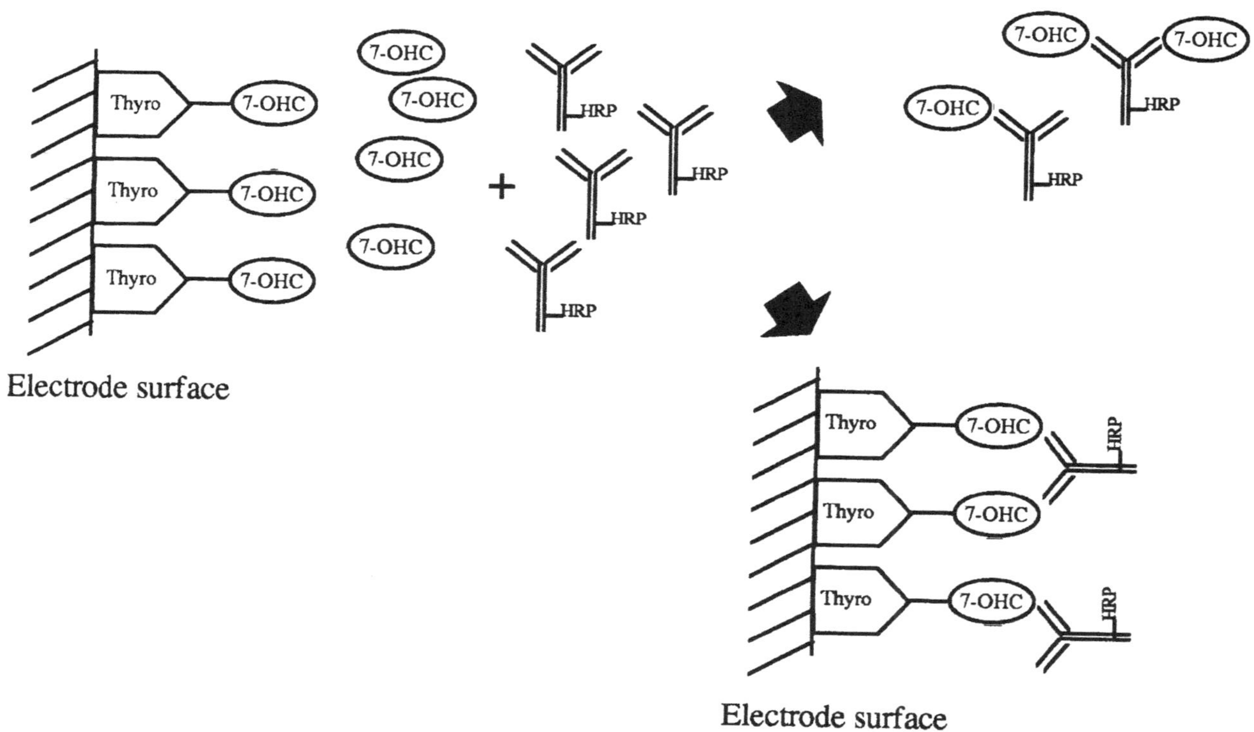

Figure 12.8. Schematic diagram of the indirect immuno-electrochemical detection of 7-hydroxycoumarin in urine[93]. Competition between the conjugated 7-hydroxycoumarin at the electrode surface and free 7-hydroxycoumarin in solution occurs for the anti-7-hydroxycoumarin horseradish peroxidase (HRP)-labelled antibodies. The electrochemical behaviour of HRP is then used for the indirect determination of 7-hydroxycoumarin.

BIAcore™ from Pharmacia Biosensor, Uppsala, Sweden[96]. The BIAcore (real-time biospecific interaction analysis) makes use of the phenomenon of surface plasmon resonance (SPR) to directly monitor biomolecular binding events without the need to label interactants[97].

The BIAcore sensor system utilises a combination of optical and continuous-flow technologies to facilitate label-free interaction analysis. Biomolecular binding is studied by immobilising one of the components on the surface of a specialised sensor chip within the biosensor, and allowing the other to flow over the surface in solution. Mass changes at the biosensor chip surface due to the binding of biological components causes a change in resonance angle directly proportional to the mass change. An indirect competitive immunoassay was developed for the determination of 7-hydroxycoumarin in serum[96].

The assay scheme is shown in Figure 12.9. The assay consists of the following steps:

(1) Immobilisation of 7-hydroxycoumarin–protein conjugate on the sensor surface.
(2) Samples/standards mixed with β-glucuronidase to deconjugate 7-hydroxycoumarin-glucuronide to 7-hydroxycoumarin.
(3) Incubation for 2 h at 37 °C.
(4) Addition of antibody (0.5 mg/ml) to samples/standards.
(5) Incubation for 5 min.
(6) Injection over sensor chip surface.
(7) Regeneration of sensor surface.

A 7-hydroxycoumarin–BSA is first immobilised on the sensor chip surface. This is done by derivitisation with *N*-hydroxysuccinimide (NHS), mediated by *N*-ethyl-N′-(dimethylaminopropyl)carbodiimide hydrochloride (EDC). The immobilisation procedure is highly reproducible, resulting in the covalent binding of about 10 000 response units (RU) of 7-hyroxycoumarin–BSA. The antibody used in the assay is a polyclonal anti-7-hydroxycoumarin purified from rabbit serum. When 20 ml of antibody–sample mixture is injected, the 7-hydroxycoumarin–BSA on the sensor surface and the 7-hydroxycoumarin in the sample compete for antibody binding. The resulting signal—due to unbound antibody binding to the drug–protein conjugate immobilised on the sensor chip—is inversely proportional to the amount of 7-hydroxycoumarin present in the sample.

A standard curve of concentration versus relative signal is prepared. In this assay, regeneration of the 7-hydroxycoumarin–BSA surface is achieved by injection of a 1 min pulse of 5 mM NaOH, a pulse of 20 mM HCl, and another pulse of 5 mM NaOH. Following the regeneration of the surface the next sample or standard is then analysed. All the steps in the assay, including the addition, and incubation, of both enzyme and antibody, are automated and carried out on the BIAcore. A typical sensogram obtained for one of the standards in the assay is shown in Figure 12.10 (page 296). The assay is suitable for the analysis of 7-hydroxycoumarin in neat serum and plasma, without the need for extraction of the drug. Neat urine samples

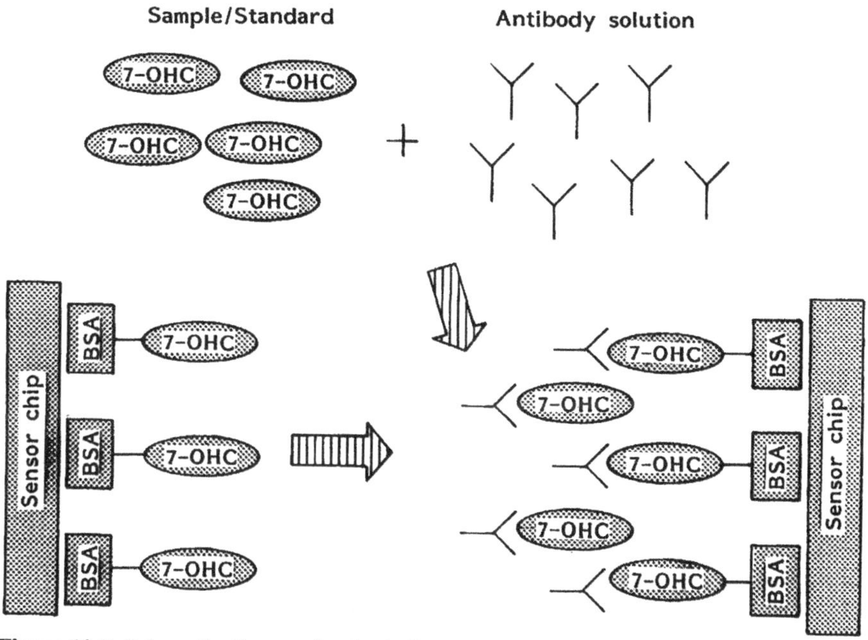

Figure 12.9. Schematic diagram for the indirect immunoassay determination of 7-hydroxycoumarin by biospecific interaction analysis (BIAcore)[96]. BSA–7-hydroxycoumarin is immobilised at the sensor chip surface. Free and immobilised 7-hydroxycoumarin compete for the binding of anti-7-hydroxycoumarin antibody. The amount of antibody available to bind at the electrode surface is dependent on the amount of free 7-hydroxycoumarin present in solution.

may also be analysed, although variations in salt content and pH are problematic. The range of the assay is from 0.5 to 80 μg/ml.

12.6 ELECTROANALYTICAL AND OTHER CHROMATOGRAPHIC TECHNIQUES

12.6.1 Electroanalytical Analysis

Cyclic voltammetry was carried out on 7-hydroxycoumarin prepared in phosphate buffer at an electrochemically pre-treated glassy carbon electrode[95]. The potential was cycled between -1.5 V and $+1.5$ V and the electrochemical characteristics of 7-hydroxycoumarin were determined. The electroanalytical assay was then developed using differential pulse voltammetry as the method of detection. It was analysed with anodic detection at 0.66 V. Samples of 7-hydroxycoumarin were prepared in urine, treated with β-glucuronidase and extracted according to Egan and

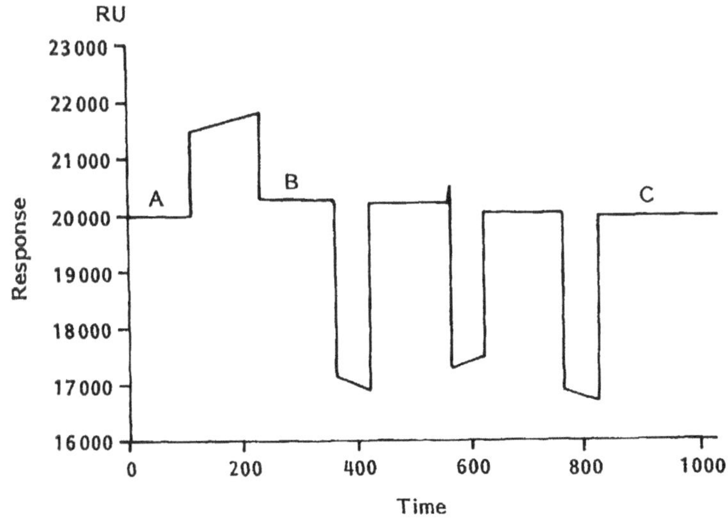

Figure 12.10. An example of a sensogram from the BIAcore 7-hydroxycoumarin competitive immunoassay[96]. (A) is the baseline after the immobilisation of BSA–7-hydroxycoumarin, (B) is the signal increase following the injection of sample–antibody mixture, and (C) is the signal after regeneration.

O'Kennedy[33] and reconstituted into methanol. The method was found to be linear between 0 and 15 mM. The method was also applied to the determination of 7-hydroxycoumarin in patient urine samples.

Differential pulse voltammetry was also used to analyse 7-hydroxycoumarin and 7-amino-4-methyl-coumarin in micellar and emulsified media[98]. Carrazon *et al.* analysed the coumarins by differential pulse voltammetry (DPV) and by linear sweep voltammetry (LSV) with a rotating and stationary glassy carbon electrode (GCE). The detection limits for 7-hydroxycoumarin by DPV were 0.87 μM for a stationary GCE, and 0.28 μM for the rotating GCE. The limit of detection for LSV at a stationary GCE was 4.9 μM, and 0.63 μM for a rotating GCE. The above values were determined in a micellar solution (0.08% Triton X-405, 0.2 M acetate buffer, pH 4.8). The samples were prepared in micellar solution or as emulsions to avoid the use of organic solvents and to allow their dissolution into aqueous media.

Polarography was used to determine coumarin in sweet clover and in preparation FIBS[99,100]. Coumarin, from samples of white and yellow sweet clover, was extracted into boiling ethanol. After distilling off the ethanol a 1 ml reconstituted sample, in ethanol, was added to 4 ml of ethanol and 1 ml of a 1% (w/v) solution of tetra-ethylammonium iodide. Polarography was carried out with a dropping mercury electrode. Coumarin is used as a stabiliser in the injection solution of FIBS and polarography is used as a selective detector to separate it from cinnamic acid, also present as a stabiliser. Samples of coumarin were analysed by polarography with cathodic polarisation in the interval 1.20–2.00 V ($E_{1/2} = 1.67$ V).

12.6.2 Other Chromatographic Techniques

Planar chromatography[101], centrifugal partition chromatography[102], and high-speed countercurrent chromatography[103] have all been used for the determination of coumarins. Betti *et al.*[101] report on a planar chromatographic method utilising a series of different stationary phases and mobile phases in order to discover the best separation technique for various compounds of interest, including umbelliferone, quercetin and psoralene. The stationary phases include C_{18}, cyano-, silica, phenyl, amino, diol, and C_2. Mobile phases include mixtures of methanol, acetic acid, acetone, toluene, ethyl acetate or formic acid.

Centrifugal liquid–liquid partition chromatography[102] relies on the separation of compounds between two immiscible phases and the application of a centrifugal force during that separation. Samples of umbelliferone, herniarin and scopoletin were separated in a light petroleum–ethyl acetate–methanol–water system. Fractions are collected and analysed by UV detection. A chloroform–methanol–water mobile phase was used for the separation of coumarins by high-speed analytical countercurrent chromatography[100] with photodiode array detection.

12.7 CONCLUSIONS

The analytical approaches used to detect coumarin and its derivatives are many and varied. The choice of method is dependent on the particular coumarin of interest, and the matrix in which it is a component. The ideal method is obviously one which will involve minimal sample preparation, short analysis time, be relatively inexpensive, and have excellent accuracy and precision for a variety of applications. Separation and detection of the coumarin in its particular matrix is dependent on many factors. These methods are dependent on whether it fluoresces or not, whether it has a strong UV absorbance or not, and whether there are antibodies available against that particular coumarin. Migration on CE, retention on a HPLC column, or its volatility for GC analysis will vary for each coumarin and determine its resolution from other species present. The selectivity of antibody-based systems may be needed to isolate the analyte of interest from complex endogenous species present that may interfere with analysis. However, it is now clear that there are a variety of methods now available with good analytical characteristics which may be applied to complex matrices and these may be modified to suit the requirements of a particular coumarin or coumarins or specific applications.

ACKNOWLEDGEMENTS

We would like to thank Forbairt (Dublin, Ireland), Schaper & Brümmer (Salzgitter, Germany), the Coumarin Research Fund, and the International Fund for Ireland (Best Centre, Dublin City University) for their financial support. We would also like to thank Andreas Engelhardt, Dr Teresa McCormack, Dr H. Stolze and Dr Brian Deasy for their help and encouragement.

REFERENCES

1 Egan D, O'Kennedy R, Moran E, Cox D, Prosser E, Thornes RD. The pharmacology, metabolism, analysis, and applications of coumarin and coumarin-related compounds. *Drug Metabolism Reviews*, 1990; **22**(5): 503–29.

2 Casley-Smith JR: Casley-Smith JR. *High-protein Oedemas and the Benzo-pyrones*. Sydney: JB Lippincott 1986.

3 Murray RD, Mendez J, Brown SA. *Coumarins. The Natural Coumarins—Occurrence, Chemistry and Biochemistry*. Chichester: John Wiley 1982.

4 Gardner KJ. Applications of chromatography in chocolate analysis. *Rev Intern Chocolat.* 1954; **9**: 33–5.

5 Ferencz V, Veres L. Fluorimetric determination of coumarin with a photoelectric colourimeter and application of this method in establishing the coumarin content of sweet clover. *Agrokémia és Talajtn*, 1953; **2**: 65–72.

6 ted Heide R and Lemmens JF. Paper chromatography in the qualitative and quantitative analysis of organic flavouring compounds. *Perfumery Essent. Oil Record* 1954; **45**: 21–3.

7 Stoll S, Bouteville Y. Characterisation of some natural constituents of vanilla. Their differentiation from ethylvanillin by chromatography. *Ann Fals et Fraudes* 1954; **47**: 183–6.

8 Nivoli M. The detection of coumarin in the presence of vanillin. *Ann Chim (Rome)* 1950; **41**: 642–4.

9 Ensminger LG. Report on vanilla extracts and imitations Colourimetric determination of vanillin and coumarin. *J Assoc Offic Agr Chemists* 1952; **35**: 264–71.

10 Patzak R, Neugebauer L. Polarographic investigation of coumarins. *Monatsh* 1952 **83**: 776–85.

11 Foffani A. Polarographic kinetic studies on coumarins. *Atti Accad Nazl Lincei, Rend* 1953; **14**: 418–23.

12 Riedl K, Neugebauer L. The paper chromatography separation of coumarins. *Monatsh* 1952; **83**: 1083–7.

13 Svedsen AB. The detection of natural coumarin in plants by paper chromatography. *Pharm Acta Helv* 1952; **27**: 44–8.

14 Swain T. Identification of coumarins and related compounds by filter paper chromatography. *Biochem J* 1953; **53**: 200–8.

15 Berlingozzi S, Fabbrini L. Paper chromatography separation of coumarin and its derivatives. *Sperimentale, Sez Chim Biol* 1954; **5**: 1–5.

16 Mitchell LC. Separation and identification of coumarin and four other vanilla-like flavouring substances by paper chromatography. *J Assoc Offic Agr Chemists* 1953; **36**: 1123–7

17 Feigl F, Feigl HE, Goldstein D. A sensitive and specific test for coumarin through photocatalysis. *J Am Chem Soc* **77**: 4162–3.

18 Glenn RA, Peake JT. Titration of phenolic esters in ethylenediamine. *Anal Chem* **27**: 205–9.

19 Wild P. Colourimetric determination of coumarin in nickel-plating baths. *Korose Ochrana Mater* 1964 **8**(4), 86.

20 Anger V, Richter HE. A new colourimetric determination of coumarin. *Mikrochim Ichnonal Acta* 1965; **5–6**: 1025–7.

21 Bucci F, Boniforti L, Cesari A. Application of gas chromatography to the analytical control of flavouring substances in foods and beverages. *Boll Lab Chim Provinciali* 1965 **16**(2): 195–201.

22 Johansen NG. Identification of vanillin in U.S.P. vanillin and the detection of various impurities by gas liquid chromatography. *J Gas Chromatogr* 1965 **3**(6): 202–3.

23 Indahl SR, Scheline RR. The metabolism of umbelliferone and herniarin in rats and by the rat intestinal microflora. *Xenobiotica* 1971; **1**(1): 13–24.

24 Sherma J, Schafer SL, Morris K. Determination of coumarin in vanilla flavourings by quantitative high performance thin layer chromatography. *J Liq Chromatogr* 1987; **10**(16): 3585–93.

25 Belay MT, Poole CF. Determination of vanillin and related flavour compounds in natural vanilla extracts and vanilla-flavoured foods by thin layer chromatography and automated multiple development. *Chromatographia* 1993; **37**(7/8): 365–73.

26 Poole SK, Daly SL, Poole CF. A thin layer chromatographic method for determining the authenticity of natural vanilla extracts. *J Planar Chromatogr* 1993; **6**(2): 129–37.

27 Cholerton SC, Idle ME, Vas A, Gonzalez FJ, Idle JR. Comparison of a novel thin-layer chromatographic-fluorescence detection method with a spectrofluorimetric method for the determination of 7-hydroxycoumarin in human urine. *J Chromatogr* 1992; **575**: 325–30.

28 Willard HH, Merritt LL, Dean JA, Settle FA, Jr. *Instrumental Methods of Analysis*. Belmont, CA: Wadsworth Publishing, 1988.

29 Meyer VR. *Practical High Performance Liquid Chromatography*. New York: John Wiley & Sons, 1988.

30 Knox JH. High Performance Liquid Chromatography. Edinburgh: Edinburgh University Press 1980.

31 Cohen AJ. Critical review of the toxicology of coumarin with special reference to interspecies differences in metabolism and hepatotoxic response and their significance to man. *Fd Cosmet Toxicol* 1979; **17**: 277–89.

32 Moran E, O'Kennedy R, Thornes RD. Analysis of coumarin and its urinary metabolites by high-performance liquid chromatography. *J Chromatogr* 1987; **416**: 165–9.

33 Egan DA, O'Kennedy R. Rapid and sensitive determination of coumarin and 7-hydroxycoumarin and its glucuronide conjugate in urine and plasma by high-performance liquid chromatography. *J Chromatogr* 1992; **582**: 137–43.

34 Sharifi S, Michaelis HC, Lotterer E, Bircher J. Determination of coumarin, 7-hydroxycoumarin, 7-hydroxycoumarin-glucuronide, and 3-hydroxycoumarin by high-performance liquid chromatography. *J Liq Chromatogr* 1993; **16**(6): 1263–78.

35 Lamiable D, Vistelle R, Trenque T, Fay R, Millart H, Choisy H. Sensitive high-performance liquid chromatography method for the determination of coumarin in plasma. *J Chromatogr Biomed Appl* 1993; **620**: 273–7.

36 Fentem JH, Garle MJ, Fry JR. Effect of various inducers of cytochrome P450 on the metabolism of coumarin by rat liver microsomes: A previously-unreported metabolite is the major component identified by HPLC. *Xenobiotica* 1992; **22**: 357–67.

37 Fentem JH, Fry JR, Whiting DA. *o*-Hydroxyphenylacetaldehyde: A major novel metabolite of coumarin formed by rat, gerbil and human liver microsomes. *Biochem Biophys Res Commun* 1991; **179**(1), 197–203.

38 Lake BG, Gaudin H, Price RJ, Walters DG. Metabolism of [3-^{14}C]coumarin to polar and covalently bound products by hepatic microsomes from the rat, syrian hamster, gerbil and humans. *Fd Chem Toxic* 1992; **30**(2), 105–15.

39 van Iersel MLPS, Henderson CJ, Walters DG, Price RJ, Wolf CR, Lake BJ. Metabolism of [3-^{14}C]coumarin by human liver microsomes. *Xenobiotica* **24**(8), 795–803.

40 Steensma A, Beamand JA, Walters DG, Price RJ and Lake BG. Metabolism of coumarin and 7-ethoxycoumarin by rat, mouse, guinea pig, Cynomolgus monkey and human precision-cut liver slices. *Xenobiotica* 1994; **24**(9): 893–907.

41 Evans RR, Relling MV. Automated high-performance liquid chromatographic assay for the determination of 7-ethoxycoumarin and umbelliferone. *J Chromatogr* 1992; **578**: 141–5.

42 Killard AJ, O'Kennedy R, Bogan DP. The analysis of the glucuronidation of 7-hydroxycoumarin by high-performance liquid chromatography. *J Pharm Biomed Anal* 1996; **14**: 1585–90.

43 Duffy CF, Bogan DP, O'Kennedy R. The determination by HPLC of the UDP-glucuronyl transferase activity for 7-hydroxycoumarin in rabbit tissue samples. *J Liq Chromatogr*. (In press)

44 Hiller DL, Cole RO. Short-column gradient elution HPLC for the rapid determination of 7-ethoxycoumarin metabolites in liver-slice incubates. *Anal Biochem* 1995; **227**: 251–4.

45 Walsh JS, Patanella JE, Halm KA, Facchine KL. An improved HPLC assay for the assessment of liver slice metabolic viability using 7-ethoxycoumarin. *Drug Metab and Dispos* 1995; **23**(8): 869–74.

46 Thompson RD, Hoffmann TJ. Determination of coumarin as an adulterant in vanilla flavouring products by high-performance liquid chromatography. *J Chromatogr* 1988; **438**: 369–82.

47 Gamache P, Ryan E, Acworth IN. Analysis of phenolic and flavanoid compounds in juice beverages using high-performance liquid chromatography with coulometric array detection. *J Chromatogr* 1993; **635**: 143–50.

48 Mazza G. Determination of coumarin in flavanoid alcoholic beverages by high-pressure liquid chromatography. *Riv Vitic Enol* 1984 **37**(6): 316–23.

49 Lander V, Wörner M, Kirchenmayer C, Wintoch H, Shreier P. Use of solid phase extraction for rapid sample preparation in the determination of food constituents. Asarone, quinine, coumarin and quassine in spirits. *Z Lebensm Unters Forsch* 1990 **190**: 410–13.

50 Nykolov N, Iossifova T, Vassileva E, Kostova I, Stoev G. Reverse-phase high-pressure liquid chromatographic analysis of hydroxycoumarins in plant extracts. Quantitative determination of hydroxycoumarins in *Fraxinus ornus Phytochem Anal* 1993 **4**: 86–8.

51 Archer AW. Determination of cinnamaldehyde, coumarin, and cinnamyl alcohol in cinnamon and cassia by high-performance liquid chromatography. *J Chromatogr* 1988; **447**: 272–6.

52 Shkarenda VV, Kuznetsov PV. Current state of the liquid chromatography of coumarins. II. High-performance liquid chromatography of coumarin derivatives. *Chem Nat Comp* 1993; **29**(2): 137–50.

53 Bogan DP, O'Kennedy R. The simultaneous determination of coumarin, 7-hydroxycoumarin and 7-hydroxycoumarin-glucuronide in human plasma and serum by high-performance liquid chromatography. *J Chromatogr.* (In press)

54 Hawthorne SB, Krieger MS, Miller DJ. Analysis of flavour and fragrance compounds using super-critical fluid extraction coupled with gas chromatography. *Anal Chem* 1988 **60**: 427–77.

55 Grundschober F. The identification of individual components in flavourings and flavoured foods. *Z Lebens Unters Forsch* **192**: 530–4.

56 Jirovetz L, Buchbauer G, Jäger W, Woidich A, Nikiforov A. Analysis of fragrance compounds in blood samples of mice by gas chromatography, mass spectrometry, GC/FTIR and GC/AES after inhalation of sandalwood oil. *Biomed Chromatogr* 1992; **6**: 133–4.

57 Christakopoulos A., Feldhusen K, Norin H, Palmquist A Wahlberg I. Determination of natural levels of coumarin in different types of tobacco using a mass fragmentographic method. *J Agric Food Chem* 1992; **40**: 1358–61.

58 Weinberg DS, Manier ML, Richardson MD, Hailbach FG. Identification and quantification of coumarin, phthalide and sesquiterone compliance markers in an umbelliferous vegetable beverage. *J Agric Food Chem* 1993 **41**: 48–51.

59 Landers JP (ed). *Handbook of Capillary Electrophoresis*. Boca Raton, FL: CRC Press 1993.

60 Li SFY. *Capillary Electrophoresis*. Amsterdam: Elsevier 1992.

61 Weinberger R. *Practical Capillary Electrophoresis*. San Diego, CA: Academic Press 1993.

62 ogan DP, Deasy B, O'Kennedy R, Smyth MR, Fuhr U. Determination of free and total 7-hydroxycoumarin in urine and serum by capillary electrophoresis. *J Chromatogr B* 1995; **663**: 371–8.

63 Bogan DP, Thornes RD, Tegtmeier M, Schafer EA, O'Kennedy R. The direct determination of 7-hydroxycoumarin and 7-hydroxycoumarin-glucuronide, in urine, by capillary electrophoresis. *Analyst* 1996 **121**: 243–7.

64 Bogan DP, Deasy B, O'Kennedy R, Smyth MR. The use of capillary electrophoresis for studying interspecies differences in coumarin metabolism in liver microsomes. *Xenobiotica* 1996 **26**: 437–48.

65 Deasy B, Bogan DP, Smyth MR, O'Kennedy R, Fuhr U. Study of coumarin metabolism by human liver microsomes using capillary electrophoresis. *Anal Chim Acta* 1995 **310**: 101–7.

66 Bogan DP, Killard AJ, O'Kennedy R. The use of capillary electrophoresis for studying the *in vitro* glucuronidation of 7-hydroxycoumarin. *J Capillary Electrophoresis* 1995; **2**(5): 241–5.

67 Schützner W, Kenndler E. Electrophoresis in synthetic organic polymer capillaries: variation of electroosmotic velocity and ζ potential with pH and solvent composition. *Anal Chem* 1992; **64**: 1991–5.

68 Morin P, Dreux M. Factors influencing the separation of ionic and non-ionic chemical natural compounds in plant extracts by capillary electrophoresis. *J Liq Chromatogr* 1993; **16**(17): 3735–55.

69 Gariel P, Gramond JP, Guyon F. Separation and determination of warfarin enantiomers in human plasma samples by capillary zone electrophoresis using a methylated β-cyclodextrin-containing electrolyte. *J Chromatogr B* 1993; **615**: 317–25.

70 Banwell CN. *Fundamentals of Molecular Spectroscopy.* London: McGraw-Hill 1983.

71 Tan HSI, Ritschel WA, Sanders PR. Determinations of coumarin and umbelliferone mixtures in whole blood by spectrofluorimetry. *J Pharm Sci* 1976; **65**: 30–3.

72 Ritschel WA, Brady ME, Tan HSI, Hoffmann KA, Yiu IM, Grummich KW. Pharmacokinetics of coumarin and its 7-hydroxy-metabolites upon intravenous and peroral administration of coumarin in man. *Europ J Clin Pharmacol* 1977; **12**: 457–61.

73 Ritschel WA, Grummich KW, Kaul S, Hardt TJ. Biopharmaceutical parameters of coumarin and 7-hydroxycoumarin. *Pharm Ind* 1981 **43**(3): 271–6.

74 Kaipainen P, Koivusaari U, Lang M. Catalytic and immunological comparison of coumarin 7-hydroxylation in different species. *Comp Biochem Physiol* 1985; **81C**(2): 293–6

75 Conway JG, Kauffman FC, Tsukada T, Thurman RG. Glucuronidation of 7-hydroxycoumarin in periportal and pericentral regions of the liver lobule. *Mol Pharmacol* 1984; **25**: 487–93.

76 Rautio A, Kraul H, Kojo A, Salmela E, Pelkonen O. Interindividual variability of coumarin 7-hydroxylation in healthy volunteers. *Pharmacogenetics* 1992; **2**: 227–33.

77 Merkel U, Sigusch H, Hoffmann A. Excretion of 7-hydroxycoumarin conjugates in human urine. *Pharmazie* 1993; **48**: 626–7.

78 Egan DA, O'Kennedy R. Spectrofluorimetric method for the quantification of 7-hydroxycoumarin in urine and plasma using both extracted and unextracted samples. *Analyst* 1993; **118**: 201–3.

79 Merkel U, Sigusch H, Hoffmann A. Grapefruit juice inhibits 7-hydroxylation of coumarin in healthy volunteers. *Eur J Clin Pharmacol* 1994; **46**: 175–7.

80 Iscan M, Rostami H, Iscan M, Güray T, Pelkonen O, Rautio A. Interindividual variability of coumarin 7-hydroxylation in a Turkish population. *Eur J Clin Pharmacol* 1994; **47**: 315–18.

81 Shanahan GJ, Boley NP, Traynor AD. Total luminescence spectroscopy in the identification and detection of coumarin (1,2–benzopyrone) in road fuels. *Anal Proc* 1991; **28**: 210–13.

82 Tijssen P (ed.). *Laboratory Techniques in Biochemistry and Molecular Biology*, Vol. XV. New York: Elsevier, 1985.

83 Killard AJ, Deasy B, O'Kennedy R, Smyth MR. Antibodies: production, functions and applications in biosensors. *Trends in Anal Chem* 1995; **14**(6): 257–66.

84 Catty D, Raykundalia C. *Antibodies: A Practical Approach.* Oxford: IRL Press, 97–155 1989.

85 Kemeny DM. *A Practical Guide to ELISA.* Oxford: Pergamon Press, 1991.

86 Egan DA. Analytical, immunological and toxicological studies of coumarin and coumarin-related compounds. PhD Thesis, Dublin City University, Dublin, Ireland, 1993.

87 Egan DA, O'Kennedy R. The production and characterisation of anti-7-hydroxycoumarin antibodies and their use in the development of an enzyme-linked immunosorbent assay. *J Ir Coll Phys Surg* 1993; **22**(2): 72.

88 Reinartz H, O'Kennedy R. Unpublished data.

89 Nolan O, O'Kennedy R. Bifunctional antibodies and their potential clinical applications. *Int J Clin Res* 1992; **22**: 21–7.

90 Cook AG, Wood PJ. Chemical synthesis of specific monoclonal antibodies: potential advantages in immunoassay systems. *J Immunol Methods* 1994; **171**: 227–37.

91 Reinartz HW, Quinn JG, Zänker K, O'Kennedy R. A bispecific multivalent antibody studied by real-time interaction analysis for the development of an antigen-inhibition ELISA. *Analyst* 1996; **121**: 757–71.

92 Dempsey E, O'Sullivan C, Smyth MR, Egan D, O'Kennedy R, Wang J. Development of an electrochemical biosensor for 7-hydroxycoumarin. *Analyst* 1993; **118**: 411–14.

93 Deasy B, Dempsey E, Smyth MR, Egan D, Bogan DP, O'Kennedy R. Development of an antibody-based biosensor for determination of 7-hydroxycoumarin (umbelliferone). using horseradish peroxidase-labelled anti-7-hydroxycoumarin antibody. *Anal Chim Acta* 1994; **294**: 291–7.

94 Lu B, Smyth MR, Quinn J, Bogan D, O'Kennedy R. A regenerable enzyme immunosensor for 7-hydroxycoumarin based on electrochemical detection. *Electroanalysis* 1996; **8**: 619–22.

95 Dempsey E, O'Sullivan C, Smyth MR, Egan D, O'Kennedy R, Wang J. Differential pulse voltammetric determination of 7-hydroxycoumarin in human urine. *J Pharm Biomed Anal* 1993; **11**(6): 443–6.
96 Keating G, O'Kennedy R. Unpublished data.
97 Fägerstam LG, O'Shannessy DJ. Surface plasmon resonance detection in affinity technologies BIAcore. In: T. Kline ed. *Handbook of Affinity Chromatography.* New York: Marcel Dekker, 1993.
98 Carrazon JMP, Vergara AG, Garcia AJR, Diez LMP. Determination of coumarins by voltammetric techniques in micellar and emulsified media. *Anal Chim Acta* 1989; **216**: 231–42.
99 Orlov YE. Determination of coumarin in sweet clover herbage by a polarographic method. *Chem Nat Comp* 1988; **24**(1): 118–19.
100 Orlov YE, Sirenko LY, Lichino OP. Polarographic determination of coumarin and cinnamic acid in the preparation FIBS. *Chem Nat Comp* 1992; **27**(5): 633–4.
101 Betti A, Lodi G, Fuzzati N. Analysis of flavanoid compounds by planar chromatography: A chemometric approach. *J Planar Chromatogr* 1993; **6**(3): 232–7.
102 Marston A, Borel C, Hostettmann K. Separation of natural products by centrifugal partition chromatography. *J Chromatogr* 1988; **450**: 91–9.
103 Schaufelberger DE. Analytical high-speed countercurrent chromatography with photodiode array detection (HSCCC-UV). *J Liq Chromatogr* 1989; **12**(12): 2263–80.

CHAPTER 13

Coumarins—Multifaceted Molecules with Many Analytical and Other Applications

Deirdre Cooke, Brian Fitzpatrick, Richard O'Kennedy, Teresa McCormack
School of Biological Sciences, Dublin City University, Dublin, Ireland

and

Denise Egan
School of Science, Regional Technical College, Tallaght, Co. Dublin, Ireland

13.1 INTRODUCTION

The purpose of this chapter is to describe the multiple uses of coumarin and coumarin derivatives not described in other sections of this book. There are very many different derivatives with a myriad of applications and this review is by no means fully comprehensive. Many aspects of coumarins and their applications have been described in the comprehensive treatise by Murray and co-authors[1]. Other reviews worth consulting are those by Feuer[2], Cohen[3], Murray[4] and Egan *et al.*[5].

What is obvious from reading this review is the importance of coumarins as labels, dyes or derivatising agents in almost every area of analytical science.

Coumarins: Biology, Applications and Mode of Action. Edited by R. O'Kennedy and R.D. Thornes.

13.2 WARFARIN-TYPE COUMARINS—MODE OF ACTION, PRESENT AND POTENTIAL APPLICATIONS

The oral anticoagulants emerged from veterinary research in the 1920s following the identification by Schofield in 1924 that a haemorrhagic disorder affecting cattle was caused by consumed spoiled sweet clover hay (*Melilotus* spp.). The toxic principle which resulted in reduced plasma prothrombin interested Dr Karl Link and his University of Wisconsin team. Subsequent research led to the identification of dicoumarol (3-3′-methylenebis(4-hydroxycoumarin)) as the offending agent. Link's group examined over 100 4-hydroxycoumarins and found that the minimal structural requirements for anticoagulant activity were an intact 4-hydroxycoumarin residue with the 3-position substituted by a carbon chain or unsubstituted. Ikawa and coworkers[6] synthesised a very potent agent called warfarin (3-(α-acetonylbenzyl)-4-hydroxycoumarin) (Figure 13.1) in 1951, with the clinical trials taking place in 1954, following which it was recommended for human use. Since the 1930s, increasing research has been directed at the molecular basis of action of this 4-hydroxycoumarin, particularly with respect to its anticoagulant function and the involvement of vitamin K in the role of anticoagulation[7].

13.2.1 Pharmacokinetics and Metabolism of Warfarin

The extent of plasma protein binding has significant effects on the pharmacological and pharmacokinetic properties of warfarin[8,9]. The rate of absorption of warfarin is dose-dependent, and once in circulation, the drug is bound almost 99% to albumin. Coumarin compounds, such as coumarin and 4-hydroxycoumarin which are almost lacking in anticoagulant activity, show only one albumin binding site with the coumarin anticoagulants showing two[10]. An inverse relationship also seems to exist between the degree of albumin binding and anticoagulant sensitivity, as the species most sensitive (humans and rat) exhibit strongest albumin binding.

The *in vivo* metabolism of warfarin has also been extensively studied[11–14] with the main metabolic routes being:

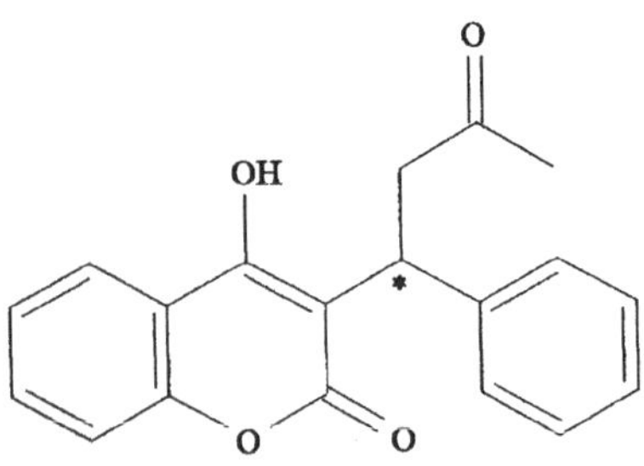

Figure 13.1. Warfarin (3-(α-acetonylbenzyl)-4-hydroxycoumarin) * Chiral centre.

(a) keto reduction to warfarin alcohols, each enantiomer yielding two diastereomeric products;
(b) dehydration of warfarin alcohol resulting in ring closure, yielding a benzopyran derivative;
(c) dehydration of 9-hydroxywarfarin to yield 9,10-dehydrowarfarin;
(d) hydroxylation reactions yielding the regioisomeric products 4, 6, 7, 8 and 9-hydroxy(benzylic)warfarin; and
(e) various conjugation reactions[15].

Substantial metabolic differences do occur between animal species, strains and organs. In humans, the main metabolites are 6- and 7-hydroxywarfarin, (*S*) and (*R*) warfarin alcohols, the (*R*) form is essentially metabolised to the 6-hydroxy metabolite, whilst the (*S*) form is metabolised to the 6- and 7-hydroxy derivatives. In rats and guinea-pigs, the major metabolites are formed by hydroxylation at positions 6, 7, 8, and at the *p*-position of the phenyl substituent, condensation to produce 3,4-dihydro-2-methyl-4-phenyl-2*H*,5*H*-pyrano[3,2-c]benzopyran-5-one and glucuronic acid conjugation. Clinically, warfarin is administered as a racemate, with the hypoprothrombinemic activity of the *S*-isomer being 2–5 times more potent than that of the *R*-isomer, and the *S*-isomer binding more strongly to human serum albumin (HSA) than the *R*-isomer[8]. Several *in vitro* studies have been carried out on warfarin compounds using the cytosolic and microsomal fractions of several species[16–20].

13.2.2 Activation of Clotting Factors

One of the most important biological functions of vitamin K is the post-translational modification of precursor proteins to form clotting factors[21], which involves the γ-carboxylation of glutamic acid residues (GLU) to γ-carboxyglutamic acid (GLA) in the presence of the vitamin and molecular oxygen (Figure 13.2). In conjunction with this carboxylation reaction, an epoxidation reaction occurs, converting the active form of vitamin K to vitamin K 2,3-epoxide, which is in turn converted back to the quinone form by vitamin K_1 epoxide reductase to complete the vitamin K cycle. It is evident, therefore, that interference with the normal functioning of the vitamin K cycle will impair the production of active clotting factors.

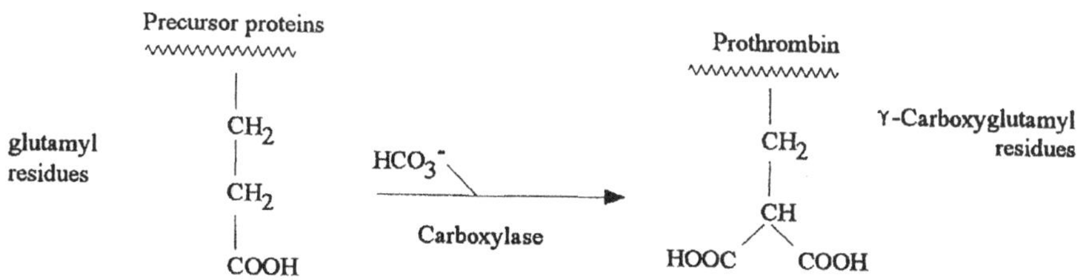

Figure 13.2. Vitamin K and molecular oxygen are required as cofactors in the reaction that converts glutamic acid residues in microsomal precursor proteins of prothrombin to γ-carboxyglutamyl residues in prothrombin, to produce active clotting factors.

13.2.3 Mechanism of Anticoagulation Action

The anticoagulant activity of the coumarins is thought to be governed by a combination of several factors including molecular shape and increased activity with 6-membered heterocyclic rings. Warfarin inhibits the activity of individual vitamin K dependent clotting factors: II, VII, IX and X. In the absence of vitamin K or during anticoagulant administration, abnormal prothrombin proteins have been found to accumulate[22]. These proteins, named PIVKAs (Proteins Induced Vitamin K Antagonism), are immunologically identical to the normal prothrombin[23] but lack its activity due to the fact that the final γ-carboxylation of glutamic acid (GLU) to yield γ-carboxyglutamic acid (GLA) has not occurred. The PIVKAs have fewer GLA residues than are normally present on the clotting factors. The GLA residues on clotting factors enhance the binding of Ca^{2+} by chelating phosphate on the phospholipid, accelerating and providing a template for the clotting mechanism.

The PIVKAs are, therefore, the precursor proteins that are carboxylated to give clotting factors. Ren *et al.*[24] have shown that there is a good correlation between prothrombin synthesis and conversion of the vitamin K 2,3-epoxide to vitamin K, supporting the hypothesis that warfarin acts by inhibiting this regeneration (Figure 13.3).

13.2.4 Clinical Uses of Warfarin

Warfarin is one of the most widely used anticoagulants in the treatment of thromboembolic disorders and has found use in cases of venous thromboembolism, threatened stroke and acute myocardial infarction. During treatment the plasma prothrombin activity should be determined frequently so that the daily dosage may be adjusted to the desired therapeutic level[25]. Drugs such as phenylbutazone may increase the anticoagulant effect by plasma protein displacement[19], whilst barbiturates may reduce its effects by promoting its metabolism. Undesirable effects

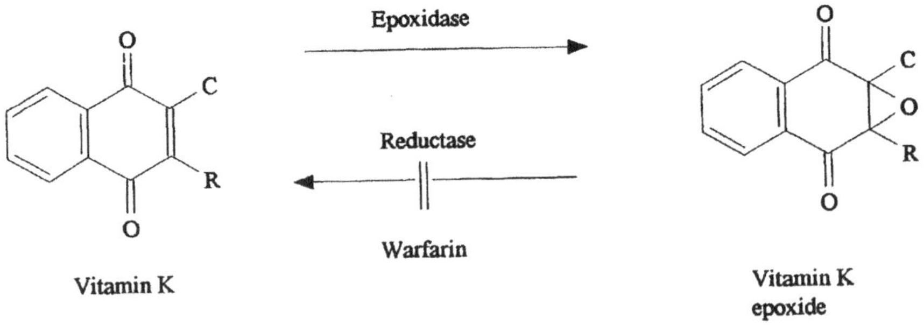

Figure 13.3. Liver microsomes contain an enzyme system that forms the vitamin 2,3-epoxide and a second system that reduces the epoxide to the vitamin. The activity of the vitamin K_1 epoxide reductase is strongly inhibited by warfarin and other coumarin anticoagulants.

of warfarin include rashes, pyrexia and it may also cause skin necrosis. Early clinical signs of overdosage are mild bleeding from the gums or elsewhere and the presence of erthyrocytes in the urinary deposit. Phytomenadione (vitamin K_1) is the most effective antidote to overdosage[25] with the administration of water-soluble vitamin K analogue being less effective.

Warfarin has also found use in the clinical setting in the treatment of a variety of cancer cell types[26–28]. Randomised clinical trials[29] have shown that warfarin improves tumour response rates and survival in patients with small cell carcinoma of the lung (SCCL)[30], a tumour cell type that is characterised by the existence of tumour cell-associated coagulation pathway[31,32]. Results obtained to date suggest that warfarin may be more useful as an aid to the management and control of tumour growth rather than its eradication.

Various types of warfarin resistance have been reported in humans and rats including[33–36]:

(a) dietary resistance caused by an excess dietary intake of vitamin K[37,38];
(b) acquired resistance shown by an initial response followed by a subsequent decrease in the response[39]; and
(c) hereditary resistance where no initial response is observed. This condition is transmitted as an autosomal dominant trait[40].

An understanding of this resistance is therefore essential for effective anticoagulant therapy.

Various studies have been carried out on the antiviral effect of warfarin and other 4-hydroxycoumarins with respect to HIV-1 replication and spread[41–45]. The most potent inhibitors examined to date are peptidomimetic compounds whose low oral bioavailability and rapid excretion has hampered their clinical potential. Warfarin, with its widely studied pharmacology and excellent oral availability, makes it an extremely promising candidate for study. Recent studies[41] illustrate that warfarin and a series of structurally related analogues exhibit a dose-dependent inhibitory effect on HIV-1 replication and cell-mediated or cell-free viral spread in MOLT-4 lymphocytes with minimal cytotoxicity. A similar study[42] examined the inhibition of the HIV-1 protease by warfarin and its analogues *in vitro*, results showing that warfarin is a competitive inhibitor at micromolar concentrations. The HIV-specificity of coumarin compounds appears equivalent or better than that of AZT, a reverse transcriptase (RT) selective nucleoside inhibitor. Another drawback with AZT is that it appears not to inhibit cell-to-cell transmission of HIV as effectively as the coumarins.

13.3 USE OF COUMARINS AS LABELLING AGENTS

The coumarin family of compounds have found widespread use in many analytical applications, in many cases as a result of their natural fluorescence. The fluorescent labelling of various molecules has been accomplished with coumarin derivatives, with 7-Amino-4-methyl-coumarin acetic acid (AMCA) a popular choice for protein labelling. AMCA is available as the *N*-hydroxysuccinamide ester which can be linked to lysine residues under relatively mild conditions. It has an excitation maximum at 350 nm with emission at 450 nm. This Stokes shift of 100 nm is advantageous compared to other labels, e.g. fluorescein isothiocyanate (FITC) which has a shift of only 30 nm. In addition, AMCA is not as susceptible to photobleaching, may have certain advantages for photographic work, and does not alter the isoelectric point of the protein being labelled. AMCA has been used to label antibodies for histochemical work[46]. Its use may also facilitate multiple colour fluorescence by flow cytometry[47]. This can be applied to analysis of cell surface antigens.

Commercially AMCA-based probes are available that may be conjugated to proteins and oligonucleotides[48]. A sulphosuccinyl ester of AMCA which is more water soluble is available and may be applied for labelling to surface proteins. AMCA derivatives with variable length spacer arms have also been produced.

Coumarin derivatives have also been used to facilitate structural analysis. For example, Forster and Hass[49] prepared three fluorescent coumarin derivatives. These were 7-carboxymethyl-amino-4-trifluoromethyl-2*H*-1-benzopyran-2-one, 7-carboxymethyl-amino-4-methyl-2*H*-1-benzopyran-2-one and 7-dimethylamino-4-formyl-2*H*-1-benzopyran-2-one. These reagents can be attached to the primary amino group of proteins and are applicable for simultaneous detection of labelled components in mixtures and for conformational studies. Coumarin-6-sulphonyl chloride may be a potential label for derivatisation of amino and phenolic groups for use in luminescence-based studies[50].

Coumarin derivatives have also been used as agents to facilitate labelling by other approaches such as iodination, e.g. coumarin-3-acetic acid-*N*-OH-succinimidyl ester can bind directly to ovalbumin via free amino groups which can in turn be radiolabelled by the chloramine-T method[51].

Other applications of coumarin-based labels for use in DNA analysis are described in Section 13.6.2.

13.4 USE OF COUMARINS IN ENZYME ASSAYS

Some of the most sensitive methods for assaying enzymes are based on the liberation of a fluorescent product from a fluorogenic reactant. The intrinsic fluorescence properties of coumarins led to the initial development of coumarin-derived enzyme substrates, in the guise of 4-methylumbelliferone. Many enzyme substrates incorporating the coumarin moiety have since been developed, and are now widely applied to research and diagnostic enzymology.

13.4.1 4-Methylumbelliferone Derivatives

In the 1950s, Mead and co-workers[52], while carrying out metabolic studies of coumarin in rabbits, observed that although umbelliferone (7-hydroxycoumarin) fluoresced strongly in ultraviolet light at pH 9–10, its conjugates showed little or no fluorescence. They also found the more easily isolated 4-methylumbelliferone (4-MU) to be equally fluorescent, with its glucuronide also practically non-fluorescent when pure. Based on their observations, they introduced a method for the fluorimetric assay of β-glucuronidase activity in various tissues. The method consisted of the enzymatic hydrolysis of 4-methylumbelliferyl β-glucuronide, followed by fluorimetric determination of the liberated methylumbelliferone ($\lambda_{ex} = 364$ nm, $\lambda_{em} = 448$ nm)[52]. Subsequently Robinson synthesised 4-methylumbelliferyl β-glucoside, and used it in the study of β-glucosidases from a variety of sources[53]. Robinson, like Mead *et al.* before him, found the fluorogenic method convenient and extremely sensitive.

Leaback and Walker evaluated the above fluorescence technique in an investigation of the hydrolysis of 4-MU-*N*-acetyl-β-D-glucosaminide by testicular *N*-acetyl-β-glucosaminidase, and considered the MU-glycoside to be advantageous over the corresponding *p*-nitrophenyl glycoside in this enzymatic determination[54]. Their work continued to derivatise 4-MU to its acetate, phosphate, sulphate and β-D-galactoside forms, and they concluded that all but the MU-acetate derivative would find application as enzyme substrates[55]. 4-MU-xyloside was soon added to the list[56,57] and many others have since been developed. An inherent advantage related to the synthesis of these 4-MU derivatives is the fact that the parent compound is inexpensive, available in large quantities and is soluble in a wide range of organic solvents. Many 4-MU compounds are now commercially available (Table 13.1) including 4-MU-acetate which is widely used for the assay of esterases ($\lambda_{ex} = 340$ nm, $\lambda_{em} = 430$ nm). The 4-methylumbelliferone compounds are widely used for the classification of many bacterial strains[58–60]. 4-Methylumbelliferyl-β-D-glucuronide has found an important application in the testing of water for coliform bacteria[61,62]. This fluorogenic β-D-glucuronidase test can also be used in conjunction with the indole capillary test as a reliable identification method for *E. coli* in foods[63]. Clinically, some 4-MU derivatives have potential application in the pre-natal diagnosis of certain diseases, such as Tay–Sachs disease and Krabbes' disease[64–66].

Table 13.1. Some commercially available 4-methylumbelliferone derivatives.

4-Methylumbelliferone derivative	Enzyme assayed	References
4-MU-Acetate (MUA)	Esterases	76
4-MU-*N*-acetyl-β-D-glucosaminide	*N*-acetyl-β-D-glucosaminidase	54
2(4-MU)-α-D-*N*-acetylneuraminic acid	Neuraminidase	67
4-MU-β-D-glucuronide (MUG)	Glucuronidase	52
4-MU-*p*-guanidinobenzoate (MUGB)	Trypsin, thrombin, Factor Xa	68
4-MU-phosphate	Phosphatases	69
4-MU-β-*N*-acetylglucosamine-6-sulphate	β-*N*-Acetylglucosaminidase isoenzyme A	70
4-MU-β-D-*N'N'N'*-Triacetylchitotriose	Lysozyme	71
4-MU-*p*-trimettylammonium cinnamate chloride (MUTMAC)	α-Chymotrypsin	68
4-MU-β-D-xyloside	Xylosidase	56,57

The 4-methylumbelliferone derivatives are widely used in the localisation of enzymes in gels following electrophoresis[72]. They are particularly used for detecting α- and β-glucosidases[73], arylsulphatase[74,75], esterases[76] and phosphatases. They are especially advantageous over corresponding chromogenic substrates, such as *p*-nitrophenol derivatives or indigogenic substrates in the case of phosphodiesterases, by virtue of their sensitivity and speed (results of zymograms are obtained in <1 h)[76,77].

13.4.2 7-Amino-4-Methyl Coumarin Derivatives

In the search for sensitive substrates for proteases many peptidyl substrates have been coupled with chromophores (e.g. peptide–*p*-nitroanilide and 4-phenylazobenzyloxycarbonyl(PZ)–peptide)[78]. Many of these chromogenic substrates, however, have the same absorption spectrum as the enzyme-liberated product, and in order to assay the product, time-consuming extraction procedures are required. Additionally, colorimetry is not a highly sensitive technique—fluorescent substrates offer greater potential for sensitive detection of enzyme activity. The search for more convenient and more sensitive substrates for proteinases/peptidases led to the development of many fluorogenic substrates, including peptide amides of 7-amino-4-methylcoumarin (AMC).

The original peptide–AMC substrates were synthesised for the sensitive assay of α-chymotrypsin[79] and an aminopeptidase[80]. These have since been widely used for the investigation of a variety of peptidases and proteases including elastase, carboxypeptidases, pyroglutamyl peptidase and Xaa-prolyl dipeptidase[81–84]. The basic principle of these highly sensitive assays using peptide–AMC substrates, is outlined for an Xaa-prolyl dipeptidase in Figure 13.4.

Improvement in the specificity of substrates for enzymes can be achieved by various methods, for example by selectively lengthening the peptide sequence—Gutierrez *et al.*[85] reported a 300-fold increase in the specificity of the pentapeptide BOC(*Butoxycarbonyl*)-Nle-Gln-Leu-Gly-Arg–AMC for CVF3b (a complement pathway C3/5 converting enzyme) over Factor Xa, when compared to the previously

Gly-Pro-MCA —***X-Pro-DAP***→ Gly-Pro + AMC (**Fluorescent compound**)

↓

Measure fluorescence

($\lambda_{ex} = 380$ nm, $\lambda_{em} = 460$ nm)

Figure 13.4. Principle of the fluorimetric assays involving 7-amino-4-methylcoumarin–peptide substrates. Gly-Pro–MCA, Gly-L-Pro–4-methylcoumaryl-7-amide; X-Pro-DAP, X-prolyl dipeptidyl-amino-peptidase; AMC, 7-amino-4-methylcoumarin.

reported tripeptide substrate BOC-Leu-Gly-Arg-AMC. Many more specific and sensitive peptide–AMC substrates have been developed for α-thrombin, plasmin, urokinase, activated protein C, Factor Xa, Factor XIIa and kallikreins[86–90]. The design of these specific substrates was accomplished by using information from the COOH-terminal sequences of the 'activation peptides', which are liberated during the conversion of plasma serine protease zymogens to their active forms. Many of these fluorogenic substrates are available commercially, and some are shown in Table 13.2.

Since cellular proteinases participate in an array of processes in biological systems—fertilisation, growth and ageing, cytoskeleton reorganisation, protein synthesis, memory, degradation of endocytosed matter and necrosis—these enzymes are the focus of research in thousands of laboratories worldwide, and AMC substrates enjoy widespread use in protease studies carried out in these programmes. The substrates are essential to the identification and classification of proteolytic enzymes in many organisms, both pro- and eukaryotic[91–95]. They are also important in the localisation of enzymes within cells (lysosomal versus cytosolic)[96,97], and feature in studies of protease secretion from cells and organisms[98,99]. They aid in the analysis of species variants of different proteases[100], and are extremely useful in protease inhibition studies[101–103]. The substrates also aid in assessing the purity of isolated cells/membranes from native tissue, and in the characterisation of cell culture lines[104].

AMC–peptide substrates can also be used to examine protease activity in cells, but modification of the AMC moiety is necessary since AMC is highly membrane permeable. This modification is accomplished by the addition of an impermeant high molecular weight dextran, as in the case of glycine–AMC–3-acetic acid–PEG–dextran[105]. This derivative has been used to measure intracellular aminopeptidase

Table 13.2. Various commercially available AMC substrates.

Enzyme	AMC substrates
Aminopeptidases, e.g. leucine aminopeptidase	H-Leu–AMC
Carboxypeptidases, e.g. carboxypeptidase Y	Suc-Ala-Ala-Pro-Phe–AMC
Chymotrypsin	Suc-Ala-Ala-Phe–AMC
	Suc-Ala-Ala-Pro-Phe–AMC
Elastase	Suc-Ala-Pro-Ala–AMC
	MeOSuc-Ala-Ala-Pro-Val–AMC
Kallikreins	H-Pro-Phe-Arg–AMC.Salt
Plasmin	BOC-Glu-Lys-Lys–AMC.Salt
	BOC-Val-Leu-Lys–AMC.Salt
Thrombin	BOC-Val-Pro-Arg–AMC.HCl
	BOC-Asp(oBzl)-Pro-Arg–AMC.HCl
Trypsin	BOC-Gln-Ala-Arg–AMC.HCl
	Bz-DL-Arg–AMC.HCl

Suc, succinyl; BOC, butoxycarbonyl; Bz, benzoyl.

activity in hepatocytes using digitised video fluorescent microscopy (DVFM), and by synthesising specific peptidyl–AMC–3-acetic acid–PEG–dextran conjugates other *in situ* peptidase activities may be examined.

The AMC substrates have also been used to localise proteases in gels[106–108], but as mentioned later there are inherent problems associated with them in this application.

13.4.3 7-Amino-4-Trifluoromethyl Coumarin Derivatives

As mentioned, the above AMC-coupled peptide substrates have proved invaluable in proteolytic enzyme research. However, one drawback that must not be neglected when using these compounds in assays involving biological fluids or tissue extracts, is that these complex matrices may contain substances which quench the fluorescence of the AMC leaving group. This problem has been somewhat overcome by the design of substrates with a 7-amino-4-trifluoromethylcoumarin (AFC) leaving group[109]. These have the advantage of being both chromogenic (yellow colour with $E_{380} = 12\,600$) and fluorogenic ($\lambda_{ex} = 400$ nm, $\lambda_{em} = 505$ nm). These AFC substrates can thus be used in both fluorimetric and spectrophotometric assays, although the sensitivity is $\sim$50-fold lower in the latter method. This novel capability of direct comparison of spectrophotometric and fluorimetric analyses of enzyme activity allows a flexibility not possible with any other system for identifying and quantifying proteinases, and is useful for inter-laboratory comparisons, as fluorescence units are often not reproducible.

The specificity of the AFC substrates is comparable to that of the AMC derivatives and AFC substrates have been used to examine peptidases in diverse research areas. For example, they have been utilised to identify some of the extracellular proteolytic enzymes associated with the development and healing of acute inflammatory lesions[110], and they are used to detect Gram-negative bacteria, streptococci group A and enterococci[111]. They also feature in studies to follow activation of immobilised plasminogen to plasmin by tissue activator[112], to observe kallikreins circulating in the blood[113,114], and in research into the expression of arginine aminopeptidase in *C. parvum* oocytes[115].

One distinct advantage of the AFC substrates is that peptide–AFCs are blue in fluorescence, while free AFC has a yellow-green fluorescence. In contrast, both free and acylated AMC fluoresce in the blue region with overlapping spectra, and are thus ineffective for the fluorescent detection of enzyme activity in cells or on electrophoresis gels, a task quite amenable to AFC substrates. This has been demonstrated effectively for the determination of erythrocyte acid phosphatase phenotype in forensic blood samples following isoelectric focusing. A trifluoromethylcoumarin substrate was used and produced a non-diffusing, highly fluorescent permanent zymogram[116].

The above advantage of AFC–peptide substrates can also be turned to their favour for assays of proteinase activities in tissue sections[117] or in intact cells as suggested by Smith *et al.*[108].

13.4.4 Quenching Coumarin Substrates

In addition to the direct monitoring of fluorescence in enzyme reactions as a result of formation of 4-methylumbelliferone, 7-amino-4-methylcoumarin or 7-amino-4-trifluoromethylcoumarin, enzyme reactions can also be followed by resonance energy transfer (RET) studies. Originally, enzyme substrates used in quenching studies required two key features: the presence of a fluorescent tryptophan residue, and a 2,4-dinitrophenol (DNP) group at the N-terminus. In the intact substrate the tryptophan fluorescence could be quenched by the adjacent DNP group, but hydrolysis of the substrate relieved this quenching and the resulting fluorescence increase served as the basis of the assay[118]. However, tryptophan has certain problems associated with its use in enzyme assays based on RET, not least the fact that it is not highly fluorescent. These problems have been overcome by replacing tryptophan with derivatives of 7-methoxycoumarin[119–121]. The 7-methoxycoumarin derivatives are highly fluorescent and efficiently quenched by dinitrophenol. For example, tryptophan has been replaced with a fluorescent amino acid derived from 7-methoxycoumarin—DL-2-amino-3-(-7-methoxy-4-coumaryl)propionic acid (AMP)—which has been used in studies of thimet peptidase[119]. This method has also been successfully applied to the sensitive, continuous assay of matrix metalloproteinases using MCA-Pro-Leu-Gly-Leu-Dpa-D-Arg-NH_2 as the fluorogenic substrate[121].

13.4.5 Medical Applications of Coumarin-based Substrates

Cellular proteinases play an important role in many diseases such as cancer, diabetes and multiple sclerosis to name a few, and the enzymatic expression changes can aid in diagnosis and prognosis of certain diseases[122]. Consequently the coumarin-based peptide substrates have received widespread application and are important in medical research (see Table 13.3).

13.5 USE OF COUMARINS IN CHEMICAL ANALYSES

13.5.1 pH Indication

It is a well-established fact that the absorption/emission spectrum of certain fluorescent dyes is sensitive to fluctuations in pH. Usually this sensitivity to $[H^+]$ is as a result of the ionisation of a phenolic moiety on the dye, and the response to protonation is usually manifested as a shift in the absorption spectrum to shorter wavelengths. Coumarin compounds used for pH indication and their corresponding pH ranges[132] are outlined in Table 13.4.

Table 13.3. Coumarin substrates in medical research.

Substrate	Enzyme(s)	Clinical application	References
Z-Arg-Arg–AMC	Cathepsin B	Breast/colorectal cancer	123,124
Glutaryl-Ala-Ala-Phe–AMC	Proline iminopeptidase	Paget's disease, prostatic bone metastases	125
Z-Arg-Arg–AMC	Cathepsin H,B metalloproteinases	Acute myocardial infarction	126
γ-Glutamyl–AMC	γ-Glutamyltransferase	Pancreatic, kidney and hepatobiliary diseases	127
Z-Phe-Arg–AMC	Cysteine proteinases and elastase	Rheumatoid arthritis, osteoarthritis	128
Gly-Pro–AMC	Dipeptidyl peptidase and PPCE	Osteoporosis	129
Glutaryl-Ala-Ala-Phe–AMC	Endopeptidase 24.11	Hypertension	130
H-Gly-Pro–AFC	Dipeptidylpeptidase IV	Autoimmune diseases	131

Table 13.4. Coumarins as pH indicators.

Indicator	From pH	To pH
Esculin	1.5 (colourless)	2.0 (blue fluor.)
β-Methylesculetin	4.0 (colourless)	6.2 (blue fluor.)
β-Methylumbelliferone	7.0 (colourless)	7.5 (blue fluor.)
Umbelliferone	6.5 (colourless)	7.6 (blue fluor.)
Coumarin	9.8 (deep green fluor.)	12.0 (light green fluor.)

13.5.2 Metal Detection

Coumarin compounds are also well established as analytical reagents in the determination of metals. Gravimetric detection of Ce, Th, Ti, U and Zr can be achieved with 3-acetyl-4-hydroxycoumarin, 6,7-dihydroxy-4-phenylcoumarin or 7,8-dihydroxycoumarin, and a variety of coumarins (e.g. 6,7-dihydroxycoumarin, 6,7-dihydroxy-4-methylcoumarin and 3-acetyl-4-hydroxycoumarin) are also available to photometrically detect Fe(III), Ti(IV), U(VI), Mo(VI), V(VI) and Co[133–139].

Some coumarins are also useful as indicators in metal ion titrations. A well-known example is Calcein Blue (4-methylumbelliferone-6-methyleneiminodiacetic acid), which exhibits a bright blue fluorescence up to a pH of 11[140,141]. However, the addition of copper ions to a solution of the free indicator results in the quenching of this fluorescence. This can be used effectively in the chelation-titration of metal ions. This analytical procedure is normally accomplished by back-titration of excess chelating agent using a standard copper solution with addition of a few drops of Calcein Blue. Since the end-point of the titration is the point where the fluorescence of Calcein Blue is quenched by copper ions, this compound can be successfully used as a metallochromic fluorescent indicator for the titrimetric determination of Ba, Ca, Co, Cu, Fe(III), Ni and Sr[140–143]. 7-Hydroxy-8-aminomethylcoumarin-diacetic acid, 4-methylcoumarin and umbelliferone can similarly be used for the detection of Ca and Cu[142,143].

13.5.3 Intracellular Ion Indicators

Detection of most of the biologically important ions, including hydrogen, calcium, magnesium, sodium, potassium and chloride intracellularly has been achieved using fluorescent dye indicators (for complete review see ref. 144). Coumarin-based molecules are among the fluorescent compounds which have been used in the development of many of these ion indicators.

As mentioned earlier, various coumarins can be used as pH sensors. Intracellularly umbelliferone has been used to measure pH in different brain studies, including studies on epilepsy[145–148]. However, the use of some of the coumarins as simple pH-sensitive dyes has been limited as they suffer such problems as dye leakage and inappropriate pK_a values with regard to their desired role as intracellular pH sensors.

Coumarin-based compounds have been utilised in the measurement of intracellular calcium (Ca^{2+})[149]. Four new potential fluorescent probes for Ca^{2+} have been described, all using the coumarin moiety as the fluorophore. The most promising derivative for intracellular use appears to be BTC, loaded into cells as its tetraacetoxymethyl ester (AM ester): on binding Ca^{2+} it exhibits an excitation shift from 462 nm to 401 nm, with an emission wavelength of 530 nm. Since this dye only responds when cytoplasmic Ca^{2+} levels exceed typical intracellular calcium concentrations, it appears to be very useful in situations (e.g. excitatory cells) where the intracellular calcium level is high enough to saturate typical Ca^{2+} indicators such as Fura-2[149].

The accurate measurement of potassium is crucial in the fields of clinical medicine and cell physiology. The techniques widely used for its measurement, such as flame photometry, atomic absorption spectrometry and NMR[150–152], are expensive and require a large sample size. In addition, due to their slow response time they are all unsuitable for the continuous monitoring of [K^+] in kinetic studies. However, several potassium-sensitive optical probes have been described which are instantaneous in their response time, and inexpensive. One such probe combines the selectivity of using a K^+-ionophore (cryptand [2.2.2.]), with the sensitivity of fluorescence to yield a highly specific K^+ probe, 6,7-(4-methyl)coumaro-[2,2,2] cryptand (MCC)[153]. This fluorescent potassium indicator allows the optical measurement of K^+ in aqueous solutions ($\lambda_{ex} = 340$ nm, $\lambda_{em} = 420$ nm). Na^+, $NH_4{}^+$, Ca^{2+} and Mg^{2+} only minimally interfere with the fluorescence of the probe. The affinity of potassium for MCC increases $\sim$160-fold in methanol, and the effect of Na^+ on MCC fluorescence is diminished. This has enabled the development of a microfluorimetric assay for potassium, useful in renal tubule transport studies, and a clinical assay for measuring K^+ in human blood and urine[153].

13.5.4 Other ions

The detection of aluminium and alkali earth metals is possible using a crown ether-linked coumarin[154]. Bourson *et al.* studied the complexation ability of coumarin 153 linked to a crown compound monoaza-15-crown-5. They found that both absorption and emission spectra undergo marked shifts on ion complexation. The compound was most sensitive to the complexation of Al^{3+}. They concluded that this new fluoroionophore would be useful as a probe for Al^{3+} which may have an application in research into Alzheimer's disease[154].

A simple, specific and sensitive determination of nitrite in foods and biological fluids is highly desirable since nitrite ion, apart from being an intermediate in the nitrogen cycle (and as such a water quality indicator), is also a precursor that forms carcinogenic and/or mutagenic nitrosamines by reaction with various amines. A simple, sensitive and reproducible method is available[155] for the fluorimetric detection of nitrite with 4-hydroxycoumarin, which exploits the following simple chemical reactions: nitrite can react rapidly with phenolic compounds (in this case 4-hydroxycoumarin) in acidic solution, to give their nitroso derivatives, which are generally non-fluorescent due to the electron-withdrawing nitroso group. Since amino groups increase the fluorescence intensity of aromatic compounds, subsequent reduction of the nitroso to an aromatic amide (3-amino-4-hydroxycoumarin) allows for the fluorimetric determination of nitrite[155]. Also, 7-amino-3-methyl-coumarin and 4,5-dihydroxycoumarin have been used for the spectrophotometric determination of small amounts of nitrite[156,157].

The ability to monitor rapid changes in inorganic phosphate (P_i) in real time, e.g. in muscle fibres, can contribute substantially to the understanding of many biological processes. Recently a sensitive probe has been developed to measure P_i release rates from phosphatases[158]. It utilises a genetically engineered phosphate binding protein which has been labelled with *N*-[2-(1-maleimidyl)ethyl]-7-(diethylamino)coumarin-3-carboxamide (MDCC). The labelled protein has an excitation maximum at 425 nm and an emission maximum at 474 nm in the absence of inorganic phosphate. This emission maximum shifts to 464 nm with a 4–5-fold increase in fluorescence when complexed with P_i at pH 7.0. The probe has recently been used for the measurement of P_i release in contracting and relaxed permeabilised muscle fibres in a millisecond time-scale, with a sensitivity in the 10 μM range, which provides new data on the relationship between ATP hydrolysis and force production[159].

13.5.5 Coumarins in Chemical Derivatisations

Aside from their uses as metal and ion indicators, the coumarins are also valuable in the determination of other biologically and chemically important compounds. 4-Bromomethyl-7-methoxycoumarin (BrMmC)[160,161] reacts with both mono- and poly-substituted carboxylic acids to produce ester derivatives of marked fluorescence (blue) according to the reaction in Figure 13.5. As can be seen, the fluorophore-labelled carboxylic acid is detected after column separation.

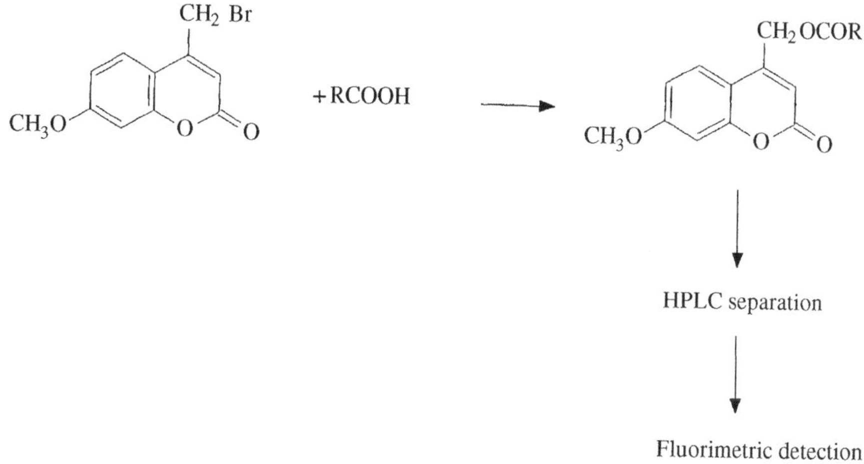

Figure 13.5. Derivatisation of carboxylic acids using 4-bromomethyl-7-methoxycoumarin.

Thus, by pre-column derivatisation of the carboxyl moieties prior to chromatographic separation Turk *et al.*[162] and then Wintersteiger and Juan[163] were able to detect prostaglandin E_2 levels with an eventual limit of detection of 100 pg. However, an important point to remember when using BrMmC for derivatisation is that the quantum yield of these derivatives is dependent on the composition of the solvent, which can limit the choice of HPLC mobile phase[164]. Despite this, the technique has been successfully applied to the determination of gibberellins in urine[165] and caprylic and valproic acids in plasma[166]. Free and conjugated bile acids in serum, bile, gastric juice, faeces and liver biopsies have also been determined by this derivatisation procedure.[167] McGuffin and Zare[168] also utilized BrMmC to derivatise prostaglandins, but in their method they combined micro-column HPLC and laser fluorimetry detection (He/Cd, 325 nm). This favoured unparallel resolution of the major classes of prostaglandins and permitted detection limits in the subfemtomolar range. Similarly, the determination of serum metabolic profiles of bile acids can be accomplished[169].

A reagent analogous to BrMmC is 4-bromomethyl-7-acetoxycoumarin (BrMaC). As the fluorescence intensities of this compound and its reaction products are unaffected by the polarity of the solvent, BrMaC is a more convenient derivatising agent than BrMmC for chromatography, because no limitations regarding the choice of mobile phase are imposed[170]. The reaction of BrMaC with a carboxylic acid is outlined in Figure 13.6. It can be seen that the detection of fluorescence is achieved differently than for BrMmC. BrMaC reacts with carboxylic acids to give ester derivatives which are separated by HPLC, but after separation each compound is hydrolysed by addition of an alkaline solution, and the resultant hydroxy compound's fluorescence is analysed. In this way, the fluorescence intensity is hardly influenced by the type of carboxylic acid being determined as can be the case for BrMmC. Furthermore, the fluorescence is measured in alkaline surroundings in which coumarin derivatives with an acetoxy group at position 7 are more fluorescent than their corresponding methoxycoumarins. Thus a higher sensitivity of detection can be achieved[170]. Tsuchiya *et al.*[170] used this reagent effectively in the determination of prostaglandins, in the range 1 nmol–5 pmol with a limit of detection of ~10 fmol, by forming stable prostaglandin esters, which were separated by HPLC and subsequently hydrolysed to the more fluorescent coumarin analogue prior to detection. This technique is also applicable to other metabolic products of arachidonic acid, including leucotrienes, thromboxanes and non-esterified fatty acids[171,172].

4-Bromomethyl-6,7-dimethoxycoumarin, 4-bromo-7,8-benzocoumarin and 4-diazomethyl-7-methoxycoumarin are other coumarin derivatives used as fluorogenic reagents in the determination of carboxylic acids[173–175].

BrMmC has also been used for the derivatisation of compounds containing imidic nitrogens of an acidic nature. This reaction has been utilised for derivatisation, either pre- or post-column, of thiouracil and thiouridine[176]. It has also been used for the determination of 5-fluoro-2′-deoxyuridine (a cancer chemotherapeutic agent), the reaction product of which displays fluorescence ($\lambda_{ex} = 320$ nm, $\lambda_{em} = 400$ nm) with a linear range of 2–10 ng/ml[177].

BrMmC and another coumarin derivative, 7-(diethylamino)-3-[4-((iodoacetyl)-amino)phenyl]-4-methylcoumarin (DCIA), have been evaluated as derivatising agents for carboxylic acid detection using peroxyoxalate chemiluminescence as the method of detection following chromatographic separation. Only DCIA derivatives of the straight-chain carboxylic acids were detected with peroxyoxalate chemiluminescence with detection limits in the low femtomole range[178].

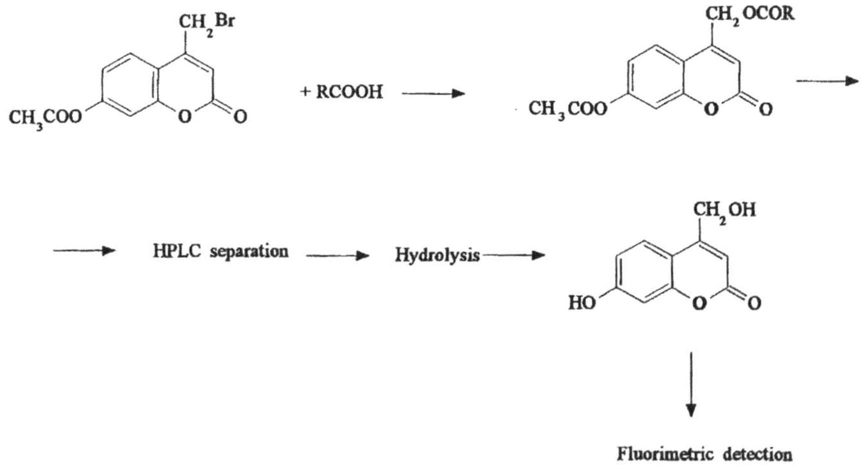

Figure 13.6. Derivatisation of carboxylic acids using 4-bromomethyl-7-acetoxycoumarin.

4-Bromomethyl-7-methoxy-6-nitrocoumarin can be used as an electrochemical label for carboxylic acids in liquid chromatography[179].

Hydroxycompounds are ubiquitous in nature—carbohydrates, steroids and prostaglandins being examples of important classes—and in order to detect them in various complex biological mixtures a high sensitivity is required. The analysis of hydroxyl moieties may be accomplished using coumarin derivatives, an example being an acid halide derivative of 4-methylcoumarin, 7-[(chlorocarbonyl)methoxy]-4-methylcoumarin (CMMC), which has been suggested for the determination of hydroxysteroids and prostaglandins[180]. Various steroid compounds, such as androsterone, tetrahydrocortisol and pregnanediol, have been derivatised using CMMC[180]. By coupling the derivatisation (depicted in Figure 13.7) with chromatographic separation and analysis of fluorescence signals ($\lambda_{ex} = 320$ nm, $\lambda_{em} = 380$ nm), these steroids and their metabolites can be quantified with detection limits between 0.3 ng and 0.7 ng, depending on the mobile phase employed (as is the case with other coumarin reagents, the quantum yield of the CMMC esters of steroids is dependent on the composition of the solvent used).

Coumarin-3-carbonylchloride has been successfully used as a derivatising agent for hydroxy compounds and has been applied to the fluorescent detection of trichothecene mycotoxins following their separation by HPLC[181].

The determination of thiol compounds is also possible using coumarin compounds. Ayers and co-workers[182] used *N*-[4-(7-diethylamino-4-methyl-3-coumarinyl)-phenyl]-maleimide (CPM) for the microfluorimetric quantitation of cellular and non-protein thiols. Their results found CPM microfluorimetry to be an accurate, specific and sensitive method for the analysis of thiols[182]. The detection of glutathione has also been accomplished using *N*-(7-dimethylamino-4-methyl-coumarinyl)-maleimide as a pre-column HPLC derivatising reagent[183].

13.6 USE OF COUMARIN IN OTHER ANALYSES

13.6.1 Cytochrome P450 Assessment using Coumarins

An inherent part of many research and safety assessment programmes is the measurement of cytochrome P450-mediated oxidative metabolism and its induction as a result of xenobiotic administration. A range of assays exist for this task but many are

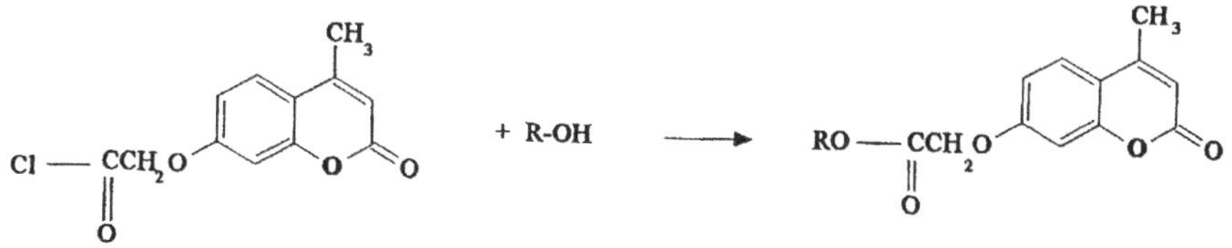

Figure 13.7. Reaction between 7-[(chlorocarbonyl)methoxy]-4-methylcoumarin and hydroxy compounds.

time-consuming and cumbersome, using limited radiolabelled substrates and/or lengthy solvent extraction protocols. Many are also relatively insensitive, which is particularly problematic when assessing *in vitro* systems such as cultured hepatocytes. One of the commonly employed assays for P450 assessment is a direct fluorimetric assay for cytochrome P450-catalysed *O*-deethylation of 7-ethoxycoumarin[184,185]. While this assay is useful, it suffers from the fact that the excitation/emission maxima (340/450 nm) of 7-ethoxycoumarin correspond to those of NADPH. As NADPH is a required cofactor for the reaction, its presence limits the sensitivity of the direct 7-ethoxycoumarin assay. This problem has been overcome though, through the use of 7-ethoxy-4-trifluoromethylcoumarin (EFC), an analogue of 7-ethoxycoumarin, which on deethylation by liver microsomes yields the fluorescent compound 7-hydroxy-4-trifluoromethylcoumarin (HFC). The production of HFC can be monitored fluorimetrically ($\lambda_{ex} = 410$ nm, $\lambda_{em} = 510$ nm), without interference from NADPH and EFC[186]. The high sensitivity of this method renders it useful for studies on both *in vitro* and *in vivo* systems[186,187].

Apart from *O*-deethylation of ethoxycoumarins, the hydroxylation of orally administered coumarin to 7-hydroxycoumarin, and its subsequent urinary excretion, can be used to assess the activity of the P450 enzyme CYP2A6 in liver microsomes. This 'coumarin test', if used in conjunction with the *N*-deethylation of lignocaine to monoethylglycinexylide (MEGX) by CYP3A4, may be an effective measure of liver function in clinical practice. Sotaniemi *et al.*[188] recently investigated the utility of monitoring serum MEGX concentration and urinary 7-hydroxycoumarin excretion to reflect liver function in patients with liver and kidney diseases, and in epileptic patients. Previously, coumarin 7-hydroxylase activity has been used in an animal study on the alteration in the cytochrome P450 profile in cocaine-induced hepatotoxicity[189], and also to examine the association of liver fluke infestation with the development of liver cancers[190].

Hepatic microsomal epoxidising activities have also been assayed using 7-vinyloxycoumarin (VOC)[191]. VOC is metabolised in the liver to 7-hydroxycoumarin and glycoaldehyde, via an unstable epoxide, and direct fluorimetry of 7-hydroxycoumarin production enables the determination of epoxidising activities in the liver.

13.6.2 DNA Analysis using Coumarin Compounds

Non-isotopically labelled probes have become popular for *in situ* hybridisation experiments recently, and coumarin-labelled oligonucleotides (with blue fluorescence) are among those which have been used as fluorescent probes. They have proved particularly appropriate for experiments requiring simultaneous multiple probing of sequences and have been used frequently in conjunction with rhodamine- and fluorescein-labelled DNA (red and green fluorescing, respectively)[192]. This three-colour *in situ* hybridisation has been applied to the study of numerical chromosome abnormalities as occur in human solid tumours, and may be applied to the prenatal diagnosis of autosomal aneuploidy[193].

Another convenient method used for examining nucleic acid structures and sequences is the use of fluorescence energy transfer studies[194]. By covalently linking oligonucleotides with different fluorophores, the primary and secondary structure of DNA fragments can be easily probed since hybridisation of two labelled oligonucleotides to adjacent nucleic acids results in the transfer of fluorescence energy between the two dyes (donor and acceptor). Many different fluorophores, including a 4-trifluoromethyl-7-aminocoumarin, have been used for this type of study[194].

13.6.3 Coumarins as Lasing Dyes

Coumarins have been used extensively to tune lasers to desired wavelengths and there are a number of coumarin dyes available for this purpose. The use of these dyes is based on their broad absorption and fluorescent spectra which allow the possibility of producing tunable lasers at a relatively low cost. One example of this use involves the laser-enhanced ionisation spectrometry of 23 elements. The elements in aqueous solution were detected by light in the UV region which was generated by frequency doubling the output from coumarin 153, the dye used[195]. Coumarin 153 has also been used by Spanner and Niessner to tune a laser between 520 and 600 nm for the non-invasive determination of blood components[196]. Coumarin 440 has been used in conjunction with other dyes in the determination of polynuclear aromatic compounds using laser-excited synchronous fluorescence[197]. The application of coumarin-based laser dyes to matrix-assisted laser desorption ionization (MALDI)-mass spectrometry has also been achieved[198]. Coumarin 2, 47, 120 and 152 were used as matrices and were evaluated in the analysis of peptides, proteins, deoxynucleotides and polymers with masses extending up to 150 000[198]. Nilsson and co-workers have used the coumarin 500 dye to set their laser at 488 nm for use in fluorescence imaging of capillary electrophoresis in 'real-time'. This was applied to DNA samples and the migration of components was monitored during the separation process with detection of fluorescence using a charge-coupled device (CCD) camera[199].

13.6.4 Coumarins in Miscellaneous Analyses

Coumarin compounds have also been applied to analysis in a diverse range of experiments. Some coumarins have been used in optically-based rapid sensing devices. Coumarin 102 has, for example, been employed as an oxygen-sensitive reagent phase encapsulated in a gas-permeable membrane at the tip of an optical fibre. This system was used for the detection of gaseous oxygen[200]. Such an approach has potential industrial, environmental and, perhaps, medical applications. It is envisaged that other sensors based on coumarin compounds will be developed in the near future.

7-Amino-4-methylcoumarin has been conjugated via a succinimide bridge to the cancer chemotherapeutic drug vinblastine[201]. The coumarin–vinblastine (cou-VBL) compound exhibits a pharmacological activity and transport kinetics identical to the native vinblastine, but due to the fluorescence of the coumarin moiety ($\lambda_{ex} = 330$ nm, $\lambda_{em} = 395$ nm), which is virtually unaffected by conjugation to vinblastine, the cou-VBL compound can be continuously monitored by fluorescence. The fluorescence of cou-VBL is quenched intracellularly and therefore the moiety can be used to study the kinetics of P-glycoprotein-controlled drug transport in multidrug resistant cells, by monitoring the increase in fluorescence over time[201].

Aqueous coumarin-3-carboxylic acid (or its succinimide ester SECCA) is converted to the highly fluorescent 7-hydroxycoumarin-3-carboxylic acid upon irradiation. The intensity of the fluorescent signal is linearly proportional to the number of 7-hydroxycoumarin-3-carboxylic acid molecules formed, which is in turn proportional to the absorbed radiation dose. This property may prove these coumarins to be useful as chemical dosimeters for radiation therapy applications[202]. The above hydroxylation of SECCA and subsequent fluorescence is the basis of many current studies on radical formation—for example, the reaction is used to monitor the formation of hydroxyl radicals in the vicinity of biomolecules (avidin, polylysine, histone) and DNA[203,204]. It has also been used *in vitro* to illustrate that benzoyl peroxide acts as a promoter of radiation-induced malignant transformation[205].

7-Amino-4-methylcoumarin and 7-amino-4-trifluoromethylcoumarin have been used in bioreactor characterisation to examine the aeration and agitation effects in fermentations by using *in situ* fluorimetry[206].

Coumarin itself is used as a chemical marker in rebated kerosene (it is added at a level of 2 mg/l). Rebated kerosene has a significantly lower excise duty than unrebated kerosene, and may be used illegally as an extender to diesel fuel. Therefore, a method is required by customs and excise officials to detect the presence of coumarin in kerosene and diesel mixtures. A simple roadside test is used in which any coumarin present is initially converted into *cis*-hydroxycinnamate, prior to conversion to the *trans*-hydroxycinnamate form—this *trans* form exhibits a green fluorescence under ultraviolet light[207]. Total luminescence spectroscopy can then be used in the laboratory to confirm the roadside test and quantify the coumarin present[208].

Coumarin has also been used in assays for adulteration of food flavourings[209], and in elucidating the storage conditions of different wines and brandies[210].

13.7 CONCLUSION

This review clearly indicates the usefulness of coumarins and coumarin derivatives in virtually every area of analysis. The inherent fluorescent properties of many coumarins is a key factor in many applications. However, it is the ingenuity of the ways in which analytical scientists have exploited the coumarins that has led to their true importance as analytical reagents.

REFERENCES

1 Murray RDH, Mendez J, Brown SA. *The Natural Coumarins—Occurrence, Chemistry and Biology*. Chichester: John Wiley, 1982.

2 Feuer G. The metabolism and biological actions of coumarins. *Prog Med Chem* 1974; **10**: 85–158.

3 Cohen AJ. Critical review of the toxicology of coumarin with special reference to interspecies differences in metabolism and hepatotoxic response and their significance to man. *Fd Cosmet Toxicol* 1979; **3**: 277–89.

4 Murray RDH. Coumarins. *Natural Prod Rep* 1989; 591–624.

5 Egan D, O'Kennedy R, Moran E, Cox D, Prosser E, Thornes RD. The pharmacology, metabolism, analysis and applications of coumarin and coumarin-related compounds. *Drug Metab Rev* 1990; **22**(5): 503–29.

6 Ikawa M, Stahmann MA, Link KP. Studies on 4-hydroxycoumarin. V. The condensation of α,β-unsaturated ketones with 4-hydroxycoumarin. *J Am Chem Soc* 1944; **66**: 902–6.

7 Suttie JW. Warfarin and vitamin K. *Clin Cardiol* 1990; **13**(4, Suppl. 6): VI 16–18.

8 Shibukawa A, Nagoa M, Kuroda Y, Nakagawa T. Stereoselective determination of free warfarin concentration in protein-binding equilibrium using direct sample injection and on-line chromatographic system. *Anal Chem* 1990; **62**(7): 712–16.

9 De Vries JX, Volker U. Determination of the plasma protein binding of the coumarin anticoagulants phenprocoumon and its metabolites, warfarin and acenocoumarol, by ultrafiltration and high-performance liquid chromatography. *J Chrom* 1990; **529**: 479–85.

10 Sutcliffe FA, MacNicholl DA, Gibson G. Aspects of anticoagulant action: a review of the pharmacology, metabolism and toxicology of warfarin and congeners. *Drug Metab Drug Interact* 1987; **5**(4): 225–72.

11 Lewis RJ, Trager WF Warfarin metabolism in man: identification of metabolites in urine. *J Clin Invest* 1970; **49**: 907–13.

12 Lewis RJ, Trager WF. The metabolic fate of warfarin: studies on metabolites in plasma. *Ann NY Acad Sci* 1971; **179**: 205–12.

13 Kaminsky LS, Dannan GA, Guengerich FP. Composition of P-450 isozymes from hepatic microsomes of C57BL/6 and DBA/2 mice, assessed by warfarin metabolism, immunoinhibition, and immunoelectrophoresis and anti-rat cytochrome P-450. *Eur J Biochem* 1984; **141**: 141–8.

14 O'Reilly RA, Aggeler PA, Hoag MS, Leong L. Studies on the coumarin anticoagulant drugs: The assay of warfarin and its biological application. *Thromb Diath Haemorrh* 1962; **8**: 82–95.

15 Testa B. Regio- and stereoselectivity in drug metabolism: two examples. *Pharm Int* 1981; **2**: 34–7.

16 Hermans JJ, Thijssen HH. Human liver microsomal metabolism of the enantiomers of warfarin and acenocoumarol: P-450 isozyme diversity determines the differences in their pharmacokinetics. *Br J Pharmacol* 1993; **110**(1): 482–90.

17 Hermans JJ, Thijssen HH. The *in vitro* ketone reduction of warfarin and analogues. Substrate stereoselectivity and species differences. *Biochem Pharmacol* 1989; **38**(19): 3365–70.

18 Rettie AE, Eddy AC, Heimark LD, Giboldi M, Trager WF. Characteristics of warfarin hydroxylation catalysed by human liver microsomes. *Drug Metab Dispos* 1989; **17**(3): 265–70.

19 Chan E, McLachan AJ, Rowland M. Warfarin metabolites: stereochemical aspects of protein binding and displacement by phenyl butazone. *Chirality* 1993; **5**(8): 610–15.

20 Ikeda M, Ullrich V, Staudinger H. Metabolism *in vitro* of warfarin by enzymic and non-enzymic systems. *Biochem Pharmacol* 1968; **17**: 1663–9.

21 Stenflo J. Vitamin K, prothrombin and γ-carboxyglutamic acid. *New Engl J Med* 1977; **296**: 624–5.

22 Hemker HC, Veltkamp JJ, Neusen A, Loeliger EA. Nature of prothrombin biosynthesis: preprothrombinemia in vitamin K deficiency. *Nature (Lond)* 1963; **200**: 589–90.

23 Stenflo J, Ganrot PO. Vitamin K and the biosynthesis of prothrombin I. Identification and purification of a dicoumarol-induced abnormal prothrombin from bovine plasma. *J Biol Chem* 1972; **247**: 8160.

24 Ren P, Stark PY, Johnson RL, Bell RG. Warfarin and the metabolism of vitamin K_1. *Biochem Pharmacol* 1974; **23**: 3355–7.

25 Lloyd JV. Uses of heparin and warfarin. *Med J Austr* 1985; **142**: 197–201.

26 Thornes RD. Warfarin as maintenance therapy for cancer. *J Irish Coll Phys Surg* 1972; **2**: 41–2.

27 Thornes RD, Edlow DW, Wood S Jr. Inhibition of cancer cells *in-vivo* by anticoagulant therapy: Effects of sodium warfarin on V2 cancer cells, granulocytes, lymphocytes and macrophages in rabbits. *Johns Hopkins Med J* 1968; **123**: 305–16.

28 Zacharski LR. Warfarin therapy for cancer. *J Irish Coll Phys Surg* 1993; **22** (Suppl 1): 20–3.

29 Zacharski LR, Henderson WG, Rickles FR *et al* Effect of warfarin anticoagulation on survival in carcinoma of the lung, colon, head, neck and prostate. Final report of the VA Cooperative study No. 75. *Cancer* 1984; **53**: 2046–52.

30 Zacharski LR, Henderson WG, Rickles F R *et al* Effect of warfarin on survival on small cell carcinoma of the lung V.A. Study No. 75. *JAMA* 1981; **245**(8): 831–5.

31 Zacharski LR, Memoli VA, Constantini V, Wojtukiewicz MZ, Ornstein DL. Clotting factors in tumour tissue: implications for cancer therapy. *Blood Coag Fibrinol* 1990; **1**: 71–8.

32 Zacharski LR, Howell AL, Memoli VA. The coagulation biology of cancer. *Fibrinolysis* 1992; **6** (Suppl 1): 39–42.

33 Kempkin SJ. Warfarin resistance caused by broccoli. *New Engl J Med* 1983; **308**: 1229–30.

34 Lee M, Schwartz RN, Sharifi R. Warfarin resistance and vitamin K. *Ann Intern Med* 1981; **94**: 140–1.

35 Hallak HO, Wedlund PJ, Modi MW *et al.* High clearance of (S)-warfarin in a warfarin resistant subject. *Br J Clin Pharmacol* 1993; **35**(3): 327–30.

36 Zallman JA, Lee DP, Jeffrey PL. Liquid nutrition as a cause of warfarin resistance. *Am J Hosp Pharm* 1981; **38**: 1174.

37 Holt RJ, Freytes CO. Familial warfarin resistance. *Drug Int Clin Pharmacol* 1983; **17**: 281–3.

38 Oren B, Shvatrzman P. Unsuspected source of vitamin K in patients treated with anticoagulants: a case report. *Family Pract* 1989; **6**(2): 151–2.

39 Talstad I, Gamst ON. Warfarin resistance due to malabsorption. *J Intern Med* 1994; **236**(4): 465–7.

40 Diab F, Feffer S. Hereditary warfarin resistance. *South Med J* 1994; **87**(3): 407–9.

41 Tummino PJ, Ferguson D, Hupe D. Competitive inhibition of HIV-1 protease by warfarin derivatives. *Biochem Biophys Res Commun* 1994; **201**(1): 290–4.

42 Bourinbaiar AS, Tan X, Nagorny R. Inhibitory effect of coumarins on HIV-1 replication and cell mediated or cell-free viral transmission. *Acta Virol* 1993; **37**(4): 241–50.

43 Thaisrivongs S, Tomich PK, Watenpaugh KD *et al.* Structure-based design of HIV protease inhibitors: 4-hydroxycoumarins and 4-hydroxy-2-pyrones as non-peptidic inhibitors. *J Med Chem* 1994; **37**: 3200–4.

44 Lunney EA, Hagen SE, Domagala JM *et al.* A novel nonpeptide HIV-1 protease inhibitor: elucidation of the binding mode and its application in the design of related analogs. *J Med Chem* 1994; **37**(17): 2664–77.

45 Thaisrivongs S, Watenpaugh KD, Howe WJ *et al.* Structure-based design of novel HIV-protease inhibitors: carboxamide-containing 4-hydroxycoumarins and 4-hydroxy-2-pyrones as potent non-peptidic inhibitors. *J Med Chem* 1995; **38**(18): 3624–37.

46 Khalfan H, Abuknesha R, Rard-Weaver M, Price RG, Robinson D. Aminomethyl coumarin acetic acid: a new fluorescent labelling agent for proteins. *Histochem J* 1986; **81**: 497–9.

47 Aubry JP, Durand I, De Paolip, Banchereau J. 7-Amino-4-methylcoumarin 3-acetic acid conjugated streptavidin permits simultaneous flow cytometry analysis of either three cell surface antigens or one cell surface antigen as a function of RNA and DNA content. *J Immunol Meth* 1990; **128**: 39–49.

48 Bioprobes 20—New products and applications from Molecular Probes Inc. (commercial data). Supplied by Cambridge Bioscience, 25 Signet Court, Newmarket Road, Cambridge CB5 8LA, UK.

49 Forster Y, Hass E. Preparation and characterisation of three fluorescent labels for proteins, suitable for structural studies. *Anal Biochem* 1993; **209**: 9–14.

50 Al-Kindy SMZ, El-Sherbini SA, Abdel-Kader MH. UV-visible absorption and fluoresence characteristics of the luminescent label coumarin-6-sulphonyl chloride in homogeneous and micellar solutions. *Anal Chem Acta* 1994; **285**: 329–33.

51 Samuel D, Amlot PL, Abuknesha RA. A new method of iodinating ovalbumin, a protein which lacks accessible tyrosine groups, by conjugation to a highly fluorescent coumarin active ester, CASE. *J Immunol Methods* 1985; **81**: 123–30.

52 Mead JAR, Smith JN, Williams JT. The biosynthesis of the glucuronides of umbelliferone and 4-methylumbelliferone and their use in the fluorimetric determination of β-glucuronidase. *Biochem J* 1955; **61**: 569–73.

53 Robinson D. Fluorimetric determination of β-glucosidase: its occurrence in the tissues of animals including insects. *Biochem J* 1956; **63**: 39–44.

54 Leaback DH, Walker PG. Fluorimetric assay of N-acetyl-β-glucosaminidase. *Biochem J* 1961; **78**: 151–6.

55 Leaback DH. Properties of some derivatives of 4-methylumbelliferone. *Biochem J* 1961; **78**: 22P–23P.

56 DeBruyne CK, Loontiens FG. A fluorogenic substrate for β-D-xylosidase. *Naturwissenschaften* 1965; **52**: 661.

57 Takagaki K, Kon A, Kawasaki H, Nakamura T, Endo M. Preparation and application of a fluorogenic substrate for endo-β-xylosidase. *Biochem Biophys Methods* 1989; **19**(2–3): 207–14.

58 Goodfellow M, Thomas EG, Ward AC, James AL. Classification and identification of rhodococci. *Int J Med Microbiol* 1990; **274**(3): 299–315.

59 Mangels J, Edvalson I, Cox M. Rapid presumptive identification of *Bacteroides fragilis* group of organisms with use of 4-methylumbelliferone-derivative substrates. *Clin Infect Dis* 1993; **16**(Suppl 4): S319–21.

60 Sepulveda JL, Stager CE, Davis JR. Rapid presumptive identification of gram-negative rods directly from blood cultures by simple enzymatic tests. *Clin Microbiol* 1990; **28**(2): 177–81.

61 U.S. Federal Register (1992); **57**: 24744.

62 Park SJ, Lee EJ, Dee DH, Lee SH, Kim SJ. Spectrofluorimetric assay for rapid detection of total and fecal coliforms from surface water. *Appl Environ Microbiol.* 1995; **61**(5): 2027–9.

63 Bulte M, Reuter G. Glucuronidase detection and indol capillary test as reliable rapid identification procedures for the detection of *E coli* in food–toxinogenic strains included. *Zentralbl Hyg Umweltmed* 1989; **188**(3–4); 284–93.

64 Grebner EE, Wenger DA. Use of 4-methylumbelliferyl-6-sulpho-2-acetamido-2-deoxy-beta-D-gluco-pyranoside for the prenatal diagnosis of Tay–Sachs disease using chorionic villi. *Prenat Diagn* 1987; **7**(6): 419–23.

65 Petushkova NA, Kozlova IK, Tsvetkova IV, Vidershain GIa. A study of the various properties of β-galactocerebrosidase from human chorion using fluorescent substrates. *Biokhimica* 1988; **53**(9): 1539–45.

66 Zhao H, Van Diggelen OP, Thoomes R *et al.* Prenatal diagnosis of Morquio disease type A using a simple fluorimetric enzyme assay. *Prenat Diagn* 1990; **10**(2): 85–91.

67 Potier M, Mameli L, Belisle M, Dallaire L, Melancon SB. Fluorimetric assay of neuraminidase with a sodium (4-methylumbelliferyl-α-D-*N*-acetylneuraminate) substrate. *Anal Biochem* 1979; **94**: 287–96.

68 Jameson GW, Roberts DV, Adams RW, Kyle WSA, Elmore DT. Determination of the operational molarity of solutions of bovine α-chymotrypsin, trypsin, thrombin and factor Xa by spectrofluorimetric titration. *Biochem J* 1973; **131**: 107.

69 Fernley HN, Walker PG. Kinetic behaviour of calf-intestinal alkaline phosphatase with 4-methylumbelliferone phosphate. *Biochem J* 1965; **97**: 95–103.

70 Bayleran J, Hechtman P, Saray W. Synthesis of 4-methylumbelliferyl-β-*N*-acetylglucosamine-6-sulfate and its use in classification of GM_2 gangliosidosis genotypes. *Clin Chim Acta* 1984; **143**: 73–89.

71 Yang Y, Hamaguchi K. Hydrolysis of 4-methylumbelliferyl N-acetyl chitotrioside catalysed by hen and turkey lysozymes. *Biochemistry* 1980; **87**: 1003–14.

72 Heeb MJ, Gabriel O. Enzyme localisation in gels. *Meth Enzymol* 1984; **104**: 416–39.
73 Swallow DM, Corney G, Harris H, Hirschorn R. Acid α-glucosidase: A new polymorphism in man demonstrable by 'affinity' electrophoresis. *Ann Hum Genet* 1975; **38**: 391–406.
74 Christomenou H, Sandhoff K. A sensitive fluorescence assay for the simultaneous and separate determination of arylsulphatases A and B. *Clin Chim Acta* 1977; **79**: 527–31.
75 Chang PL, Ballantyne SR, Davidson RG. Detection of arylsulfatase A activity after electrophoresis in polyacrylamide gels: problems and solutions. *Anal Biochem* 1979; **97**: 36–42.
76 Hawley DM, Tsou KC, Hodes ME. Preparation, properties and uses of 2 fluorogenic substrates for the detection of 5′- (venom) and 3′ spleen nucleotide phosphodiesterases. *Anal Biochem* 1981; **117**: 18–23.
77 Lo KW, Aoyagi S, Tsou KC. A fluorogenic method for the demonstration of human serum 5′-nucleotide phosphodiesterase isozymes after polyacrylamide gel electrophoresis. *Anal Biochem* 1981; **117**: 24–7.
78 Aswanikumar S, Radhakrishnan AN. Purification and properties of a peptidase acting on a synthetic collagenase substrate from experimental granuloma tissue in the rat. *Biochim Biophys Acta* 1972; **276**: 241–9.
79 Zimmerman M, Yurewicz EC, Patel G. A new fluorogenic substrate for chymotrypsin. *Anal Biochem* 1976; **70**: 258–62.
80 Kanaoka Y, Takahashi T, Nakayama H. A new fluorogenic substrate for aminopeptidase. *Chem Pharm Bull* 1977; **25**: 3126–8.
81 Zimmerman M, Ashe B, Yurewicz EC, Patel G. Sensitive assays for trypsin, elastase and chymotrypsin using new fluorogenic substrates. *Anal Biochem* 1977; **78**: 47–51.
82 Kunugi S, Fukuda M, Hayashi R. Action of serine carboxypeptidases on endopeptidase substrates, peptide-4-methyl-coumaryl-7-amides. *Eur J Biochem* 1985; **153**: 37–40.
83 Fugiwara K, Tsuru D. New chromogenic and fluorogenic substrates for pyrridonyl peptidase. *Biochemistry* 1978; **83**: 1145–9.
84 Kato T, Hama T, Kojima K, Nagatsu T, Sakakibara S. Excretion of X-prolyl dipeptidyl-aminopeptidase in human urine as determined with a new fluorogenic substrate. *Clin Chem* 1978; **24**: 1163–6.
85 Gutierrez JC, Gotze O, Caporale L.H. An improved fluorogenic substrate for the alternative complement pathway C3/5 converting enzyme CVF3b. *Biochem Biophys Acta* 1983; **744**: 276–80.
86 Morita T, Kato H, Iwanaga S, Takada K, Kimura T, Sakakibara S. New fluorogenic substrates for α-thrombin, factor Xa, kallikreins and urokinase. *Biochemistry* 1977; **82**: 1495–8.
87 Ohno Y, Kato H, Morita T *et al.* A new fluorogenic peptide substrate for Vitamin K dependent blood coagulation factor, bovine protein C. *Biochemistry* 1981; **90**: 1387–95.
88 Kato H, Adachi N, Ohno Y, Iwanaga S, Takada K, Sakakibara S. New fluorogenic peptide substrates for plasmin. *Biochemistry* 1980; **88**: 183–90.
89 Sugo T, Kato H, Iwanaga S, Takada K, Sakakibara S. Kinetic studies on surface mediated activation of bovine factor XII and prekallikrein. *Eur J Biochem* 1985; **146**: 43–50.
90 Kawabata S, Muira T, Morita T *et al.* Highly sensitive peptide-4-methylcoumaryl-7-amide substrates for blood-clotting proteases and trypsin. *Eur J Biochem* 1988; **172**, 17–25.
91 Smyth M, O'Cuinn G. Alanine aminopeptidase of guinea-pig brain: a broad specificity cytoplasmic enzyme capable of hydrolysing short and intermediate length peptides. *Int J Biochem* 1994; **26**(10–11): 1287–97.
92 Cromlish JA, Seidah NG, Marcinkiewicz M, Hamelin J, Johnson DA, Chretien M. Human pituitary tryptase: molecular forms, NH_2-terminal sequence, immunocytochemical localisation, and specificity with prohormone and fluorogenic substrates. *J Biol Chem* 1987; **262**(3): 1363–73.
93 McGinty A, Moore M, Halton DW, Walker B. Characterisation of the cysteine proteinases of the common liver fluke *Fasciola hepatica* using novel, active site directed affinity labels. *Parasitology* 1993; **106**(5): 487–93.
94 Robichon A, Kuks P. Proteolysis in rat hypothalamic neurosecretory granules: characterisation of an alpha-chymotrypsin-like activity in the pathway of intracellular processing of prohormones. *Endocrinology* 1991; **128**(4): 1974–80.

95 Meyer J, Jordi R. Purification and characterisation of X-prolyl-dipeptidyl-aminopeptidase from *Lactobacillus lactis* and from *Streptococcus thermophilus*. *J Dairy Sci* 1987; **70**(4): 738–45.
96 Sohar I, Wolz RL, Bond JS. Soluble metalloendopeptidase (MMP-7ase) activity in mouse kidney cytosol. *Acta Biol Hung* 1991; **42**(1–3): 265–74.
97 Khalfan HA. Study of the thiol proteases of normal human skin fibroblasts. *Cell Biochem Funct* 1991; **9**(1): 55–62.
98 Drake LJ, Bianco AE, Bundy DA, Ashall F. Characterisation of peptidases of adult *Trichuris muris*. *Parasitology* 1994; **109**(5): 623–30.
99 Sklar LA, McNeil VM, Jesaitis AJ, Painter RG, Cochrane CG. A continuous, spectroscopic analysis of the kinetics of elastase secretion by neutrophils. The dependence of secretion upon receptor occupancy. *J Biol Chem* 1982; **257**(10): 5471–5.
100 Mason RW. Species variants of Cathepsin L and their immunological identification. *Biochem J* 1986; **240**(1): 285–8.
101 Falase EA, Storey BT, Teuscher C. Purification and preliminary characterisation of guinea pig testicular proteinase inhibitors. *Mol Reprod Dev* 1991; **29**(1): 29–39.
102 North MJ, Robertson CD, Coombs GH. The specificity of trichomad cysteine proteinases analysed using fluorogenic substrates and specific inhibitors. *Mol Biochem Parasitol* 1990; **39**(2): 183–93.
103 Billings PC, Carew JA, Keller-McGandy CE, Goldberg AL, Kennedy AR. A serine protease activity in C3H/10T1/2 cells that is inhibited by anticarcinogenic protease inhibitors. *Proc Natl Acad Sci USA* 1987; **84**(14): 4801–5.
104 Blackmon DL, Watson AJM, Montrose MH. Assay of apical membrane enzymes based on fluorogenic substrates. *Anal Biochem* 1992; **200**: 352–8.
105 Bronk SF, Powers SP, Gores GJ. Dynamic measurements of intracellular aminopeptidase activity in hepatocytes using multiparameter digitized video fluorescent microscopy. *Anal Biochem* 1993; **210**: 219–25.
106 Board PG. The use of fluorogenic peptide substrates for the detection of coagulation factors II and X after electrophoresis. *Ann Hum Genet* 1982; **46**(4): 293–8.
107 Sinha P, Gossrau R. Isoelectric focusing (IEF) and band detection with fluorogenic protease substrates. *Histochemistry* 1984; **81**(2): 167–9.
108 Smith RE, Bissell ER, Mitchell AR, Pearson KW. Direct photometric or fluorimetric assay of proteinases using substrates containing 7-amino-4-trifluoromethylcoumarin. *Thromb Res* 1980; **17**: 393–402.
109 Tchoupe JR, Moreau T, Gauthier F, Bieth JG. Photometric or fluorimetric assay of cathepsin B, L and H and papain using substrates with an aminotrifluoromethylcoumarin leaving group. *Biochim Biophys Acta* 1991; **1076**: 149–51.
110 Higuchi K, Kajiki A, Nakamura M *et al.* Proteases released in organ culture by acute dermal inflammatory lesions produced *in vivo* in rabbit skin by sulfur mustard: hydrolysis of synthetic peptide substrates for trypsin-like and chymotrypsin-like enzymes. *Inflammation* 1988; **12**(4): 311–34.
111 Agban A, Ounanian M, Luu-Duc C, Monget D. Synthesis of new fluorogenic substrate derivated of 7-amino-4-trifluoro-methylcoumarin. Detection of gram-negative bacteria. Streptococci group A and Enterococci. *Ann Pharm Fr* 1990; **48**(6): 326–34.
112 Silverstein RL, Nachman RL, Leung LL, Harpel PC. Activation of immobilised plasminogen by tissue activator. Multimolecular complex formation. *J Biol Chem* 1985; **260**(18): 10346–52.
113 Harpel PC, Lewin MF, Kaplan AP. Distribution of plasma kallikrein between C1 inactivator and alpha 2 macroglobulin in plasma utilising a new assay for alpha-2-macroglobulin-kallikrein complexes. *J Biol Chem* 1985; **260**(7): 4257–63.
114 Garrett JR, Chao J, Proctor GB *et al.* Influences of secretory activities in rat submandibular glands on tissue kallikrein circulating in blood. *J Exp Physiol* 1995; **80**(3): 429–40.
115 Okhuysen PC, Dupont HL, Sterling CR, Chappell CL. Arginine aminopeptidase, an integral membrane protein of the *Cryptosporidium parvum* sporozoite. *Infect Immun* 1994; **62**(10) 4667–70.
116 Budowle B, Gambel AM. An alternative, effective substrate for erythrocyte acid phosphatase phenotype determinations. *J Forensic Sci* 1988; **33**(4): 915–20.

117 Gossrau R. Fluorescence histochemical detection of hydrolases in tissue sections and culture cells. *Histochemistry* 1983; **79**(1): 87–94.

118 Stack MS, Gray RD. Comparison of vertebrate collagenase and gelatinase using a new fluorogenic substrate peptide. *J Biol Chem* 1989; **264**: 4277–81.

119 Knight CG. A quenched fluorescent substrate for thimet peptidase containing a new fluorescent amino acid, DL-2-amino-3-(7-methoxy-4-coumaryl) propionic acid. *Biochem J* 1991; **274**(1): 45–8.

120 Tisljar U, Knight CG, Barrett AJ. An alternative quenched fluorescent for Pz-peptidase. *Anal Biochem* 1990; **186**: 112–15.

121 Knight CG, Willenbrock F, Murphy G. A novel coumarin-labelled peptide for sensitive continuous assays of the matrix metalloproteinases. *FEBS Lett* 1992; **296**(3): 263–6.

122 Bond JS, Butler PE. Intracellular proteases. *Ann Rev Biochem* 1987; **56**: 333–64.

123 Dengler R, Lah T, Gabrijelcic, Turk V, Fritz H, Emmerich B. Detection of cathepsin B in tumour cytosol and urine of breast cancer patients. *Biomed Biochim Acta* 1991; **50**(4–6): 555–60.

124 Kolar Z, Krepela E, Kovarik J, Kohoutek M, Rypka M, Vicar J. Clinical experience with the testing of serum proteinase activity by cathepsin B-like sensitive amino acid derivatives of 7-amino-4-methylcoumarine. *Neoplasm* 1987; **34**(5): 601–8.

125 Bouvier M, Colson F, Tebib J, Mathieu M, Vianey-Liaud C. Détermination de l'activité proline imino-peptidase dans le serum à l'aide d'un substrat fluorescent dans la maladie de Paget et les metastases osseuses condensantes prostatiques. Résultats preliminaires. *Pathol Biol* 1986; **34**(4): 235–9.

126 Laszlo A, Sohar I, Sagi I, Kovacs J, Kovacs A. Activity of cathepsin H, B and metalloproteinase in the serum of patients with acute myocardial infarction. *Clin Chim Acta* 1992; **210**(3): 233–5.

127 Sacchetti L, Castaido G, Fortunado G, Salvatore F. Improved procedure for measuring gamma-glutamyltransferase isoenzymes in serum. *Clin Chem* 1988; **34**(2): 419–22.

128 Huet G, Flipo RM, Richet C *et al.* Measurement of elastase and cysteine proteinases in synovial fluid of patients with rheumatoid arthritis, spondylarthropathies and osteoarthritis. *Clin Che.* 1992; **38**(9) 1694–7.

129 Gotoh H, Hagihara M, Nagatsu T, Iwata H, Muira T. Activity of dipeptidylpeptidase IV and post-proline cleaving enzyme in sera from osteoporotic patients. *Clin Chem* 1988; **34**(12) 2499–501.

130 Yandle T, Richards M, Smith M, Charles C, Livesey J, Espiner E. Assay of endopeptidase 24.11 activity in plasma applied to *in vivo* studies of endopeptidase inhibitors. *Clin Chem* 1992; **38**(9) 1785–91.

131 Smith RE, Reynolds CJ, Elder EA. The evolution of proteinase substrates with special reference to dipeptidylpeptidase IV. *Histochem J* 1992; **24**(9) 637–47.

132 Dement J. (1995) Fluorescent indicators. In: *CRC Handbook of Chemistry and Physics*, 76th edn, DR Lide (ed). Boca Raton, FL: CRC Press, 8-19–8-20.

133 Bhat AN, Jain BD. Separation and determination of uranium and thorium with 3-acetyl-4-hydroxy-coumarin. *Talanta* 1960; **4**: 13–16.

134 Bhat AN, Jain BD. Analytical applications of 3-acetyl-4-hydroxycoumarin—Spectrophotometric determination of iron. *Talanta* 1960; **5**: 271–5.

135 Jain BD, Singh HB. Gravimetric determination and separation of titanium(IV) from iron(III) using 4-phenyl-7,8-dihydroxycoumarin. *Ind J Chem* 1963; **1**: 317–18.

136 Jain BD, Singh HB. Molybedenum(VI) complexes with *o*-dihydroxycoumarins. *Ind J Chem* 1963; **1**: 369–71.

137 Jain BD, Singh HB. Spectrophotometric determination of niobium using *o*-dihydroxycoumarins. *Anal Chim Acta* 1967; **37**: 135–8.

138 Katyal M, Singh HB. Analytical applications of hydroxycoumarins. *Talanta* 1968; **15**: 1043–54.

139 Sane RT, Honavar PP, Deshpande BR. Simultaneous spectrophotometric determination of vanadium and uranium with 4-methyldaphnetin. *Ind Chem Soc* 1978; **55**: 290.

140 Eggers JH. Umbellikomplexon und xanthokomplexon—ein beitrag zur kenntnis komplexome trisches fluoreszensindikatoren. *Talanta* 1960; **4**: 38–43.

141 Wilkins DH. Calcein Blue—a new metalfluorochromic indicator for chelatometric titrations. *Talanta* 1960; **4**: 182–4.

142 Huitnik GM, Diehl H. Methylcalcein blue and other analogues of Calcein Blue. *Talanta* 1974; **21**: 1193–202.

143 Huitnik GM, Poe DP, Diehl H. On the properties of Calcein Blue. *Talanta* 1974; **21**: 1221–9.

144 Haugland R. Intracellular ion indicators. In: *Fluorescent and Luminescent Probes for Biological Activity*, WT Mason (ed) London; Academic Press, London, 34–43 1993.

145 Sundt TM, Anderson RE. Intracellular brain pH and the pathway of a fat soluble pH indicator across the blood–brain barrier. *Brain Res* 1980; **186**: 355–64.

146 Csiba L, Paschen W, Hossmann KA. A topographic quantitative method for measuring brain tissue pH under physiological and pathophysiological conditions. *Brain Res* 1983; **289**: 334–7.

147 Tomlinson FH, Anderson RE, Meyer FB. Panoramic imaging of brain pH and CBF during penicillin and metrazole induced status epilepticus. *Epilepsy Res* 1992; **13**(1): 49–58.

148 Hohn-Berlage M, Okada Y, Kloiber O, Hossmann KA. Imaging of brain tissue pH and metabolites. A new approach for the validation of volume selective NMR spectroscopy. *NMR Biomed* 1989; **2**(5–6) 240–5.

149 Iatridou H, Foukaraki E, Marais EM, Haugland RP, Katerinopoulos HE. The development of a new family of intracellular calcium probes. *Cell Calcium* 1994; **15**(2) 190–8.

150 Emelianov NA. Use of the flame photometer of the PFM type for determination of sodium and potassium in biological materials. *Lab Delo* 1978; 697–9.

151 Burguera JL, Burguera M, Gallignani M. Direct determination of sodium and potassium in blood serum by flow injection and atomic absorption spectrometry. *An Acad Bras Cienc* 1983; **55**: 209–11.

152 Adam WR, Koretsky AP, Weiner MW. Potassium adaptation ^{39}K-NMR evidence for intracellular compartmentalisation of K^+. *Am J Physiol* 1988; **254**: F401–F406.

153 Golchini K, Mackovic-Basic M, Gharib SA, Masilamani D, Lucas ME, Kurtz I. Synthesis and characterisation of a fluorescent probe for measuring potassium. *Am J Physiol* 1990; **258**: F438–F443.

154 Bourson J, Borrel M-N, Valeur B. Ion-responsive fluorescent compounds. Part 3. Cation complexation with Coumarin 153 linked to monoaza-15-crown-5. *Anal Chim Acta* 1992; **257**: 189–93.

155 Ohta T, Arai Y, Takitani S. Fluorimetric detection of nitrite with 4-hydroxycoumarin. *Anal Chem* 1986; **58**: 3132–5.

156 Zhou JY, Dauphin C, Hamon M. Utilisation de la 7-amino-4-methylcoumarine pour le dosage spectrophotométrique des nitrites. *Ann Pharm Fr* 1993; **51**(2) 94–100.

157 Nakamura M, Murata A. Spectrophotometric determination of nitrite using 4,5-dihydroxycoumarin. *Analyst* 1979; **104**: 985–8.

158 Brune M, Hunter JL, Corrie JET, Webb MR. Direct real-time measurement of rapid inorganic phosphate release using a novel fluorescent probe and its application to actomyosin subfragment I ATPase. *Biochemistry* 1994; **33**: 8262–71.

159 Ferenczi MA, He ZH, Chillingworth RK *et al.* A new method for the time-resolved measurement of phosphate release in permeabilised muscle fibres. *Biophys J* 1995; **68**(4 Suppl.): 191S–193S.

160 Dunges W. 4-Bromomethyl-7-methoxycoumarin as a new fluorescence label for fatty acids. *Anal Chem* 1977; **49**: 442–5.

161 Wolf JH, Korf J. 4-Bromomethyl-7-methoxycoumarin and analogues as derivatisating agents for high performance liquid chromatography determinations: a review. *J Pharm Biomed Anal* 1992; **10**(2–3): 99–107.

162 Turk J, Weiss SJ, Davis JE, Needleman P. Fluorescent derivatives of prostaglandins and thromboxanes for liquid chromatography. *Prostaglandins* 1978; **16**: 291–309.

163 Wintersteiger R, Juan H. Prostaglandin determination with fluorescent reagents. *Prostaglandins, Leukotrienes Med* 1984; **14**: 25–40.

164 Lloyd JBF. Fatty acid esters of 4-hydroxymethyl-7-methoxycoumarin: fluorescence characteristics and their variation in high performance liquid chromatographic solvents. *J Chromatogr* 1979; **178**: 249–58.

165 Crozier A, Zaerr JB, Morris RO. Reversed- and normal-phase high performance liquid chromatography of gibberellin methoxycoumarin esters. *J Chromatogr* 1982; **238**: 157–66.

166 Wolf JH, Veenma-van der Duin L, Korf J. Automated analysis procedure for valproic acid in blood, serum and brain dialysate by high performance liquid chromatography with 4-bromomethyl-4-methoxycoumarin as fluorescent label. *J Chromatogr* 1989; **487**(2) 496–502.

167 Guldutuna S, You T, Kurts W, Leuschner U. High performance liquid chromatographic determination of free and conjugated bile acids in serum, liver biopsies, bile, gastric juice and feces by fluorescence labeling. *Clin Chim Acta* 1993; **214**(2): 195–207.

168 McGuffin VL, Zare RN. Femtomole analysis of prostaglandin pharmaceuticals. *Proc Natl Acad Sci USA* 1985; **82**: 8315–19.

169 Andreolini F, Beale SC, Novotny M. Determination of serum metabolic profiles of bile acids by microcolumn liquid chromatography/laser-induced fluorescence. *J High Resolut Chromatogr Chromatogr Commun* 1988; **11**: 20.

170 Tsuchiya H, Hiyashi T, Naruse H, Takagi N. High performance liquid chromatography of carboxylic acids using 4-bromomethyl-7-acetoxycoumarin as fluorescence reagent. *J Chromatogr* 1982; **234**: 121–30.

171 Kelly RA, O'Hara DS, Kelley V. High performance liquid chromatographic separation of femtomolar quantities of endogenous carboxylic acids, including arachidonic acid metabolites, as 4-bromomethyl-7-acetoxycoumarin derivatives. *J Chromatogr* 1987; **416**: 247–54.

172 Tsuchiya H, Hayashi T, Narube H, Takagi N. Sensitive high performance liquid chromatographic method for prostaglandins using a fluorescence reagent, 4-bromethyl-7-acetoxycoumarin. *J Chromatogr* 1982; **231**: 247–54.

173 Farinotti R, Siard P, Bourson J, Kirkiacharian S, Valeur B, Mahuzier G. 4-Bromomethyl-6,7-dimethoxycoumarin as a fluorescent label for carboxylic acids in chromatographic analysis. *J Chromatogr* 1983; **269**: 81–90.

174 Takadate A, Tahara T, Fujino H, Goya S. Synthesis and properties of 4-diazomethyl-7-methoxycoumarin as a new fluorescent labeling reagent for alcohols and carboxylic acids. *Chem Pharm Bull* 1982; **30**: 4120–5.

175 Marr JC, McDowell LM, Quilliam MA. Investigation of derivatisation reagents for the analysis of diarrhetic shellfish poisoning toxins by liquid chromatography with fluroescence detection. *Nat Toxins* 1994; **2**(5): 302–11.

176 Iwamoto M, Yoshida S, Hirose S. High performance liquid chromatographic determination of pyrimidine nucleobases and 5-fluorouracil labelled with a fluorescent label. *Yakugaku Zasshi* 1984; **104**(12) 1251–6.

177 Michealis HC, Foth H, Kahl GF. Determination of 5-fluoro-2′-deoxyuridine in human plasma by high performance liquid chromatography with pre-column fluorimetric derivatisation. *J Chromatogr* 1987; **416**: 176–82.

178 Grayeski ML, DeVasto JK.Coumarin derivatising agents for carboxylic acid using peroxyoxalate chemiluminescence with liquid chromatography. *Anal Chem* 1987; **59**: 1203–6.

179 Kubab N, Farinotti R, Mahuzier G. 4-Bromomethyl-6-nitro-7-methoxycoumarin as an electrochemical label for carboxylic acids in liquid chromatographic detection. *Analysis* 1986; **14**: 125–30.

180 Karlsson KE, Wiesler D, Alasandro M, Novotny M. 7-[(Chlorocarbonyl)methoxy]-4-methylcoumarin: A novel fluorescent reagent for precolumn derivatisation of hydroxycompounds in liquid chromatography. *Anal Chem* 1985; **57**: 229.

181 Cohen H, Boutin-Muma B. Fluorescence detection of trichothecene mycotoxins as coumarin-3-carbonylchloride derivatives by high performance liquid chromatography. *J Chromatogr* 1992; **595**: 143–8.

182 Ayers FC, Warner GL, Smith KL, Lawrence DA. Fluorometric quantitation of cellular and non-protein thiols. *Anal Biochem* 1986; **154**: 186–93.

183 Obata H, Imai K, Inoue N, Umebayashi M. A highly sensitive quantitative determination of glutathione in plant roots by high performance gel filtration chromatography and fluorometric detection after precolumn derivatisation with *N*-(7-dimethylamino-4-methyl-coumarinyl-maleimide). *Phytochem Anal* 1994; **5**: 239–42.

184 Greenlee WF, Poland A. An improved assay of 7-ethoxycoumarin *O*-deethylase activity: induction of hepatic enzyme activity in C57BL/6J and DBA/2J mice by phenobarbital, 3-methylcholanthrene and 2,3,7,8-tetrachlorodibenzo-*p*-dioxin. *J Pharmac Exp Ther* 1978; **205**: 596–605.
185 Ullrich V, Weber P. The *O*-dealkylation of 7-ethoxycoumarin by liver microsomes. *Hoppe-Seyler's Z physiol Chem* 1972; **353**: 1171–7.
186 DeLuca JG, Dysart GR, Rasnick D, Bradley MO. A direct highly sensitive assay for cytochrome P450 catalysed *O*-deethylation using a novel coumarin analog. *Biochem Pharmacol* 1988; **37**(9): 1731–9.
187 Buters JT, Schiller CD, Chou RC. A highly sensitive tool for the assay of cytochrome P450 enzyme activity in rat, dog and man. Direct fluorescence monitoring of the deethylation of 7-ethoxy-4-trifluoromethylcoumarin. *Biochem Pharmacol* 1993; **46**(9): 1577–84.
188 Sotaniemi EA, Rautio A, Backstrom M, Arvela P, Pelkonen O. CYP3A4 and CYP2A6 activities marked by the metabolism of lignocaine and coumarin in patients with liver and kidney diseases and epileptic patients. *Br J Clin Pharmacol* 1995; **39**: 71–6.
189 Pellinen P, Steinback F, Raunio H, Pelkonen O, Pasanen M. Modification of hepatic cytochrome P450 profile by cocaine-induced hepatotoxicity in DBA/2 mouse. *Eur J Pharm* 1994; **292**: 57–65.
190 Kirby GM, Pelkonen P, Vatanasapt V, Camus AM, Wild CP, Lang MA. Association of liver fluke (*Opisthorchis viverrini*) infestation with increased expression of cytochrome P450 and carcinogen metabolism in male hamsters. *Mol Carcinog* 1994; **11**(2): 81–9.
191 Sone T, Isobe M, Takabatake E, Ozawa N, Watabe T. 7-ethenyloxycoumarin as a new substrate for fluorophotometric assay of hepatic microsomal epoxidising activities. *J Pharmacobiodyn* 1989; **12**(3): 149–58.
192 Wiegant J, Wiesmeijer CC, Hoovers JM *et al.* Multiple and sensitive fluorescence *in situ* hybridisation with rhodamine-, fluorescein- and coumarin-labeled DNAs. *Cytogenet Cell Genet* 1993; **63**(1): 73–6.
193 Nederlof PM, Robinson D, Abuknesha R *et al.* Three-colour fluorescence *in situ* hybridisation for the simultaneous detection of multiple nucleic acid sequences. *Cytometry*1989; **10**(1): 20–7.
194 Mergny JL, Boutorine AS, Garestier T *et al.* Fluorescence energy transfer as a probe for nucleic acid structures and sequences. *Nucl Acid Res* 1994; **22**(6): 920–8.
195 Axner O, Magnusson J, Petersson J, Sjostrom S. Investigation of the multielement capability of laser enhanced ionisation spectrometry in flames for analysis of trace elements in water solutions. *Appl Spectrosc* 1987; **41**: 19–26.
196 Spanner G, Niessner R. A photoacoustic laser sensor for the non-invasive determination of blood contents. *Analyt Meth Instrum* 1993; **1**(4): 208–12.
197 Stevenson CL, Tuan V-D. Analysis of polynuclear aromatic compounds using laser excited synchronous fluorescence. *Anal Chim Acta* 1995; **303**: 247–53.
198 Perera IK, Kantartzoglou S, Dyer PE. Coumarin laser dyes as matrices for matrix assisted UV laser desorption/ionization mass spectrometry. *Int J Mass Spectr Ion Processes* 1994; **137**: 151–71.
199 Nilsson S, Johansson J, Mecklenburg M *et al.* Real-time fluorescence imaging of capillary electrophoresis. *J Cap Elec* 1995; **002**(1): 46–52.
200 Li PYF, Narayanaswamy D. Oxygen-sensitive optical fibre transducer. *Analyst* 1989; **114**: 1191–4.
201 Bornmann WG, Roepe PD. Analysis of drug transport kinetics in multidrug-resistant cells using a novel coumarin-vinblastine compound. *Biochemistry* 1994; **33**: 12665–75.
202 Collins AK, Makrigiorgos GM, Svensson CK. Coumarin chemical dosimeter for radiation therapy. *Med Phys* 1994; **21**(11): 1741–7.
203 Makrigiorgos GM, Baranowska-Kortylewicz J, Bump E, Sahu SK, Berman RM, Kassis AI. A method for detection of hydroxyl radicals in the vicinity of biomolecules using radiation-induced fluorescence of coumarin. *Int J Radiat Biol* 1993; **63**(4): 445–58.
204 Makrigiorgos GM, Bump E, Huang C, Baranowska-Kortylewicz J, Kassis AI. A fluorimetric method for the detection of copper-mediated hydroxyl free radicals in the immediate proximity of DNA. *Free Radic Biol Med* 1995; **18**(4): 669–78.
205 Kennedy SH, Manevich Y, Biaglow J. Benzoyl peroxide acts as a promoter of radiation induced malignant transformation *in vitro*. *Biochem Biophys Res Commun* 1995; **212**(1): 118–25.

206 Scheper T, Schuegerl L. Bioreactor characterisation by means of *in situ* fluorimetry. *Chem Ing Tech* 1986; **58**(5) 433.
207 Institute of Petroleum, *Methods for Analysis and Testing of Petroleum Products*, no. IP374. Institute of Petroleum, London.
208 Shanahan GJ, Boley NP, Traynor AD. Total luminescence spectroscopy in the identification and detection of coumarin (1,2-benzopyrone) in road fuels. *Anal Proc* 1991; **28**: 210–13.
209 Thompson RD, Hoffmann TJ. Determination of coumarin as an adulterant in vanilla flavouring products by high-performance liquid chromatography. *J Chromatogr* 1988; **438**: 369–82.
210 Salagoity-Auguste M-H, Tricard C, Sudraud P. Dosage simultané des aldehydes aromatiques et des coumarines par chromatographie liquide haute performance. Application aux vins et eaux-de-vie viellis en fut chenes. *J Chromatog.* 1987; **392**: 379–87.

CHAPTER 14

Suggested Modes of Action of Coumarins and Some Comments on their Significance

Richard O'Kennedy
School of Biological Sciences, Dublin City University, Dublin, Ireland

and

R.D. Thornes
Colman K. Byrnes Research Centre, Beaumont Hospital, Dublin, Ireland

14.1 THE ACTIONS OF COUMARIN

A biologically active substance such as coumarin, with numerous metabolites, widespread in nature, produces a complex picture that is difficult to analyse when mode of action is being considered. At a meeting of the Biochemical Society in Liverpool (April, 1996) it was suggested that bioflavanoid compounds have a range of actions. These include:

(1) Inhibition/induction of enzymes
(2) Modulation of immune cells
(3) Antiviral effects
(4) Antixodidant properties
(5) Endocrine effects
(6) Potential anticarcinogenic effects.

Coumarins: Biology, Applications and Mode of Action. Edited by R. O'Kennedy and R.D. Thornes.

Coumarins may not have all of these functions. However, the above list illustrates the complexity involved in unravelling the mode of action of such compounds. It is interesting, however, to note that we ingest relatively large amounts of coumarin-derived compounds in foodstuffs (e.g. in tea, apples, onions, etc.) and that such compounds may have effects on our health.

It is, perhaps, best to choose a topic in human medicine where coumarin (1,2-benzopyrone) has been observed to have a role. John Casley-Smith and his wife Judith, together with Neil Piller, have spent many years unravelling the role of coumarin in lymphoedema (Chapters 7 and 8). They showed that the main action of coumarin is through the reticuloendothelial system, and macrophages in particular, where they produce enzymes that break down the protein in the oedema fluid and thereby facilitate its removal. How coumarin interacts with the macrophage and the immune system presents another problem (Chapter 5).

The action of warfarin, a 4-hydroxycoumarin derivative, in growth control of tumour is also macrophage-dependent. Its action on lymphoedema around transplanted tumours can be seen directly using Sumner Wood's rabbit ear chamber (Chapter 11), where activated polymorphonuclear leucocytes and macrophages migrate in, resulting in reduction of oedema.

The action of coumarins is dose-dependent. In plants, low doses stimulate while high doses inhibit germination and growth (Chapter 1). Piller (Chapter 8) found the optimum dose of coumarin to treat oedema in the rat to be 25 mg/kg body weight and estimated that the human dose should be 125 mg given daily as a single dose. Zlabinger likewise found that there was an optimum dose of coumarin, 7-hydroxycoumarin and 4-hydroxycoumarin that induced monocytes to produce interleukins (Chapter 5). The induction of phagocytosis by macrophages is also dose-dependent (Chapter 7) and this varies from species to species. The long-term use of stimulators of immune responses may lead to the production of inhibitors, hence the necessity to use the lowest effective dose. This makes the metabolism of coumarins and their derivatives important (Chapter 3). It is now very clear that for most individuals, coumarin is rapidly converted to its 7-hydroxy derivative. However, some individuals lack the enzyme responsible, and this may have significance in relation to how they metabolise drugs and other compounds. 7-Hydroxycoumarin is converted mainly to its glucuronide derivative (Chapter 3). However, for most *in vitro* work using cell lines, 7-hydroxycoumarin was shown to be more effective as an inhibitor of cell growth. The precise significance of this in the *in vivo* situation is not clear. What is clear is that results extrapolated from *in vitro* studies have not necessarily taken into account the actual levels of the effective form of the drug which may occur *in vivo*. Indeed, for coumarin to be effective when taken orally, methods to reduce the rate of its metabolism and elimination may be necessary. Current studies show that for most individuals, up to 60% may be eliminated within 4–8 h. Virtually no free coumarin can be detected in the blood when taken orally. In the clinical sphere, if coumarin is being used as an immune stimulant, it is essential to ensure that the patient has an intact immune system and that it responds to stimulation.

There is still no defined picture as to exactly how coumarin or 7-hydroxycoumarin bind to cells or pass across the cell membrane and exert their effects. While coumarin may simply pass through the cell membrane because of its non-polar nature, the process for 7-hydroxycoumarin may be more complex. Efforts are in progress to try to answer at least some of these problems, and it is suggested that cell surface receptors (Chapter 5) and gap junctions may be involved (Chapter 6).

The process by which coumarins act at the molecular level is still unclear. Studies by Yevdokimov, O'Kennedy, Egan and others using sensor-based methods have indicated that some coumarins do not, in fact, interact with DNA although direct interaction with DNA has also been described[1]. Coumarins may, however, modulate the effects of other drugs on DNA or interfere with repair mechanisms[1]. Some coumarins may also induce apoptosis. Coumarin also has an effect on oncogene expression (Chapters 4 and 6) and its main human metabolite, 7-hydroxycoumarin, inhibited oncogene-induced transformation of murine fibroblasts. Tseng and his colleagues showed that the cell-cycle progressions of MTV EJ *ras* cells is markedly reduced and *ras* expression is decreased in both G and S phase by coumarin. The production of p21 *ras* protein is inhibited by coumarin, thus it may have a possible role in control of prostate cancer (Chapter 10).

14.2 USE OF COUMARINS IN ANALYSIS

Chapter 13 shows the major role and importance of coumarin and coumarin derivatives in a whole range of analytical areas and these are expanding at an ever increasing rate. Areas where coumarins are widely used include estimation of enzymatic activity, labelling of proteins, antibodies, DNA and lipids, derivatising agents in chromatography, dyes for tuning lasers in ion analysis, intracellular ion indicators, pH and gas detection, measurement of drug/ion transport, studies on bioreactor characterisation, chemical markers in kerosene, food adulterant detection and in sensors.

14.3 FUTURE POTENTIAL USES OF COUMARIN

Reactive oxygen species react with a series of coumarin derivatives in both aqueous and hydrophobic environments so they should play a role as protective agents where oxidative stress is present. They may be useful in preventing oxidative stress and apoptosis in HIV infection[2]. The role of coumarin in virus infection is not covered in detail in this book but cell-to-cell communication is important in the phenomenon. Coumarin can prevent the cell-to-cell transfer of HIV *in vitro*[3] and 4-hydroxycoumarin blocks the entry of HIV into CD4 cells[4] by inhibiting HIV-associated proteases[4]. This is an exciting area for further development.

Through their effects either directly or indirectly, and as a result of their inhibitory effect in HIV infection, the coumarins offer a gateway to the understanding and long-term control of the entire cellular immune system.

The presence of many coumarin-containing compounds in foods and their positive properties indicates that this will be an expanding area of research. The same can be said for the use of coumarins in analysis. There is great potential for future research and development to increase our understanding of their mode of action. We hope that this book may stimulate and assist such progress.

REFERENCES

1 Grigg GW. Genetic effects of coumarins, *Mutation Res* 1978; **47**: 161–81.

2 Greenspan HC, Aruoma OI. Inhibitory effects of coumarins on HIV-1 replication and cell-mediated or cell free viral transmission, *Immunol Today* 1994; **14**: 209–13.

3 Bourinbaiar AS, Tan X, Nagorny, R. Oxidative stress and apoptosis in HIV infection: A role for plant-derived metabolites with synergistic anti-oxidant activity, *Acta Virol* 1993; **37**: 241–50.

4 Thaisrivongs S, Tomich PK, Watenpaugh, KD *et al.* Structure-based design of HIV protease inhibitors: 4-hydroxy-coumarins and 4-hydroxy-2-pyrones as non-peptidic inhibitors, *J Med Chem* 1994; **37**: 3200–4.

Index

Index compiled by Liza Weinkove

Printed and bound by CPI Group (UK) Ltd, Croydon, CR0 4YY

18/09/2026

14965009-0003